OPPORTUNITIES, USE, AND TRANSFER OF SYSTEMS RESEARCH METHODS IN AGRICULTURE TO DEVELOPING COUNTRIES

Systems Approaches for Sustainable Agricultural Development

VOLUME 3

Scientific Editor
F.W.T. Penning de Vries, CABO-DLO, Wageningen, The Netherlands

International Steering Committee
D.J. Dent, Edinburgh, U.K.
J.T. Ritchie, East Lansing, Michigan, U.S.A.
P.S. Teng, Manila, Philippines
L. Fresco, Wageningen, The Netherlands
P. Goldsworthy, The Hague, The Netherlands

Aims and Scope
The book series *Systems Approaches for Sustainable Agricultural Development* is intended for readers ranging from advanced students and research leaders to research scientists in developed and developing countries. It will contribute to the development of sustainable and productive systems in the tropics, subtropics and temperate regions, consistent with changes in population, environment, technology and economic structure.

The series will bring together and integrate disciplines related to systems approaches for sustainable agricultural development, in particular from the technical and the socio-economic sciences, and presents new developments in these areas.

Furthermore, the series will generalize the integrated views, results and experiences to new geographical areas and will present alternative options for sustained agricultural development for specific situations.

The volumes to be published in the series will be, generally, multi-authored and result from multi-disciplinary projects, symposiums, or workshops, or are invited. All books will meet the highest possible scientific quality standards and will be up-to-date. The series aims to publish approximately three books per year, with a maximum of 500 pages each.

The titles published in this series are listed at the end of this volume.

Opportunities, use, and transfer of systems research methods in agriculture to developing countries

Proceedings of an international workshop on systems research methods in agriculture in developing countries, 22-24 November 1993, ISNAR, The Hague

Edited by

PETER GOLDSWORTHY
International Service for National Agricultural Research, The Hague, The Netherlands

and

FRITS PENNING DE VRIES
DLO Research Institute for Agrobiology and Soil Fertility, Wageningen, The Netherlands

SPRINGER-SCIENCE+BUSINESS MEDIA, B.V.

A C.I.P. Catalogue record for this book is available from the Library of Congress

ISBN 978-0-7923-3206-0   ISBN 978-94-011-0764-8 (eBook)
DOI 10.1007/978-94-011-0764-8

Printed on acid-free paper

Contents

vi

SECTION C. NARS NEEDS AND PRIORITIES FOR ADDRESSING NATURAL RESOURCES ISSUES

SECTION D. EXPERIENCE AT A NATIONAL LEVEL OF USE OF SYSTEMS ANALYSIS METHODS

SECTION E. TRAINING IN THE USE OF SYSTEMS APPROACHES

SECTION F. THE USE OF SYSTEMS METHODS IN RESEARCH AT AN INTERNATIONAL LEVEL

viii

Foreword

In December 1993, ISNAR, in collaboration with International Consortium for Application of Systems Approaches, organized a three-day workshop on systems approaches and modelling for agricultural development. Sponsored by the Dutch Ministry for Development Cooperation, the workshop was attended by participants from 12 national agricultural research systems (NARS), nine international agricultural research centers (IARCs), and five advanced research organizations (AROs).

Although application of systems approaches in agricultural research and resource management is a rather new field, there is already increasing demand for implementation of these approaches. This will require a critical mass of specialists in the NARS and IARCs. Before this critical mass can be obtained, however, the experience that has been gained in this area needs to be evaluated, further possibilities need to be explored, and new objectives and targets need to be set.

This book, which contains the papers presented at the workshop, assesses the state of the art of systems approaches in agricultural research, resource management, and rural planning. It also gives an impression of the evolution of this interdisciplinary field and its use in national and international research centers.

Another, less tangible, outcome of the workshop was its contribution toward strengthening the network of NARS, IARCs, and AROs. It gave participants and organizers a chance to develop contacts, and provided an opportunity to make the first proposals for collaborative programs. Special thanks are due to Peter Goldsworthy and Luc Boerboom for their crucial role in making the workshop a success in this regard.

C. Bonte-Friedheim
R. Rabbinge

Editorial note

These proceedings follow the same structure as the workshop. The papers and the discussions that accompany them are ordered in four sections. Readers who are relatively new to the subject of natural resource management, and unfamiliar with a systems approach to research, will find **Section A** of the proceedings helpful. It introduces the concepts of sustainable growth in agricultural production, explains why a systems approach is required, and how this differs from the more familiar production-oriented, commodity-based approach to agricultural research. **Section B** contains papers that describe some of the systems methods available, and how they are used to support decision making at different levels of organization in the agricultural technology sector. In **Sections C** and **D**, some developing countries give their perceptions of their needs and priorities for the management of natural resources, with accounts of the institutional changes and the addititional information needs that these changes imply. Those that have applied systems approaches in their agricultural research give an account of their experience. **Section E** concerns some of the needs and opportunities for training in the use of systems methods for scientists and managers from developing countries. **Section F** contains accounts by the international agricultural research centres of the CGIAR of the systems approaches that they have developed and applied. The accounts are a useful milestone of progress since the last occasion when the centers gathered to discuss the subject of agricultural environments at an inter-center workshop in Rome in 1986.

The discussions have been condensed and reordered from the notes of the 12 working-group sessions and the concluding panel discussion. The editors have done everything possible to ensure that the reports reflect accurately what was said.

The editors wish to thank the contributors for their generous cooperation. We also owe our gratitude to Gisela Soffner for tracking the documents so that nothing was lost and for assisting in preparing the manuscript, and to Jacobine Verhage for helping with the graphical work. We are particularly grateful to Pablo Eyzaguirre and Willem Janssen for their assistance in compiling the discussion notes, and to Kathleen Sheridan and Jan van Dongen for their sustained assistance with the editing and preparation of the manuscript for this publication.

In memoriam C.T. de Wit

C.T. de Wit, known as Kees to many, died on 8 December 1993, nearly 70 years of age.

Since the late 1950s, Kees de Wit has provided enormous stimuli to agricultural research and contributed substantially to the solid scientific basis that is now supporting the broad field of "production ecology". His scholarly work yielded the now classical papers "Transpiration and crop yields", "Photosynthesis of leaf canopy", and "On competition". These papers, and his introduction of simulation techniques in crop ecology and his contributions to the development of explorative techniques for strategic policy making, all provided milestones that found, and continue to find, many followers. Kees received many distinctions, including the prestigeous Wolf price. Directly and indirectly, he influenced research in several CGIAR institutes.

Photo courtesy WAU

Although he remained based in The Netherlands, Kees de Wit was strongly interested in problems of developing countries ever since his appointment in post-war Burma. This interest grew when he led research in the Near East and the Sahelian zone. From 1986 to 1991, Kees de Wit was a critical but enthusiastic member of TAC. A method for setting priorities for research in a transparent manner was one of his last contributions.

Kees de Wit participated briefly in this ISNAR/ICASA meeting in The Hague. Although his weak health became more and more of a burden, he still loved the dialogue on international agricultural research, and contributed wise comments on exploiting rather then suppressing genotype x environment interactions.

An excellent scientist has left us. Original, eloquent, uncompromising, he showed how a strong social conviction stimulates the use of science to address practical problems. He has left us many thoughts to explore and examples to follow.

P. Goldsworthy
F.W.T. Penning de Vries

SECTION A

Agricultural sustainability and systems approaches

Sustainable growth in agricultural production: the links between production, resources, and research

JOHN K. LYNAM
Rockefeller Foundation, International House, P.O. Box 47543, Nairobi, Kenya

Key words: equity, food production, food demand, genetic diversity, natural resource management, population, research, sustainability concepts, systems level

Abstract
This paper outlines some conceptual issues concerning the sustainability of agricultural production, and surveys the emerging natural resource management issues. It examines the implications for agricultural research, which is probably the most critical factor in determining whether sustainable growth in agricultural production will be maintained.

Introduction

Agriculture is central to the concerns about sustainable development. Science has found no alternative to humankind's dependence on plants and animals for food and sustenance. And yet, with the continued growth in the world's population, we must command more of the earth's resources or use them more efficiently in order to feed ourselves. Moreover, today's husbandry of these same agricultural resources has direct bearing on the world's future ability to meet its food needs. A deep grounding in moral philosophy is, thus, not required to accept that food is a basic human right or entitlement (Sen 1981). The Bruntland report's dictum—to satisfy the moral imperative of meeting today's food requirements without jeopardizing the ability for a much larger future population to feed itself (World Commission on Environment and Development 1987)—may be more difficult to resolve in the agricultural sector than in any other sector, except possibly the energy sector.

This largely moral dilemma can be embedded within a number of striking characteristics of agricultural systems.

First, agriculture is both a biological/ecological activity and an economic activity. In agriculture, biological systems are designed to meet particular economic objectives. As agriculture is brought more into the market sphere, its demands on the resource base increase.

Second, agriculture is a land-use system, competitive with other land-use systems. Because of its scale, it plays a major role in determining the quantity and quality of other natural resources, such as fresh water, forests, and grasslands, as well as the undomesticated plant and animal life on our planet. As a human activity, agriculture is both extremely decentralized in terms of how agricultural resources are managed, and it is very extensive in terms of the proportion of the earth's land resources devoted to it.

3

P. Goldsworthy and F.W.T. Penning de Vries (eds.), Opportunities, use, and transfer of systems research methods in agriculture to developing countries, 3 - 27.
© 1994 *Kluwer Academic Publishers.*

4

Third, agriculture is a hierarchial system, as illustrated in figure 1. This has led to a number of different definitions of sustainable agricultural systems (table 1), which primarily differ in terms of the systems level. Meeting the world's food needs under conditions of increasing competition for resources leads to a definition of sustainable agricultural systems borrowed from Crosson (1993), who describes it as a production system that can indefinitely meet demands for food, fiber, and fuel at socially acceptable economic and environmental costs. The ability of agricultural systems to produce the needed increases in food will be a function of costs, both in economic terms, which reflects the scarcity of resources and the ability of the population to afford its food basket, and in environmental terms, in the rate at which natural resource capital is used to produce sufficient food. Starting with this definition, the paper will outline some conceptual issues underlying the sustainability of agricultural production, then survey emerging natural resource management issues, and finally draw implications for agricultural research, probably the most critical factor in determining whether sustainable growth in agricultural production systems will be maintained into the future.

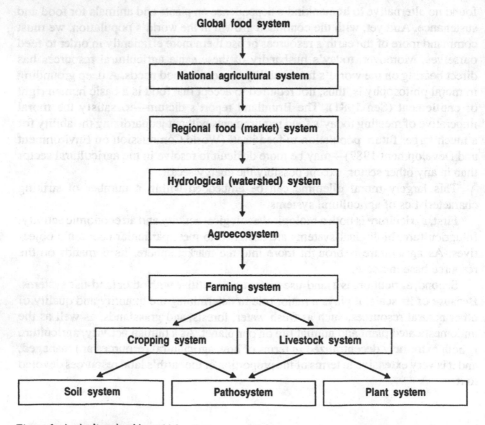

Figure 1. Agriculture is a hierarchial system

Table 1. Definitions of sustainable agricultural systems

System level	Definition
Global, national, regional	A system that can indefinitely meet demands for food, fiber and fuel at socially acceptable economic and environmental costs (Crosson 1993)
Agroecosystem	The dynamic equilibrium between national inputs and outputs, modified by external events such as climatic disasters (Fresco and Kroonenberg 1992)
Farming and cropping systems	A system that has a non-negative trend in total factor productivity (Lynam and Herdt 1989)

Conceptual and practical issues in sustainability

Designing an operational definition of sustainability that enjoys universal acceptance remains elusive. Nevertheless, this elusiveness has not dampened the exponential increase in literature on the subject in the past years. This paper will venture into this minefield by exploring both problematic and conceptual elements that together make up the debate on sustainable growth in agricultural production.

Food demand and the economic dimensions of sustainability
The increasingly stronger pressures that are being put on the management of natural resources derive largely from the growing demand for food and agricultural products. Both population and income growth determine the overall pattern of growth in demand for agricultural products, with population the relatively more important factor in the next half century. World population has grown from 2.5 billion in 1950 to 5.3 billion in 1990, 77 percent of whom live in developing countries. UN medium-growth estimates project world population to more than double again by 2050 to around 11 billion (United Nations 1992). This projection is supported by recent evidence that the drop in population growth rates, which started in the 1970s, has stagnated in the 1980s (Horiuchi 1992). Moreover, the distribution of this projected growth is highly skewed. Almost all of it will come in developing countries, with the bulk of the increase in Asia, where almost 60 percent of the world's population currently resides, and in Africa, where growth rates continue to remain the highest.

This growth in population is usefully considered in relation to the resource base. In Asia, the potential to expand cropland is extremely limited today. Cropland is expected to drop to 0.09 hectares per capita by 2025 (WRI 1992), and although urban growth rates will be high, more than half the population will continue to reside in the rural sector by that time. In Africa, the population is already growing faster than the increase in food production. While cropland can be expanded, the expansion is distributed unevenly among countries. Population today is calculated to exceed the carrying capacity in 14 African countries, primarily the Sahel and the East African highlands (Ho 1990). Moreover, Africa lacks the irrigation potential of Asia, nor does it have an industrial base that is as rapidly expanding. The question whether food production will keep pace with population growth into the next half century is relevant in particular in South Asia, North Africa, and West Asia (Tutwiler et al. 1991).

The other major forces acting on world food demand are income growth and urbanization. The theoretical interplay between population and economic growth on food demand is set out in table 2. This shows the ballooning in food demand in middle-income countries that are experiencing rapid economic growth, such as Southeast Asia and parts of Latin America. Moreover, income growth and urbanization affect the structure of demand, with an increase in demand for vegetable oils, horticultural products and livestock products, relative to cereal grains, root crops, and grain legumes. A food basket made up of vegetable oils, horticultural products, and livestock products requires more land resources than do diets of mainly starchy staples. This also applies to the diversion of grains to animal feed, given the less than unitary conversion efficiencies of grain to meat. In the developed world, changes in demand have virtually run their course. In the developing world, food consumption is still heavily skewed toward basic caloric staples.

Given plausible population and income-growth projections, Crosson and Anderson (1992) estimate that cereal production in the developing world will have to triple by 2030 to meet demand. While taking into account increased feedgrain demand, the estimates leave out the increasing horticultural production, especially in Asia, or oilseed production, in particular in Latin America and South Asia. Nevertheless, the prospect of a tripling of the cereal production in 40 years in the developing world raises a number of questions as to how this target will be achieved.

In 1950 only 17 percent of the population of the developing world lived in urban areas. By 1990 this percentage had grown to 33 and by 2025 it is expected to increase to 53 (WRI 1988). Such high urbanization rates suggest that an increasing share of

Table 2. Comparison of growth of demand for agricultural commodities, at different stages of development (hypothetical cases)

Level of development	Percent of population in agriculture	Rate of population growth	Rate of per capita income growth	Income elasticity of demand	Rate of growth of demand
Very low income	70	2.5	0.5	1.0	3.0
Low income	60	3.0	1.0	0.9	3.9
Medium income	50	3.0	3.0	0.7	5.1
High income	35	2.0	4.0	0.5	4.0
Very high income	10	1.0	3.0	0.1	1.3

(*Source:* Mellor and Paulino 1989)

the developing world's population depends on others to produce their food, and that a growing proportion of the world's food production is put on the market. This trend brings agriculture increasingly into the economic sphere, causing production to depend on the calculus of cost, profit, and price, and linking production to demand through the market-price mechanism. Countries strong in industrial exports need not be self-sufficient in agricultural production but can rely on income from trade to meet their food needs. In fact, given the locations of population growth and resource constraints sketched above, trade by developing countries with resource-surplus areas may be a way to reduce pressure on resources, were it not that South Asia and sub-Saharan Africa contain some of the poorest countries in the world.

The ability of an agricultural system to meet demand indefinitely can thus be defined at various levels, from the farming system to the regional, national, or common market, or the world agricultural economy. A point of some controversy has been whether the food security of the system is improved by the incorporation of, usually subsistence, farming systems into the market economy and the increased reliance on trade (Kennedy et al. 1992). Sen (1981), in evaluating the 1943 Bengali famine, argues that the famine was precipitated by the poor being priced out of the market, not necessarily by insufficient food supplies. Even at the level of national agricultural policies, governments are loathe to depend too heavily on imports to meet the domestic demand for basic food staples, given the political importance of food prices. Nevertheless, the sustainability of food systems will inevitably rely increasingly on how food distribution and marketing systems are structured and managed.

Production, distribution, and equity in sustainability
Is the sustainability of agricultural systems determined more by the biological or the economic properties of such systems? Of course, both properties interact in creating sustainable systems, but the question is important because the sustainability debate has shifted the balance of the relative importance of these two factors in both the research and policy spheres. The issue is considered at various levels, demonstrating the shifts in the foundations of the discussion within agricultural institutions. The debate also illustrates in some detail how economic considerations interact with the biological basis of sustainable systems, and it demonstrates in particular how equity concerns form a significant part of sustainable food systems.

At the level of the farming system and below, the science of ecology has had a major impact on both research and policy institutions in changing the status of farms from production systems to agroecosystems. Farm management has shifted its focus from input-output relationships to enhancing an ecological balance and system resilience through such avenues as increased diversity, component interactions, and enhanced biological subsystems. This approach has also resulted in a shift from externally purchased inputs—principally fertilizer and agrochemicals—to a reliance on more efficient nutrient cycling, biological control of pests and diseases, and genetic solutions to managing environmental stresses. These techniques are brought together in low external-input sustainable agricultural (LEISA) systems. However, in much of the developing world, the actual management strategies of farmers are moving toward greater specialization and input use. This move takes place because there is a pressure to increase yields—at a time when the average farm is becoming ever smaller—and because farming systems are becoming more commercialized (Lynam, Sanders, and Mason 1986). The relatively recent shift in the research community towards biological solutions should, therefore, be tempered by and designed within the wider forces acting upon farming systems and affecting farmers' management strategies.

The sustainability of agricultural production is defined in terms of the ability of the food system to meet current and future demand. The food system as a whole integrates production with distribution, in the sense that the ability of a food system to meet demand depends not only on production but also on how that production is distributed, especially to the lower-income strata of the population. During the 1970s and early 1980s, ensuring equitable access to food was a major concern of agricultural price and research policy. Meeting the food requirements of the whole population meant increasing the production and lowering the prices of food staples important in the diets of the poor, as well as designing policy and public food-rationing programs (von Braun et al. 1992). The emergence of sustainability principles has subtly shifted the research and policy agenda to a sole focus on the production or farming-systems level, with a move away from commodities to management of the natural resource base, and to ensuring that future rather than current food demand is met. Part of the reason for the shift in focus is that the agricultural sustainability debate takes place at the production or farming-system level rather than at the food-system level. Only at the higher systems level are marketing, price decisions, and trade seen as important components of sustainability of the system. Nevertheless, if meeting demand is to be part of the definition of sustainable agriculture, then equity considerations can and must be integrated with production considerations.

Even within production, sustainability has expanded, if not shifted, the domain of agricultural research and development. The impetus here has come from the donor community, who have found in sustainable development a framework for building a domestic constituency for maintaining foreign aid. Global warming and climate change, rainforest destruction, loss of biodiversity, and desertification are problems that reach across borders, and while many of these problems occur in the developing world, their impact is felt just as strongly in the developed world. However, the increasing environmental concerns of aid agencies come at a time when aid funds are being frozen and even cut. Funding of research and policy work on these problems has, therefore, resulted in a drop in support to more traditional areas of agricultural research and development and therefore in a shift in relative priorities.

The interests in limiting the destruction of natural ecosystems, however, may not be shared evenly between developing and developed countries. Antle (1993) presents preliminary evidence that the demand for environmental services, such as the conservation of rainforests, varies with the income level of the country, with higher-income countries demanding such services much more. Poorer nations would prefer to substitute such services for services that improve agricultural production. Because of the transborder effects of rainforest destruction and the value that developed countries attach to these sources, the benefits of conservation could be of greater importance for developed countries. This suggests that they should be prepared to foot a large part of the conservation bill. Yet, developing countries with increasingly higher income levels may develop a growing interest in environmental conservation and, hence, in environmental services (Howarth and Norgaard 1990).

Preserving the agricultural resource base for future generations introduces intergenerational equity into the discussion of sustainable agricultural production. For renewable resources, the attainment of intergenerational equity could be defined as a husbandry principle, whereby each generation bears the costs of maintaining the quality—not drawing down the stock—of both natural and manmade capital employed in agricultural production. However, as these resources are required to be ever more productive, husbandry costs for resources such as land and water will be expected to rise. For farmers with limited incomes, there may be an increased propensity to draw down the capital stock. Market prices may, in fact, not justify investments in husbandry, which means that protecting the interests of future generations will almost certainly require public-sector intervention.

For resources that can be depleted beyond recoverable thresholds, such as aquifers and fisheries, or for resource management decisions that are irreversible, such as the conversion of a tropical forest, intergenerational equity becomes even more difficult to define and much harder to achieve. In essence, what property rights do future generations have over resources that today could be consumed rather than conserved? The problem extends not so much to the existing agricultural resources, where husbandry criteria apply, but rather to the conversion of natural ecosystems, such as the transformation of tropical forests into agricultural land. The problem is primarily one of valuation, which involves not only determining the values of environmental services, but also theoretically extending the concept of the values associated with natural resources. Such natural resource values include in particular so-called non-use values (Cicchetti and Wilde 1992)—resources valued even by non-users for their mere existence or the possibility of bequesting them to future generations. However, again, such valuation is not applied in terms of the profit-cost evaluation of slash-and-burn cultivators, but in terms of public policy and investments.

In summary, sustainable agricultural development does not only involve economic considerations, but also considerations where incomplete or non-existent markets, equity, uncertainty, and externalities impinge on a society's ability to achieve sustainable agricultural production. Under such circumstances, market prices calculated to achieve private profitability fail to effect an allocation or use of resources that maximizes social welfare. Improving the relationship between market prices and social valuation of natural resources remains one of the key challenges in achieving sustainable management and conservation of such resources. Moreover, even in the public sphere, economic considerations must be combined with political, social justice, or "sustainability" considerations to represent the interests of future generations (Markandya and Pearce 1991; Howarth and Norgaard 1990).

Scale and systems level in sustainability

Sustainability can be evaluated or measured only as a system state or response (Lynam and Herdt 1989). As such, evaluation of sustainability requires the system under study to be closely defined. Agriculture is a highly hierarchial system, with its foundations on biological and edaphic sub-systems and with higher levels that are defined largely in economic and social terms. Any discussion of sustainable agriculture requires the systems level to be specified, which also then largely defines the spatial and temporal scales over which sustainability of the system can be analyzed (Fresco and Kroonenberg 1992).

The sustainability of a farming system is a measure of its internal organization, management and performance over time, and it may be described by such "health" indicators as rates of soil erosion or nutrient depletion, crop yield trends, pest and disease build up, or profitability. However, if one thinks of farming systems in an evolutionary framework, unsustainable systems will by definition disappear. The current concern about sustainable farming systems arises from the rapid change in forces acting on these systems, the increasing limits put on an adaptive response, and the geographic scope of the problem, leading to widespread resource degradation (Oldeman et al. 1990). Solutions to this situation are either to attack the problem at the level of the farming system by developing more options for farmers to maintain the productivity of their systems, or to try to solve the problem at a higher systems level by addressing the forces that lead to unsustainable systems.

With rural populations rapidly increasing, as in the highlands of East Africa, a trajectory of adaptive responses has been charted (Boserup 1981; Pingali and Binswanger 1984). In response to the declining farm size, there is both an internal reorganization of the farming system, visible especially in a drop in the use of fallow and a move to more permanent cropping, and an exploitation of other land resources. The latter moves the analysis up to another scale. Reliance on cultivation of the mid slopes (which are usually more fertile and more easily cultivated) gives way to the exploitation of the upper slopes, at the expense of forests, of the steeper slopes, leading to increased soil erosion, and of the wetlands in the valley bottoms (Loevinsohn and Wangati 1993).

Given such dynamics, much of the research on agricultural resource management is designed at the scale of the watershed. There are yet no global criteria to track the sustainability of a watershed. The focus has rather been the understanding of the various processes, including soil erosion and the deposition and movement of pesticides in the water system. In addition, studies were aimed at the impact of land-use change—especially deforestation and wetland conversion—hydrology, and the effects of wetland conversion on the epidemiology of vector-borne human diseases. Such processes effect the sustainability of farming systems across the toposequence in various ways. Understanding these process is necessarily part of designing the interventions to maintain the sustainability of watersheds that undergo rapid land-use change.

A potential intervention in these watersheds might be a tree crop such as coffee, but this is a decision made at the regional or country level, which defines the overall food and agricultural system. The interesting question at this level is the role that market incentives and trade policy play in, first, the sustainability of individual farming systems and, second, the sustainability of the overall food system—the ability of food production to meet growing demand. The possibility of trade through the market allows individual farming systems to specialize, potentially in agricultural activities that may be more sustainable than subsistence food crops (such as tree crops on hillsides or in the humid tropics). Trade also provides the cash to purchase inputs, especially fertilizer, which may be critical to sustainable increases in production.

It is a practical question whether trade between nations leads to more sustainable systems per se. A country's agricultural trade based on comparative advantage is a function of costs of production as determined by input and market factors. However, market prices do not reflect social and environmental costs. *If* these costs were incorporated into the market price, the market would serve as a vehicle for investments in sustainable systems or the reorientation of production to regions with low environmental costs. However, market failures would have to be rectified by all producers in all countries. Otherwise, producers or countries that do not pay for environmental services would produce at a lower cost than those who do pay, thus potentially exacerbating the environmental problem (Harold and Runge 1993). The complex of environmental, economic and trade policies thus provides the background for achieving sustainable growth in agricultural production.

Sources of and limits to growth of food production

By the middle of the next century world agricultural production will have to be triple the current production to keep up with the increasing demand. Such an increase in agricultural production is bound to put pressure on non-agricultural resources and the environment. This pressure leads to several important questions. First, what are the sources of growth for such increases in world food supplies? Second, how can societies most effectively assess the trade-offs between expanding agricultural production and the increase in environmental costs that such expansion will precipitate? Finally, as Taylor (1993) asked, what political and social adjustments does the world have to consider if "exponential output growth will cease to be feasible in the decades to come." Some research (Homer-Dixon, et al. 1993) has suggested that resource scarcity is already contributing to an increase in civil strife and political instability in some parts of the world. Some of these issues will be sketched briefly in this section.

Land resources and soil
Agriculture occupies land and utilizes its properties, particularly nutrient and moisture supplies. The transition from shifting cultivation to permanent field systems (Ruthenberg 1980) transforms the natural ecosystem into an agroecosystem that is subjected to the objectives of farmers and, through the market, of society. Humans also exploit, to various degrees, most other major natural ecosystems, such as the savannah and forest areas. One estimate suggests that man now appropriates directly

or indirectly 40 percent of the terrestrial net primary productivity, that is the direct products of photosynthesis on which all biotic food chains depend (Vitousek et al. 1986).

Agriculture has now appropriated most of the fertile and well-watered lands. From 1700 to 1980, the area under cultivation expanded by 466 percent (Meyer and Turner 1992), primarily at the expense of forest. From 1970 to 1990, the total area planted for grains grew by less than 10 percent, and stagnated in the last decade (WRI 1992). As a result, the cultivated area per capita dropped, especially in the highly-populated developing countries. The growth in the world food production has increasingly depended on increasing the crop yields.

With the growing demand for food, problems in the sustainable use of the land resource arise from the search for higher yields on existing cultivated lands (the intensive margin) and the increasing environmental costs from conversion of natural ecosystems to cultivation (the extensive margin). The problems of rainforest destruction, biodiversity loss, and desertification are defined at the extensive agricultural frontier, where agriculture is expanding through migration and the cultivation of new areas. This expansion comes both at an increasing environmental cost and at a lower return in agricultural production. Ehrlich and Wilson (1991) note that "if current rates of clearing continue, one-quarter or more of the species on Earth could be eliminated in 50 years,... and for the first time in geological history, plants are being extinguished in large number."

There is a public policy question as to how society balances the preservation of these natural areas against the increase in agricultural production. As noted earlier, conservation is a typical developed-country issue. Conversion, on the other hand, is subject to the preferences usually of developing countries, and especially the farmers in those countries, where societal needs are skewed toward food provision. Moreover, it is difficult to come up with solutions that preserve the natural ecosystem when there is little governmental infrastructure and significant migration pressure, or ways to develop sustainable farming systems in situations of nutrient or moisture stress.

Three brief points may be made about the difficulties in defining policy interventions to achieve the conservation and the conversion objectives. First, although land-use planning uses common frameworks, it is difficult to implement because it relies on law enforcement in areas where police powers are weak, on community structures that are difficult to build in a frontier area, and on economic incentive structures that are difficult to target on land-use matters. Second, economic incentives to increase the profitability of agriculture at the frontier can speed up the conversion process, as the subsidies for cattle ranching in Brazil illustrated (Binswanger 1991). However, where subsidies do not aid the process, such as in West Africa, economic policies have had little impact on reducing the conversion process, especially since markets are underdeveloped in such areas. Finally, migration is the drive behind conversion. Much of the pressure to migrate comes from the inability to access resources (constraints to improving rural equity), few employment alternatives in urban areas, and limited progress in increasing productivity in the intensive agricultural margin. Sustainable increases in agricultural production without the loss of forests and marginal lands may come more from increasing the productive capacity of existing agricultural land and developing the industrial sector.

Sustainable agricultural change in the intensive margin is achieved through increasing land productivity. Increases in yield are achieved through optimizing the plant-growth environment and a shift to the most productive component, usually the starchy staple crop. Such specialization into most productive crop can be further strengthened by providing a market for it.

Sustainable use of existing agricultural lands involves three major challenges.

First, most agricultural areas display rich patterns of land use, including arable lands, pastures, forests or woodlands, and floodplains. These patterns become more pronounced as the terrain becomes more varied. The permanent characteristics of the terrain serve as the major regulators of the local hydrology by maximizing rainfall infiltration, controlling run-off across the area, optimizing deep percolation into aquifers and watertables, and normalizing discharge into stream channels. The declining farm size, the reduction of common lands, and the increasingly expansive markets shift land use to arable cropping at the expense of pasture and forest, with usually unknown effects on hydrology and the longer-term productivity of the watershed. Agroforestry is one research area which attempts to expand crop cultivation, while preserving the advantage of permanent cover and deep rooting.

Second, the search for higher yields through land intensification has raised questions about how to modify the biology and physical structure of the soil and the pest and disease complex necessary to allow continued increases in plant yields. These issues in turn raise questions as to the limits to crop yields and whether these limits are set more by the genetics of crop plants or by other intensification limits within the agroecosystem. In subsistence systems in Africa, the increasingly higher rates of nutrient depletion have caused problems, the development of profitable fertilizer responses has become increasingly more difficult, pests and diseases have become difficult to handle at the low levels of soil fertility, and the soils have become increasingly more acidified. Declining crop cover causes soil to erode more seriously, which in turn causes land degradation to spiral. On the other hand, in the intensive rice and wheat cultures of Asia, where rates of yield increases are dropping, there is some evidence that rates of fertilizer application are increased only to maintain existing yield levels. In experimental stations where high yields have been produced for a long time, yields are actually declining (Pingali, in press). Unlike the focus on soil chemistry of nutrient supply in the past, soil management under the present intensification conditions requires a more integrated assessment of the biology of the soil, its physical structure, and its chemistry. Under continued intensification, management of the existing soil resources is one of the principal challenges in the sustainable increase of agricultural production.

Finally, the above discussion of sustainable soil management under intensification should reinforce the point that increased food production will critically depend on supplies of fertilizer. Agriculture is an extractive production activity, principally in terms of soils nutrients. The shift in the post-war period to meeting the increasing food demand through increasing the yields was achieved through a more than proportionate increase in fertilizer production. World production of nitrogenous fertilizers, which are considered critical to grain production, increased almost 25-fold between 1950 and 1988 (Smil 1991). This growth may be due to the engineering advances made in the Haber-Bosch process of ammonia synthesis. Smil estimates that of the annual use of between 140 and 210 million metric tons of nitrogen, fertilizers currently account for two-fifths of all inputs (table 3), and the percentage is increasing. However, the efficiency of the conversion from nitrogen fertilizer to crop production is low, at between 40 percent to 50 percent. Even the most efficient systems fail to reach an efficiency of more than 66 percent (Smil 1993).

With increasing population growth, food production will depend even more on fertilizers. The issue for the future is cost and limits on supplies. Supplies of nitrogen are potentially unlimited, but costs are directly linked to the price of petroleum and electricity (the best ammonia plants require 35 million joules per kg of nitrogen [Smil 1993]). Over the last two decades, real nitrogen and rice prices have fallen (Rosegrant and Svendsen 1993), partly because of a continuous drop in energy prices. The question is when, not if, energy prices will begin to rise. The question is also how the increasingly closer link between energy and prices will affect the cost of food. But increasing fertilizer use in the tropics may well be expected to have an even greater potential impact on the other macronutrients: phosphorus and potassium, which are derived from mining geologic sources (table 4). The highly leached oxisols and ultisols dominate the humid tropics (National Research Council 1993) and are low in particular in phosphorus. Other soils of the tropics have very high sorption

Table 3. Estimated ranges of annual nitrogen fluxes in global croplands (in million tons)

Flux	Minimum	Average	Maximum
Inputs	140	175	210
Synthetic fertilizers	70	75	80
Mineralization	20	30	40
Animal wastes	10	15	20
Biotic fixation	25	30	35
Atmospheric deposition	10	15	20
Crop residue recycling	5	10	15
Outputs	140	175	210
Crops (including residues)	75	80	85
Erosion	30	45	50
Denitrification	15	20	35
Volatilization	10	15	20
Leaching	10	15	20

(*Source:* Smil 1991)

Table 4. Fertilizer consumption by nutrient, 1990

	Nitrogen ('000 tons N)	Phosphorus ('000 tons P_2O_5)	Potash ('000 tons K_2O)
Africa	2,146	1,164	501
Latin America	3,880	2,441	2,077
Asia	35,929	6,982	4,464
Developed countries	37,123	28,429	19,915
World	79,078	39,016	26,957

(Source: World Bank 1991)

capacities for phosphorus. As Loomis (1984) notes, "sources of high-grade ores are limited to the United States and Africa, low-grade sources are more plentiful, and recyclable sources (sewage, oceans) will require large amounts of energy to process." Moreover, in particular tropical crops such as bananas, tree crops, and root crops, especially cassava, use a lot of potassium, while soils poor in potassium are becoming increasingly more frequent, especially as little fertilizer is used on these crops and potassium is seldom available in many developing countries. Future nutrient supplies in developing countries will be available only at increasing costs.

Rainfall and water

Locally, regionally, and globally, agriculture interacts with the water cycle in significant ways. First, irrigation for agricultural purposes is by far the largest consumer of water on a global scale, in particular in Asia. Second, since it affects land cover and soil structure, agriculture influences local and regional hydrology through its impact on infiltration, run-off, and aquifer recharge. Third, there is a hypothetical causal link between the conversion of tropical forest to agriculture and regional rainfall patterns (Salati and Vose 1984). Finally, agriculture and land-use changes have had major impacts on the global carbon budget, creating a feedback loop through global warming, shifting rainfall patterns, and the local agricultural production potential. Thus, while exerting a direct and indirect impact on the availability of water, agriculture is the most critical force when it comes to managing the quantity and quality of water resources for growing future populations.

It is not surprising that the explosion in world population from about the year 1750 (Coale 1974) coincides with a similar exponential increase in irrigated area from 0.08 million square kilometers to between 2.00 and 4.58 million (depending on the estimate) (Meyer and Turner 1992). An estimated 17 percent of the world's arable land is now irrigated, reaching 34 percent in Asia (Rosegrant and Svendsen 1993). In the past two decades, the irrigated areas in the developing countries have been the centers of the increasing yields, with one estimate suggesting that these areas have been responsible for 80 percent of the total increase in cereal production in Asia (Seckler 1993). However, assessments of potentially irrigable land suggest that in the developing world a further increase of 54 percent is possible (Crosson and Anderson

1992), with potential areas well below this percentage in critical areas such as North Africa and the Near East. Although expansion of irrigable areas is already being achieved, real capital costs per hectare have increased rapidly. Rosegrant and Svendsen (1993) show that costs have risen from 40 percent in Thailand to over 100 percent in India and Indonesia since the late 1960s. This trend can only continue, at a time when an estimated 0.3 to 1.5 million hectares of irrigated land is lost annually to salinization and waterlogging, caused by insufficient investments in drainage (Rosegrant and Svendsen 1993).

The ability to increase irrigable land further will also be affected by other demands on water and potential increases in the price of that water. Irrigation already accounts for about 70 percent of the total consumption of water (WRI 1992), which in turn represents 37 percent of total available freshwater runoff (Liniger 1992). With continued industrialization and urbanization in the developing world, non-agricultural demand for water is expected to increase rapidly—since 1940 the worldwide consumption of water has quadrupled. The positive side of these trends is that water, traditionally considered a virtually costless input, will increasingly be seen as a limited resource, which in turn will introduce a search for improved efficiency in water resource management, from the field to the river basin. Seckler (1993), for one, sees substantial potential for meeting the increasing consumption through the increased recycling of the water within the river basin, what he terms water multipliers. This, however, puts an even greater stress on a better integrated planning of capital investments and allocative mechanisms for water resources.

Agriculture is not only the major determinant of water demand but also a major factor in the building up of water supplies at the local, regional, and global levels through its impact on land cover. Agriculture also plays a major role in the forming of pathways in the hydrological cycle, particularly groundwater recharge, surface run-off, and atmospheric flux. These pathways are not well understood as they are mediated by land cover, soil management, and land-use change. The effects of these mediating factors can been seen in various ways, ranging from local watersheds to the hydrology of river basins to regional rainfall patterns. The effects on regional rainfall patterns are especially hypothetical, but may be illustrated in the long-term decline in precipitation in the Sahel starting in the early 1970s. This decrease may have its origin in the substantial deforestation along coastal West Africa.

Finally, agriculture (including tropical deforestation) contributes by almost 25 percent to global greenhouse gas emissions, in terms of the equivalent effect of carbon dioxide on global warming potential (Rubin et al. 1992). Rice cultivation, ruminant production, and nitrogen fertilizer volatilization contribute as much to global warming as does tropical deforestation, and all are expected to increase. The effect of the global climate change on agriculture has become a major field of research (Parry et al. 1988). Future climate changes will require food and agricultural systems to be more resilient and adaptive. Paradoxically, world agriculture is aiming at higher productivity through more specialization.

Genetic diversity and genetic potential
Whether the ideal balance between food and population in the future is defined by technology optimists or resource-constraint pessimists may well depend on the limits imposed by the genetic yield potential of the basic food crops. The food-population equation outlined above will require increasing crop yields in the developing countries. In particular, a continued lifting of the yield ceiling on wheat and rice in the irrigated areas of Asia, where yields have already been rising rapidly for two decades, will be needed. In addition, a broad attack on increasing crop yields in rainfed agriculture in Africa is needed. Expanded food production will continue to rely on major investments in crop breeding, where progress will depend on an increase in the exploitation of variability in the gene pool. Some are concerned that the size and variability of that gene pool will be reduced through the loss of land races and wild relatives, and potentially expanded by the ability to exploit inter-species variability through biotechnology. Certainly, managing the world's genetic resources becomes more crucial as food production increasingly relies on rising yields.

The increase in yield in Asian irrigated agriculture has been brought about by specialization, including genetic specialization. The genetic base of rice in Asia has narrowed substantially, with a possible contraction of the gene pool itself. Moreover, there are increasing indications that rice yields in Asia are leveling off. The growth in rice yields has slowed substantially (Rosegrant and Svendsen 1993). On experimental stations, rice yields have stagnated, and may even be declining (Pingali, in press). This yield stagnation, together with the expected increasing demand, has prompted a major investment in biotechnology research on the crop (Khush and Toenniessen 1991). So far, however, much of this research has been directed at biotic constraints and quality characteristics. Progress on the multi-genic characters to overcome inherent and physiological yield constraints has been limited. Achieving positive results in this area remains a challenge that still faces biotechnology applications. In addition, pushing physiological yield barriers may well require a simultaneous enhancement of nutrient and moisture supply systems—as the increased economic yield shifts from altering the harvest index to increasing the overall biological yield (Evans 1980)—and this simultaneous enhancement may in turn have a negative effect on the efficiency of nutrient and moisture utilization, putting further pressure on these resources.

Raising yields in rainfed agriculture in the developing world, especially Africa, may well require expanding the use of genetic variability. It has been difficult to make progress in breeding for rainfed crops in the center of origin, such as beans, cassava, and even maize in Latin America and sorghum in Africa. Part of the reason for this may be biotic constraints (Jennings and Cock 1977), but may also be the excellent adaptation of local cultivars to micro and agroclimatic niches. Since much of the food base of Africa depends on crop species that were domesticated elsewhere, there should be substantial potential for better varietal-agroclimatic matching. Such matching, however, will require a significant change in crop breeding. A move would have to take place from the broad adaptation and release of elite materials to a substantial genetic diversity, which farmers could then evaluate (Voss 1992). This means that the significant effort at ex situ and in situ conservation of the world's germplasm should be matched by the design of mechanisms to better exploit that diversity at the level of farmers' fields. This diversity may be the key to improving both yields and production stability in rainfed agricultural systems.

The more pervasive and global problem includes the wider loss of biodiversity, due to the loss of habitat and ecosystem integrity by the advancing agricultural frontier, environmental pollution, and potential climate changes. The maintenance of an as yet ill-defined percentage of natural terrestrial ecosystems is critical to supporting the world agricultural system. Ecosystem services that sustain agriculture are best summarized by Ehrlich et al. (1993), who include "maintenance of the gaseous composition of the atmosphere, moderation of climate, control of the hydrological cycle, recycling of nutrients, control of the great majority of insects that might attack crops, pollination, and maintenance of a vast 'genetic library' containing many millions of kinds of organisms, from which humanity has 'withdrawn' the crop and livestock species on which civilization was built, and which potentially could (if preserved) provide enormous benefits in the future." No matter how humankind evaluates these costs, either by developing substitutes for these services or by accepting the costs of extinction, his relation to his continued need for food lies at the heart of the challenge of sustainable development.

Research and sustainable agricultural production

Sustainable increases in agricultural production to meet the demand in the next 50 years will require more productive use not only of land resources, but also of water, nutrient, energy, genetic, and forest resources. For many of these resources, such as water, phosphorus, and potassium, there are virtually no substitutes. Moreover, the exponential yield increases in the post-war period may have resulted from an historical confluence of significant technological breakthroughs, especially the Haber-Bosch process for ammonia production, but also hybrid technology, dwarf rice, and wheat varieties, all aided by relatively inexpensive energy, water, and nutrient resources. Efforts by the CGIAR system have also contributed significantly to increasing yields. Future agricultural research and technology development will have to search for productivity gains across a wider range of limiting resources than just land, as well as reduce the environmental costs of agricultural intensification. Also, to slow or stop further expansion of the agricultural frontier, research will have to shift the yield frontier faster than the growth in demand. The task for agricultural research is nothing short of herculean, with increases in food production becoming increasingly dependent on the flow of technology from research stations. Yet, the 1990s have seen a continued tightening of research budgets and increasing restrictions on public funds for agricultural research.

The incorporation of the sustainability agenda into agricultural research has two principal dimensions. The first is an expansion in the research domain, which under budget stringency has resulted in a shift in research priorities. This shift is best reflected in a move from commodity research to natural-resource management research, from green-revolution technologies to those that are more environmentally friendly, and from research on the intensive agricultural margin to a greater focus on the extensive margin. The process is best reflected in the current restructuring within the CGIAR, which will soon be felt within the national research systems, and fueled by shifts in donor funding. The shifts in research problems have led in turn to attempts to reorganize the research within and across research institutions. In particular, there has been a shift from a commodity program approach to a more conceptual approach, such as natural resource management (NRM) programs, agroecologies, and matrix management. This change in the problem focus has not yet resulted in a new organization model for agricultural research, and it will be difficult to find new forms that are as effective as or more effective than multidisciplinary commodity teams (Lynam 1992).

To deal with the significantly different problem structure, the sustainability agenda has also resulted in a change in research methodology. The search for increased productivity across a range of limiting resources has required a move from breaking a system down into manageable components to researching the system as a whole. Such systems research considers the component interactions to be as important as the direct component effects. It searches for multiple avenues to address a productivity problem, such as pests, and attempts to integrate these different inter- ventions within the system, as in integrated pest or soil management. This systems research has resulted in a focus on basic processes rather than the reductionist approach, which tests how individual changes in one factor affect system perform- ance. In addition, natural resource management and concerns about environmental costs and externalities have meant that research focuses on higher levels than just the field or the farm. This section touches on emerging issues in the development of more appropriate research methodologies to meet the requirements for sustainable agricul- tural systems.

Characteristics of NRM research
Research on sustainability has been implemented primarily, though not exclusively, through a greater focus on natural resource management. This includes management of the underlying soil and water resources critical to crop growth, as well as the natural resource systems that support crop and livestock systems, such as tree-forest, grass- savannah, and fish-aquatic resources. Besides the system focus and the importance of more basic research on processes, there are other characteristics of the problem structure that determine the methodological approach. First, both the temporal and spatial scaling of research are expanded. By necessity, such research is more expen- sive than short-term and experimental trials normally carried out in commodity research. Methods are required that use limited operational research budgets effi- ciently. More upfront work has to be done in designing long-term and large-scale experimental trials. Second, and related to the last point, NRM research in most respects embodies a higher degree of location specificity than commodity research, at least that of the green-revolution type. This is because NRM lacks the equivalent of the specific inputs of commodity research to buffer the cropping system against

variation in environmental stresses. In addition, issues such as the causes, rates, and effects of soil erosion, the effects of pesticide on the environment, and the hydrological impact of forest clearing require detailed specification of initial biophysical, agroclimatic, and economic conditions to evaluate such processes. Monitoring these processes require well-developed protocols, the instrumentation of which is usually quite expensive. Third, there is a much stronger interplay between the economic and biophysical domain in NRM research than in commodity research due to, first, the higher level at which the research takes place; second, the increased importance of market failures and non-market valuation; third, the increased disjunction between research and farmer evaluation criteria (economic yield provides a high level of congruence between farmer and commodity researchers' evaluation criteria); and fourth, the high farmer-management component in any system intervention or technology. All these characteristics add to the complexity and expenses of NRM research.

Pattern, process, system level, and scale
There has been much debate over how to define sustainability and how to employ it as an evaluation criterion in agricultural research (Lynam and Herdt 1989; Harmsen and Kelley 1993). To give sustainability scientific significance, the concept of sustainability has to be limited and made precise, primarily in terms of the four parameters of pattern, process, system level, and scale. Harmsen and Kelley go so far as to state that "sustainability cannot be measured; however, one can quantify the processes occurring in a particular agro-ecosystem, the interactions between them, relate these processes and interactions to environmental and social conditions, and try to predict the behavior of such a system over time." If sustainability is to be a measure of system performance, then by the time that the system, scale, and processes are sufficiently specified, one is working with already existing evaluation criteria, such as total factor productivity in the case of cropping systems (Lynam and Herdt 1989). The strength of the notion of sustainability is that its importance is shared across a significant spectrum of the policy, scientific, and donor communities. Its weakness, however, is that it is interpreted differently by everybody. All three communities describe a set of production, development, environmental, and natural resource management issues, around which directed action can be initiated. These four issues lie at the core of a more sophisticated research methodology, which is being developed in an attempt to integrate them.

The global slash-and-burn initiative (Palm, Izac and Vosti 1993) is an example of this research methodology. The objective of the initiative is to develop technological and policy options for the expanding slash-and-burn agriculture in forest ecosystems. The problem structure is defined at various system levels, ranging from understanding migration within the national economy to understanding nutrient cycling in alternative production systems. Different processes are of interest at different system levels. These processes are deduced or monitored through patterns of empirical data, which, depending on the system level, defines the level of data resolution or scale. Palm and Izac have defined a pattern-process ladder across different system levels and scales (figure 2). Pattern analysis provides hypotheses about operative processes, which can then be studied in more detail within a sampling frame defined by that pattern structure. The system levels are critical to defining the research framework, in that the higher system levels determine the constraints or boundary conditions on the lower system levels. In modelling terms, characterization of the higher system levels are critical to initializing the systems at lower levels. In turn, understanding the processes operating at the lower systems levels allows extrapolation of the results through the patterns established at the higher levels.

This methodological structure provides significant flexibility in representing the particular characteristics of any site while at the same time extrapolating results to higher levels. Additional tools deepen the methodology. First of all, a system is not a single process but a set of interacting processes that in turn describe the performance of the system, subject to conditions imposed by the external environment. Modelling has become a tool that integrates process, performance, and environment. Examples range from soil organic-matter dynamics to crops and livestock to farming systems to agricultural markets, including world commodity trade, to global climate change. Such system levels can be linked through models. To link different system levels through empirical research and modelling remains highly complex, however, and work on it is limited. Moreover, the modelling is only as good as the current understanding of underlying processes. The effective use of modelling tools critically relies on a close interaction with the empirical side of the research process. Yet, the two are usually very separate enterprises, and one of the organizational/institutional challenges in natural resource management research is to better integrate the two.

The other tools that support this research methodology are relational databases, geographic information systems (GIS), and remote sensing. The types of research envisioned under natural resource management research are data intensive. They require data that can be represented at different levels; they need a capacity for monitoring system change across different temporal scales, and they require a framework for integrating data from different sources and in different formats. Georeferenced databases of standardized variables, whether collected from surveys or from remote sensing devices, can be spatially portrayed and interpolated through GIS packages. When linked to models, where the databases provide the initializing variables, GIS provides a powerful tool for extrapolating research results and linking different scales.

Technology and the economics of NRM
A potential major shortcoming of natural resource management research is that it lacks the imperative to improve technologies. The focus on the understanding of processes is research of a more basic and strategic kind, not necessarily with an avenue to technological solutions. Moreover, the research objectives focus on more than increasing the crop yields, and the technology aims at more than improving varieties, inputs, and crop-management techniques. The incorporation of NRM into agricultural research institutes will require special vigilance in clearly defining research problems, evaluating their relative priority, designing solutions, and testing

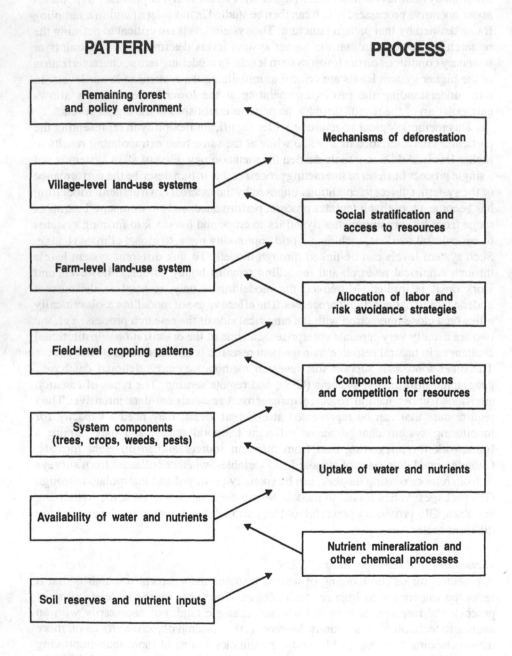

PATTERN

PROCESS

- Remaining forest and policy environment
- Mechanisms of deforestation
- Village-level land-use systems
- Social stratification and access to resources
- Farm-level land-use system
- Allocation of labor and risk avoidance strategies
- Field-level cropping patterns
- Component interactions and competition for resources
- System components (trees, crops, weeds, pests)
- Uptake of water and nutrients
- Availability of water and nutrients
- Nutrient mineralization and other chemical processes
- Soil reserves and nutrient inputs

Figure 2 Pattern process at different system levels (*Source* Palm, Izac, and Vostı 1993)

such solutions in appropriate contexts. This section will explore briefly some of the challenges facing technology development in NRM research.

First, the technological output of NRM research is heavily biased toward improved management practices. The increased knowledge of organic-matter dynamics and the timing of nutrient release to plants (Seward and Woomer 1992) is embodied in the choice of farmers of organic-matter materials, in the manipulation of litter particle size, its incorporation in the soil, the timing of incorporation, and the placement relative to the plant. Selection of the most appropriate management practices will be both site and system specific, depending on the availability of organic-matter resources, soil type, temperature and rainfall conditions, and the crop being produced. Although understanding these processes allows the researcher to design appropriate management strategies across different sites, the real challenge is how to implement these strategies at the farm level. Given the variation in agroclimatic conditions across the whole of the farm populations, the fact that the new knowledge must be imparted directly to the farmer rather than embodied in inputs, and that such technologies heavily depend on system differences such as tillage technology, cropping patterns, and farm size, the "extension" will have to become much more sophisticated as to how recommendations are made. Even more important, some have raised questions about whether there is any diffusion potential for such technologies and thus, whether there are any potential economies of scale in the technology-transfer process.

Second, many technologies to address resource management problems fail to become reality until the technology is brought to scale. Technology per se is implemented at the field or farm levels. However, many of the resource management impacts are measured at higher levels, when the technology has been deployed sufficiently widely. Biological control of pests, reducing pesticide pollution, controlling deforestation, and decreasing sediment loads all imply a certain minimum scale of technology deployment, and often even a spatial and temporal pattern to the deployment. IITA's biological control of the cassava mealybug in Africa is only one example of the logistical and infrastructural requirements to rear and distribute sufficient predator populations to achieve the required level of control. The program was even comparatively easy here, since there was no need to change farming management of cassava cropping systems. Bringing a truly integrated pest-management (IPM) program to scale requires another order of cost and sophistication.

Third, system technologies such as integrated soil or pest management, watershed management, or the replacement of slash-and-burn cultivation systems are multiple-component technologies. Deploying such technologies at the farm level remains the singular challenge in their development. To date, these technologies have focused on the deployment of a singular component. Successful IPM strategies have focused on the predator in the case of the cassava mealybug or on reduced pesticide sprayings in the case of pest control in Indonesian rice systems. For soil management, technology deployment programs focus on fertilizer, contour strips, agroforestry, and green manures, but always as single components. The challenge remains how to integrate these components at the farm level in order to achieve the full benefit of such system technologies.

The characteristics of NRM technologies also highlight that socioeconomic "technologies" will be just as important as the management or biological components of NRM technology. Such social technologies include institutional reforms, land tenure and community control mechanisms to state regulatory institutions, economic policies and incentives, mechanisms that compensate for market failures or lack of market valuation, and methodological innovations in the deployment of management or biological technologies. The design of such socioeconomic technologies will generate its own research agenda, as well as significant problems in implementing them. NRM problems will necessitate a greater parity between socioeconomic and biological disciplines in agricultural research institutes than did commodity research. This parity requires the support role of socioeconomics in commodity research to be changed into a more basic and socioeconomic research agenda. The danger in this is the potential loss of multidisciplinary teams in addressing NRM problems. The organization and institutional arrangements for NRM research programs must prevent this loss.

The NRM technology "problem" will put pressure on the economic methodology to address these issues. The simple assumption that risk leads to profit-maximizing behavior will not be sufficient to deal with the complexities in farmers' decision making of resource management and the adoption of NRM technologies, as it was in the adoption of commodity technologies. These complexities will need more explicit study of farmer decision-making, rather than rely on theory only. How farmers manage their soils, handle pest-control options, use their land, or exploit common property resources, are embedded decisions. They are conditioned by decisions at a higher economic level, where farmers decide on their choice of production activities and techniques. Yet, the key to grasping these higher-level management issues involves understanding lower-level decisions on how farmers understand and manage the biophysical aspects of the resources. Researching farmer NRM decisions will require a closer integration of economic and biophysical processes. This introduces more location specificity to the analysis and requires data that is sufficiently disaggregated to match the economic and biophysical interactions (Antle and Just 1992).

In turn, there will be an increased dependence on empirical parameter estimates than on purely theoretical deduction in the behavioral modelling of resource management (Antle 1993). Moreover, the modelling of biophysical and economic systems within an optimizing framework has had to employ more complex methodologies such as control theory. In the process, the assumption that the modelling represents actual farmer decision making has become increasingly unrealistic, and the results are useful only for public policy formulation. It is, therefore, worth to underline that progress in the economics of agricultural resource management will depend as much on solid empirical work as on theoretical development, and maybe even more so.

Conclusions

Sustainable growth in agricultural production in the developing world in the next half century faces two interrelated challenges: to increase production sufficiently to keep pace with rapidly growing demand, and to reduce as much as possible the impact this production increase will have on the extent and quality of the natural resource base. Of course, agricultural research and agricultural policy have been dealing with the first objective for years, but the immense number of policy and research topics to be addressed in the second objective is becoming apparent only now. The sustainability agenda requires new conceptual and theoretical work, new methods for empirical research, an exponential increase in data needs, the search for different organizational structures. At the same time, there is substantial uncertainty in how this agenda will be implemented. The danger is that the understanding of the critical interdependence of the two objectives is being lost, and that in the process, the balance in the allocation of resources to the traditional and NRM agendas has been shifted too far to the NRM side. The NRM agenda can potentially absorb a vast amount of public resources, with little capacity for evaluating relative priorities or potential impact. However, it is clear that the imperative to triple food production in the next fifty years remains, probably under an increasing scarcity of resources. The current uncertainty—some would say muddle—must soon give way to a clear set of priorities and strategies that will lead to sustainable agricultural growth.

Acronyms

CGIAR	Consultative Group on International Agricultural Research
GIS	geographic information systems
IITA	International Institute of Tropical Agriculture
IPM	integrated pest management
LEISA	low external-input sustainable agricultural
NRM	natural resource management

References

Antle J M (1993) Environment, Development, and Trade between High- and Low-Income Countries American Journal of Agricultural Economics, 75 784-788

Antle J M, Just R E (1992) Conceptual and Empirical Foundations for Agricultural-Environmental Policy Analysis Journal of Environmental Quality 21 307-316

Binswanger H P (1991) Brazilian Policies that Encourage Deforestation in the Amazon World Development 19 821-829

Boserup E (1981) Population and Technological Change A Study of Long-Term Trends University of Chicago Press, Chicago, USA

Von Braun J, Bouis H, Kumar S, Pandya-Lorch R (1992) Improving Food Security of the Poor Concept, Policy, and Programs International Food Policy Research Institute, Washington DC, USA

Cicchetti C J, Wilde L (1992) Uniqueness, Irreversibility, and the Theory of Nonuse Values American Journal of Agricultural Economics 74 1121-1125

Coale A J (1974) The History of the Human population In The Human population Scientific American Freeman, San Francisco, USA

26

Crosson P (1993) Sustainable Agriculture A Global Perspective Choices 38-42

Crosson P, Anderson J R (1992) Resources and Global Food Prospects Supply and Demand for Cereals to 2030 World Bank Technical Paper no 184 The World Bank, Washington DC, USA

Ehrlich P R, Ehrlich A H, Daily G C (1993) Food Security, Population, and Environment Population and Development Review 19 1-32

Ehrlich P R, and Wilson E O (1991) Biodiversity Studies Science and Policy Science 253 758-762

Evans L T (1980) The Natural History of Crop Yield American Scientist 68 388-397

Fresco L O, Kroonenberg S B (1992) Time and Spatial Scales in Ecological Sustainability Land Use Policy March 155-168

Harold C, Runge C F (1993) GATT and the Environment Policy Research Needs American Journal of Agricultural Economics 75 789-793

Harmsen K, Kelley T (1993) Natural Resource Management Research for Sustainable Production In Report of the Joint TAC/CDC Working Group on Ecoregional Approaches to International Agricultural Research

Ho T J (1990) Population Growth and Agricultural Productivity In Acsadi G T F, Johnson-Acsadi G, Bulatao R A (eds) Population Growth and Reproduction in Sub-Saharan Africa Technical Analyses of Fertility and Its Consequences The World Bank, Washington DC, USA

Homer-Dixon T F, Boutwell J H, Rathjens G W (1993) Environmental Change and Violent Conflict Scientific American February 38-45

Horiuchi S (1992) Stagnation in the Decline of the World Population Growth Rate during the 1980s Science 257 761-765

Howarth R B, Norgaard R B (1990) Intergenerational Resource Rights, Efficiency, and Social Optimality Land Economics 66 1-11

Jennings P R, Cock J H (1977) Centers of Origins of Crops and Their Productivity Economic Botany 31 51-54

Kennedy E, Bouis H E, Von Braun J (1992) Health and Nutrition Effects of Cash Crop Production in Developing Countries A Comparative Analysis Social Science and Medicine 35 689-697

Khush G S, Toenniessen G H (1991) Rice Biotechnology CAB International, Wallingford, UK

Liniger H P (1992) Water Endangered Regional Water Degradation and Solutions mimeo Environment and Development Group, University of Berne, Switzerland

Loevinsohn M, Wang'ati F (1993) Integrated Natural Resource Management Research for the Highlands of East and Central Africa ICRAF, Nairobi, Kenya

Loomis R S (1984) Traditional Agriculture in America Annual Review of Ecology and Systematics, 15 449-478

Lynam J K (1992) Sustainability Challenges for International Agricultural Research Paper for Workshop on Social Science Research and the CRSPs University of Kentucky, USA

Lynam J K, Herdt R W (1989) Sense and Sustainability Sustainability as an Objective in International Agricultural Research Agricultural Economics 3 381-398

Lynam J K, Sanders J H, Mason S C (1986) Economics and Risk in Multiple Cropping In Francis C C (ed) Multiple Cropping Systems Macmillan Publishing Company, New York, USA

Markandya A, Pearce D W (1991) Development, the Environment, and the Social Rate of Discount The World Bank Research Observer 6 137-152

Mellor J W, Paulino L (1989) Food Production Needs in a Consumption Perspective In Davis K, Bernstam M S, Sellers H M (eds) Population and Resources in a Changing World Current Readings Morrison Institute for Population and Resource Studies, Stanford, California, USA

Meyer W B, Turner II B L (1992) Human population Growth and Global Land-Use/Cover Change Annual Review of Ecology and Systematics 23 39-61

National Research Council (1993) Sustainable Agriculture and the Environment in the Humid Tropics National Academy Press, Washington DC, USA

Oldeman L R, Hakkeling R T A, Sombroek W G (1990) World Map of the Status of Human-Induced Soil Degradation UNEP, Nairobi, Kenya

Palm C, Izac A-M N, Vosti S (1993) Procedural Guidelines for Characterization and Diagnosis Alternatives to Slash and Burn Project mimeo ICRAF, Nairobi, Kenya

Parry M L, Carter T R, Konjin N T (eds) (1988) The Impact of Climatic Variations on Agriculture 2 vols Kluwer Academic Publishers, Dordrecht, The Netherlands

Pingali P (in press) Technological Prospects for Reversing the Declining Trend in Asia's Rice Productivity. In Agricultural Technology: Policy Issues for the International Community, CAB International, Wallingford, UK.

Pingali P, Binswanger H P (1984) Population Density and Agricultural Intensification: A Study of the Evolution of Technologies in Tropical Agriculture. Discussion Paper Report no. ARU 22. The World Bank. Washington DC, USA

Rosegrant M W, Svendsen M (1993) Asian Food Production in the 1990s: Irrigation Investment and Management Policy. Food Policy, February:13-32.

Rubin E S et al (1992) Realistic Mitigation Options for Global Warming. Science 257:148-49, 261-266.

Ruthenberg H (1980) Farming Systems in the Tropics. Oxford University Press, Oxford, UK.

Salati E, Vose P B (1984) Amazon Basin: A System in Equilibrium. Science 225:129-138.

Seckler D (1993) Designing Water Resource Strategies for the Twenty-First Century. Center for Economic Policy Studies Discussion Paper no. 16. Winrock International, Arlington, USA.

Sen A (1981) Poverty and Famines: An Essay on Entitlement and Deprivation. Clarendon Press, Oxford, UK.

Seward P, Woomer P L (1992) The Biology and Fertility of Tropical Soils: Report of the Tropical Soils Biology and Fertility Programme. TSBF, Nairobi, Kenya.

Smil V (1991) Population Growth and Nitrogen: An Exploration of a Critical Existential Link. Population and Development Review 17:569-601.

Taylor L (1993) The World Bank and the Environment: The World Development Report 1992. World Development 21:869-881.

Tutwiler R N, Van Schoonhoven A, Bailey E (1991) Agricultural Growth and Sustainability: Conditions for Their Compatibility in Semi-Arid Tropics (Middle East). In Vosti S A, Reardon T, Von Urff W (eds). Agricultural Sustainability, Growth, and Poverty Alleviation: Issues and Policies. DSE, Feldafing, Germany.

United Nations (1992) Long-range World Population Projections: Two Centuries of Population Growth 1950-2150. United Nations, New York, USA.

Vitousek P M et al (1986) Human Appropriation of the Products of Photosynthesis. Bioscience 36:368-373.

Voss J (1992) Conserving and Increasing On-Farm Genetic Diversity: Farmer Management of Varietal Bean Mixtures in Central Africa. In Moock J L, Rhoades R E (eds) Diversity, Farmer Knowledge, and Sustainability. Cornell University Press, Ithaca, USA.

World Bank/FAO/UNIDO/Industry Fertilizer Working Group (1991) World and Regional Balances for Nitrogen, Phosphate, and Potash 1989/90 - 1995/96. World Bank Technical Paper 44. World Bank, Washington DC, USA.

World Commission on Environment and Development (1987) Our Common Future. Oxford University Press, Oxford, UK.

World Resources Institute (1988) World Resources 1988-89. Basic Books, New York, USA.

World Resources Institute (1992) World Resources 1992-93. Oxford University Press, New York, USA.

A case for setting common objectives for natural resource management

H. ZANDSTRA

Director General of the International Potato Center (CIP), Apartado 5969, Lima, Peru, and presently chairman of the subcommittee on sustainability and environment of the Consultative Group on International Agricultural Research (CGIAR)

Key words biodiversity, equity, natural resource management, productivity, research priorities, sustainable agriculture

Abstract

The payoffs from developing a more accurate objective for NRM are likely to increase as scientists enhance their ability to improve the accuracy of natural resource models and simulations Better definitions of objectives will also help policymakers and researchers work more effectively in widely different ecologies and with different natural resource user groups

This paper provides a brief review of several types of natural resources and the manner in which they are affected by change The relationship between population growth and environmental conservation is highlighted The paper also examines the beneficial effects that a better use of natural resources can be expected to have on income levels and on the ability to provide basic necessities for existing populations

It also explores the mechanisms available to achieve a sustainable use of natural resources Efficient use of natural resource is described as a necessary condition for sustainability The paper concludes that productivity and sustainability are by definition essential components of a common objective aimed at efficient use of natural resources It is also suggested that such an objective must provide for the protection of unique environmental niches and the biodiversity they contain The pursuit of equity vis a vis NRM objectives is also reviewed

NRM involves multiple objectives—sustainability, biodiversity, productivity, and equity Although pursuit of multiple objectives applies to all resource-use processes, the relative importance of any component objective will influence the response of users to different resource endowments To achieve NRM objectives, neighboring countries may have to recognize a shared objective, to avoid the "import" or "export" of environmental problems across their borders

Finally, the paper examines research needs for NRM and it describes priority areas that require special attention These areas of research include the development of methods to compare resource-use efficiencies of different resource options, the evaluation of different policy options, and methods for conducting participatory research and for resolving conflict

Introduction

Although Agenda 21 stresses the need for sustainable development, the Rio Summit took place at a time when structural adjustments were taking place in many developing countries. The increasing liberalization of markets for trade, finance, labor and land markets, which is a main feature of these adjustments, is accompanied by a diminished role of the state in public enterprises.

While a reduced role helps to avoid public-sector inefficiencies, it excludes lower-income groups from access to public and private financing. It also weakens any regulatory capabilities set up to protect natural resources, and it tends to widen the gap between the incomes of urban and of rural populations. A reduction in public

29

P Goldsworthy and F W T Penning de Vries (eds), Opportunities, use, and transfer of systems research methods in agriculture to developing countries, 29 - 39
© 1994 *Kluwer Academic Publishers*

sector intervention may be effective in stimulating economic growth, but protecting the environment and the growth of the economy at the expense of the poor cannot be considered sustainable development (Serageldin 1993). Structured adjustments must be accompanied by policies that support the sustainability of natural resource use and of rural production systems.

This paper discusses the scope, nature, and general characteristics of natural resource management (NRM) objectives. It focuses on the agricultural sector, which uses the most valuable parts of the earth's land resource. The importance of cross-sectoral planning and the influence of the policy environment on the attainment of NRM objectives is also discussed. The paper then examines the challenge presented by the information requirements of NRM research, involving spatial and temporal dimensions of climate, soils and land-use patterns. It also examines the special contribution of social science in providing methodologies that link field and policy issues at the community, regional, and national levels.

Natural resources

Natural resources come in many forms. Some are renewable, others are not. Some are physical (mineral or geological), others biological. Some can be recycled, others can only be consumed.

Maintenance of biodiversity is essential to the continuing ability of the world to respond to the biological and physical challenges that it is exposed to. For the same reason, we need species diversity in cropping systems and genetic diversity within a species, as represented by varieties and land races.

Changes in the natural resource base can be caused by pollution, by the accumulation or depletion of resources, and by the destruction of biodiversity. Natural events such as typhoons, forest fires, and the processes of geological formation also change the quality and distribution of natural resources. A study of these processes shows that the natural resource base is constantly changing, even if left undisturbed by man. Any NRM strategy must therefore take into consideration the artificial and natural processes of environmental change.

Biological factors may also play a part in determining the environment. The ability of a species to compete for resources frequently determines the environment for other organisms. Evolution continues to influence the fitness of species to adapt to changes in the physical and biological environments. Some of these changes are caused by human intervention. For example, many insecticides have a less adverse impact on insects that regenerate quickly than on the slower breeding predators. Similar changes caused by pollution, salinization, and increased ultra-violet radiation are occurring all the time. The introduction of dominant improved varieties has led to less genetic diversity within crop species.

NRM objectives

Since the publication of the Brundtland Commission report in 1987, the sustainable use of natural resources has become a widely accepted management objective. The sustainable use of natural resources is frequently translated to mean the sustainability of a specific resource use, such as agriculture or fisheries. While there are many definitions of sustainable agriculture (Ruttan 1991), the Consultative Group on International Agricultural Research (CGIAR) accepted the following:

"The successful management of resources for agriculture to satisfy changing human needs while maintaining or enhancing the quality of the environment and conserving natural resources." (TAC/CGIAR 1987)

The advantage of this definition is that it does not specify a process for using natural resources. Indeed, the sustainability of natural resources use is best defined in terms of keeping options open on how best to use natural resources for the well-being of the planet and its inhabitants (Zandstra 1993).

Recent emphasis on sustainability modifies the previous preoccupation with productivity as the major NRM objective. To many observers, the objective of increased productivity (simply to survive) contradicts the objective of sustainability. Although some reconciliation has taken place (University of Florida and Cornell University 1993), an uneasy disagreement in views among environmentalists and survivalists remains. In my opinion, however, both views are too narrow. NRM objectives must seek to improve the resource base and at the same time increase the benefit that people can derive from responsible use of resources. This combined objective, however, fails to protect unique habitats that are relatively unproductive but that harbor valuable plant and animal diversity. It could be argued that the objective should include provision to secure the unique ecological niches required for the preservation of biological diversity. In reality, there are multiple objectives for NRM. Thus, for example, equity may be another objective in addition to sustainability, productivity, and maintaining biodiversity.

The scope for pursuing a given objective is another consideration. Environmental problems are currently imported and exported. Rich countries are polluting their environment often at the cost of the poor countries (through e.g., global climatic effects), while at the same time poor countries must deplete theirs to survive. Arriving at common objectives for managing natural resources is therefore an important means of actually achieving an objective.

While this may be so in theory, applying common objectives can be difficult to achieve in practice. Siamwalla (1991) asked whether the cost-increasing application of environmental standards is sufficient justification for "greener" countries to introduce import tariffs against other countries that implicitly subsidize exports by ignoring environmental consequences. The experience with the Common Agricultural Policy within the European Community (EC) and beyond is a case in point.

The importance of improving the natural resource base becomes evident when we consider the relationship between the quality of the resource and the carrying or production capacity of the land. A high-quality resource base provides greater choice and greater returns to additional labor and material inputs (De Wit 1992). However, when high extraction rates cause natural resources to be degraded, incomes drop and the resource base continues to decline. This, in turn, frequently results in a downward spiral in human welfare and in the type of resource depletion that is common in poorer countries (Murqueito 1992). Poverty and high birth rates cause production demands to outgrow the sustainability of an increasingly weaker resource base. The effects of this phenomenon on the natural resource base are well documented (Lipton 1991, 1985).

Mechanisms for achieving sustainability objectives

Unfortunately, the public often considers high productivity and sustainability to be mutually exclusive goals. The industrialized countries, which are all too aware of the detrimental effects on the environment of too many inputs, have come to doubt the benefits of being able to access resources. And they are often unaware of the devastation that a lack of inputs can cause (figure 1). For instance, crops need nutrients to produce vigorous ground cover. If they are deprived of nutrients, serious soil erosion may result. Similarly, the lack of soil phosphorus limits root development of vegetation in the Sahel, and can lead to the ineffective use of scarce rainfall.

The most sustainable resource-use system (corresponding to the lowest portion of the curve in figure 1) is therefore one that avoids the limitations imposed by the deficiency or excesses of one factor on the efficiency of another. The efficiency in the use of water increases with improved crop nutrition, thus the efficiency of use of nitrogen and water is adversely affected by phosphorus deficits.

A rational strategy for NRM is to maintain or improve the natural resource base and the balance among its components. In a properly functioning land-use system this often implies that the natural resources replaced through natural regeneration or

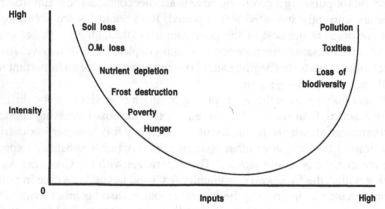

Figure 1. The occurrence of different causes for reduced sustainability at extremely low and high input levels in agricultural systems

inputs must equal or exceed those removed by the natural resource use system. An exception is the removal of toxic elements from the system.

A balance among the different components of the natural resource base and a balance between the rate at which resources are removed and returned to the system can be achieved by employing biological processes, by recycling non-renewable resources, or by substituting non-renewable resources with renewable ones (e.g., ethanol for fossil oil). Because resource flows occur between fields, communities, watershed components, and regions, quantifying these flows presents complex problems. Methodologies for measuring resource flows need to be improved.

Within the agricultural sector, the development of improved methods will require a more thorough knowledge of production systems, including a better understanding of resource transfers between crops and between crop and livestock enterprises. We will also need a better catalogue of water availability, soil physical and chemical characteristics, temperatures, and day-length regimes. In addition, the resources required for different enterprises must be better defined. Opportunities for substitution of enterprises (much influenced by food storage, processing and transport capabilities) must be explored to satisfy market demands for different products in different production environments.

Efforts to balance resource flows have often been unsuccessful, usually because of the imperfect pricing of production factors. Polluters do not bear the cost of pollution control in relation to the degree to which they cause pollution. In a market with overpriced crops, pollution is often caused by the excessive use of inputs to produce those crops. This is most evident in the EC and Japan, where subsidies have led to excessive use of pesticides (Netherlands Scientific Council for Government Policy 1992).

Conversely, the price of the crops leaving the Andean region is well below what it takes to replace their "natural resource content" and still provide appropriate returns to labor and cash required for their production. This leads to the mining of resources common in mountain agriculture. Sustainable land use would be better served if fertilizer application were universally set at the point of maximum response per unit input and at a minimum sufficient for the maintenance of soil fertility (CGIAR 1993). It is evident that the measurements and prices used in assessing NRM strategies must take into account all the resources used.

NRM objectives in the agricultural sector

Agriculture is the largest user of land resources and a major source of environmental concerns. Land areas dedicated to agriculture are usually heavily populated and they are the principal source of soil erosion and water pollution. It is therefore essential that environmentalists focus their attention on agricultural regions as well as relatively unused natural resource units such as forests.

The unit of production and the time scale in which sustainability concerns and input-output balance are measured merit careful consideration. Transfers of crop nutrients between fields or within rotations are an accepted means of optimizing resource use. In mixed farms, transfers between crops, livestock and tree enterprises have to be taken into account. The same applies to transfers between farms or regions.

Land-use systems are generally quite complex, often involving many different commodities and enterprises. They vary in their response to changes in natural resource endowments and to economic and social factors brought about by policy changes (Zandstra et al. 1981). There are interactions among enterprises and com-modity systems (e.g., through substitution and through the effects of crop rotation) and among agroecologies (e.g., through differences in harvest date and the effects on the incidence of pests). For purposes of planning and monitoring sustainable resource use, a number of different scales need to be considered. These include field or production system, farm, community, watershed, sub-national, national, and regional scales.

Time is also a concern. Given the dynamic nature of events that influence resource transfer and transformation, conclusive documentation on the impact of land-use systems is difficult to achieve in a short time. For example, changes in the level of accumulation of toxins from irrigation water that has been affected by mine spills, or a gradual shift in nutrient levels, may be hard to measure over a short period because they occur slowly. Computer simulation of the consequences of the different land-use systems can be a valuable instrument to predict the long-term effects of changes in resource use patterns.

Time also separates those who bear the costs of NRM and those who reap the benefits. Some of these costs may be indirect, for example in the form of lower profits from producing with fewer chemicals, but they can also be the direct costs of improvement of land productivity. The benefits of both kinds of costs may only provide returns after many years and they may accrue to individuals other than those who incurred the cost.

To achieve a more sustainable and efficient use of natural resources, interventions must take place at the farm level. This may require changes at the field level (e.g., land preparation, straw handling, use of hedge rows or waste utilization), at the plant level (e.g., acid-soil tolerance, phosphorus-use efficiency, pest resistance), or at the animal level (e.g., product choice, feeding system, waste recycling).

The scope and opportunity for such interventions are often determined by the access to community resources, by input costs and product prices, and by established patterns of waste management. These factors are in turn determined by the intrinsic characteristics of the resource-use process, such as the intensity of labor use and the natural resource requirements. In addition, local traditions and policies strongly influence technological opportunities for improving efficiency and sustainability.

Cross-sectoral planning

Land types and natural resource units differ greatly in the comparative advantage they represent for different natural resource use systems. The efficient use of resources depends on the selection of an appropriate enterprise mix for any one type of land. The best match between natural resource endowments and resource use is more likely to be achieved where there is cross-sectoral land-use planning. The planning process needs to compare the advantages of different agricultural enterprises, as well as non-agricultural land uses such as forestry or tourism. The value attached to biodiversity should be included in this kind of assessment for it is important not only to protected conservation areas, but also to systems that combine agricultural activities conservation areas, and eco-tourism. Rural areas can benefit substantially from the tourism industry. If the benefits can be combined with additional benefits derived from measures to maintain biodiversity, then together they may compete economically with alternatives in which only a single objective is pursued. It is the total benefits of a particular land-use system that count.

However, social and economic objectives must be considered, as well as the physical and biological aspects of resource-use efficiency. Equity consideration may take precedence over other objectives and may therefore modify land-use options which provide greater efficiency and sustainability. These are policy choices, and clear policy guidelines are required to protect long-term sustainability.

A hypothetical example of an equity objective would be to favor the production of potatoes in the mountains of Peru rather than in the coastal regions, by imposing planting restrictions on the coast or price subsidies in the mountain regions. These measures would raise the cost of potatoes for urban consumers, at least until a road system for quick transport from the mountains is developed. Such a policy would benefit the poorest areas of the country (equity and preventation of migration objectives) and increase the likelihood that farming practices in the mountains will evolve towards a more sustainable management of land resources.

Policies that change land-use patterns can be formulated to have an effect at an international level, through import restrictions or subsidies, and at national or provincial levels, through price manipulations. At the local community level, ownership patterns and decisions about the use of common resources, such as water, play an important role. Policies formulated at the community level usually address a wider range of issues than those directly associated with agriculture. Social and economic policies strongly influence resource-use sustainability. For example, access to education, the availability of health services, family-planning programs, the state of roads and other infrastructure, and the regulation of human settlement, often affect the way people manage land. Therefore, when the relationship between natural resource use and policies is at issue, it is necessary to consider the whole policy system in the analysis. This does not mean that policies directed toward education or equity are necessarily the origin of the causes of resource degradation or that they should be employed to achieve sustainability objectives. It is more effective to deal with resource issues through resource policies and not through policies dealing with the externalities that influence the resource-use process (see for the case of trade policies Siamwalla 1991 and McCalla 1991).

Research for NRM

To identify where particular production systems can be located most safely and where they have the greatest comparative advantage, it is important to compare their performance in terms of efficiency and sustainability in different ecological situations. This requires knowledge of the resource requirements of the production system over time. It also requires reliable data on the characteristics of the possible ecological situations with respect to land, water, and climate as well as existing production methods, product prices and demands. As discussed earlier in the section on NRM objectives (p. 31-32), many factors operate at different levels in the decision-making hierarchy. To avoid unnecessary data collection, careful thought should be given to what minimum data set is needed in land-use planning. An outstanding example of such work is "Ground for Choices," a study on goals and alternatives for land use in the EC, conducted by the Netherlands Scientific Council for Government Policy (1992).

Priorities for research should be set on the basis of the kind of interventions in land-use systems that can be anticipated. In this respect, much is known about the soil-plant interactions and the physical and chemical processes that determine the productivity of a production system. The challenge appears to be in putting together this knowledge in a way that allows the performance of a production systems to be compared under different resource situations and under different policy constructs. More work is needed to integrate what we know about the relations between technical, social, and economic factors at different levels of system hierarchy and how they affect the productivity of the systems. Some of these relations are illustrated in figure 2. For example, decisions that are made about a community resource (e.g., an irrigation reservoir) can hinder the adoption of a more sustainable or productive land-use system. In one particular example, village leaders had to be persuaded to abandon their practice of using water for irrigation only when the reservoir had been filled to the spillway. Once they had been persuaded, it was possible to introduce a modification to their normal practice, by substituting direct seeding of rice in place of transplanting. This change in practice resulted in an increase in the water-use efficiency of the rice crop and, in addition, it often enabled the villagers to plant a legume crop after the rice.

The introduction of NRM concerns into the agricultural research agenda gives rise to questions of methodology (Zandstra 1993). Farming-systems research methods offer much in terms of understanding the performance of existing production systems and the behavior of new technologies. Such farm-level research should include regional comparisons of the performance of different land-use systems. These comparisons must be made over a period that is sufficiently long to permit measurement of the effects of different technologies on natural resources. The results must be viewed on a geographical scale that will reveal policy aspects of land-use choices.

The broader research agenda has to pay particular attention to agronomy and soil-science issues, including nutrient cycling, maintenance of soil productivity, management of soil organic matter and its interaction with inorganic fertilizer (in

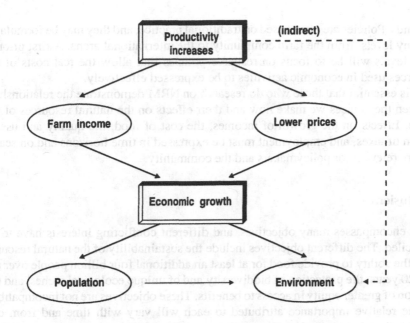

Figure 2. Schematic representation of the relation between population, productivity (income), and environmental quality

particular phosphorus), and appropriate use of land modifications for water and nutrient management. Other areas that deserve special attention are the application of integrated pest management, and multiple-cropping systems in which use is made of the complementarities between crops with different canopy structures.

The contribution that social sciences can make to NRM is even greater than to farming systems research. A better understanding is required of community decision making. This calls for more participation by the community in setting the research agenda and evaluating the outcomes. As the world population continues to increase, resolution of the issue of common-property management becomes progressively more urgent, and more work is needed on conflict management. The consequences of specific uses of natural resources will no doubt give rise to conflicts similar to those that occur commonly in association with water rights. Conflicts between short-term benefits or political expediency and the long-term maintenance of natural resources will have to be resolved.

Supporting policies

Appropriate policies are powerful instruments for changing the ways in which we use land and other resources. They are also difficult to formulate, because there are often conflicts between urban and rural interests, between the interests of producers and sellers, and between people who live in the watershed and those who are in the lowlands. Policies are often based on traditional practice, and they may be formulated at many levels, from the farm community to the international arena. A first priority at all levels will be to focus on resource policies that allow the real costs of the resources used in economic activities to be expressed effectively.

It is essential that those who do research on NRM demonstrate the relationships between the choices we make now and their effects on the natural resources of the future. Effects on the growth of incomes, the cost of food, the quality and use of burden of taxes, and employment must be expressed in time horizons and on scales that are relevant for policymakers and the community.

Conclusions

NRM encompasses many objectives, and different conflicting interests have to be reconciled. The different objectives include the sustainability of the natural resource base, the ability to provide food for at least an additional four billion people over the next 20 years, the protection of biodiversity and of unique ecological niches, and the creation of greater equity in access to benefits. These objectives are not incompatible, but the relative importance attributed to each will vary with time and from one ecological region to another. They will have to be applied as widely as possible to avoid the import or export of environmental problems.

Management choices must take into consideration induced as well as natural courses of change in the natural resource base. In this respect, measures to enhance naturally occurring factors that favor the long-term productivity of natural resources merit the attention of researchers.

Studies of NRM involve consideration of resource use at different levels (the plant, crop, farm, community, watershed, national, and regional levels), and the interactions between them. The performance of different systems for using natural resources must be understood in terms of the long-term effects on the resource base and the total productivity of the resources used. This understanding should begin at the highest possible level of natural resource use (e.g., an agroecology or land type) and should be of a form that will allow the knowledge to be extrapolated to other geographically separate but similar environments, where appropriate.

The main research challenge is the development and application of appropriate decision support methods for NRM. Methods are needed to compare the sustainability and productivity of different resource-use systems in different situations. The methodologies required would provide an understanding of the cross-sectoral relationships within land-use systems and the probable effects of the prevailing policy environment. They would provide an analysis in which values are attributed to all resources used in the management system.

Research results should represent time and spatial dimensions that are relevant to the needs of policymakers for economic and natural resource planning. High priority should be given to national and local resource-use policies that reflect the real costs of production, including the social costs of undesirable environmental consequences.

While some biological and physical factors merit particular research attention, the greatest challenge appears to be to find ways to deal with the complexities of resource management at different levels within any system of resources use. Measures are needed to define minimum datasets and standards for data collection for use in models that predict over long-term periods, the performance of resource-use systems for a range of environments and for different assumptions about the economic and social policy environment.

The social sciences can contribute greatly by contributing to more effective participatory research methods and with approaches to resolving conflicts over natural resources at the community, national, and international level.

Acronyms

CIP   Centro Internacional de la Papa
CGIAR   Consultative Group on International Agricultural Research
EC   European Community

References

CGIAR Secretariat (1993) Task Force Report. CGIAR Response to UNCED's Agenda 21. International Centers Week, Washington DC, USA.

Lipton M (1991) Accelerated resource degradation by third world agriculture: Created in the commons, in the west, or in bed? Pages 213-241 in Agricultural sustainability, growth and poverty alleviation: Issues and policies. DSE, IFPRI, Feldafing, Germany.

Lipton M (1985) Coase's theorem versus prisoner's dilemma, in Matthews R C O (ed.) Economy and democracy. Macmillan, London, UK.

McCalla A (1991) Commentary on The relationship between trade and environment, with special reference to agriculture. Pages 116-117 in Agricultural sustainability, growth and poverty alleviation: Issues and policies. DSE, IFPRI, Feldafing, Germany.

Murqueito E (1992) Sistemas sostenibles de producción agropecuaria para campesinos. En Agroecología y Desarrollo. CLADES. Número Especial 213. Santiago, Chile.

Netherlands Scientific Council for Government Policy (1992) Ground for choices: Four perspectives for the rural areas in the European Community. The Hague, The Netherlands.

Ruttan V (1991) Sustainable growth in agricultural production: Poetry, policy and science. Pages 13-28 in Agricultural sustainability, growth and poverty alleviation: Issues and policies. DSE, IFPRI, Feldafing, Germany.

Serageldin I (1993) Agriculture and environmentally sustainable development. World Bank 13th Agricultural Symposium. Washington DC, USA.

Siamwalla A (1991) The relationship between trade and environment, with special reference to agriculture. Pages 105-115 in Agricultural sustainability, growth and poverty alleviation: Issues and policies. DSE, IFPRI, Feldafing, Germany.

TAC/CGIAR (1987) CGIAR priorities and future strategies. TAC Secretariat. FAO, Rome, Italy. 246 pp.

University of Florida and Cornell University (1993) Reconciling Sustainability with Productivity Growth: A Workshop. Gainesville, Florida, USA.

De Wit C T (1992) Resource use efficiency in agriculture. Pages 125-151 in Agricultural Systems, Vol 40.

Zandstra H (1993) Preserving the options. World Bank 13th Agricultural Symposium. Washington DC, USA.

Zandstra H, Price E C, Litsinger J A, Morris R A (1981) A methodology for on-farm cropping systems research. IRRI, Los Baños, Philippines.

Discussion on Section A: Agricultural sustainability and systems approaches

The concept of sustainability of agricultural environments, *maintaining the productivity of the resource base while meeting the resource needs of present and future populations*, encapsulates many different objectives, some of which have been referred to in these first two papers. These various and at times contradictory objectives pose problems in transforming the concept of sustainability into operational research activities and programs. In developing countries, sustainability includes two main goals: increasing productivity and conservation of the environment. The relative emphasis on production or conservation receives depends on the level of institutional development, the productivity of the economy, and societal values. The industrialized countries, burdened with food surpluses, are in a position to place increased emphasis on conservation. Given the divergent goals, implementing Agenda 21 is a particularly difficult task for national research systems in developing countries.

The papers refer to systems and systems levels, but at this point a brief explanation is needed of why systems approaches are essential in addressing sustainability and environmental concerns in agriculture. First, these approaches serve to identify the agroecological production systems that are characteristic of different environments, including the social, cultural, and economic components of those systems. Agroecological systems theory distinguishes a hierarchy of system levels, and therefore serves to clearly define the geographic scale and the temporal dimension of the problem to be solved. Second, systems approaches call for greater attention to the relationships between production and the environment, with an inventory of the environmental resources and better understanding of how they are used now, and why. The purpose of an agroecological approach is also to increase knowledge of the potential uses of resources and of how they can be developed. Third, a systems perspective applied to NRM research reveals the different institutional issues that arise at the various systems levels. These issues include, for example, the need for interinstitutional linkages and which institutions are likely to have a comparative advantage to work at a given scale or to achieve a particular objective.

Intersectoral perspective
Linking research on agricultural productivity with conservation and the overall productivity of natural resources requires an intersectoral perspective. In most developing countries, agriculture is the most extensive and important user of land, water, and biological resources. As such, it impinges on other resource users and sectors. The use of natural resources in agricultural production must therefore be weighed against non-agricultural uses. By focusing on the relations within agroecological systems, research aimed at increasing productivity can also include NRM

41

objectives. However, there are important institutional and policy issues that arise when implementing such an approach.

The origins of many environmental problems in agriculture are cultural, social, and economic, not technical. Population growth, pricing policies and subsidies, and land use planning are often the root causes. For example, failure to account for the true environmental costs of chemical fertilizer and for the cost of fertilizer subsidies has favored fertilizer-based technologies over others. There is therefore a need for NARS to understand the policy environment, and to advise governments when a need for policy changes is indicated. But solutions to some of the problems are likely to come from measures taken to develop non-agricultural opportunities (e.g., reforestation of hillsides and development of ecotourism) and changes in economic development policies. While agricultural research institutions cannot bear the main burden for solving all the NRM problems that fall within the domain of agriculture, a systems perspective can be used to clearly delineate the levels at which problems can be tackled, and to identify the key areas in which agricultural research could take the lead. NARS can make important contributions by identifying the prospects for alternative land-use technologies as well as improving the productivity and resource efficiency of existing production systems.

The research balance between environment and agriculture
The move by NARS into NRM research is not without risks. NRM research can be a costly and uncertain endeavour, particularly for research agencies already overwhelmed with the demand for increased production. Demonstrating the impact of research may be more difficult given the long time frames and scales at which improved NRM can be measured. Increasingly in developing countries, sustainability and productivity are seen as complementary research objectives. The success of one depends to some extent on the success of the other. There is in fact a great deal that remains to be done to improve the efficiency of resource use in agriculture. The use of fertilizers, water, organic residues in soils and as by-products are areas where research needs to be done to improve total factor productivity.

Research on production and NRM have long-term and short-term objectives that must be balanced. To do so involves three tasks:
i) identifying the natural resources concerned and the consequences entailed in their use;
ii) measuring the costs and benefits over time of the resources used in production, as well as their impact on the environment, including the use of resources that are not priced or are undervalued;
iii) understanding why producers adopt or reject resource-management practices.
The final balance will depend on how society values immediate productivity gains relative to the conservation of resources for the benefit of future generations.

The role of systems approaches as an aid to policy decisions
and as practical tools for resource management

Supporting agricultural research policy and priority decisions: an economic-ecologic systems approach

S. WOOD and P.G. PARDEY

International Service for National Agricultural Research (ISNAR), P O Box 93375, 2509 AJ The Hague, The Netherlands

Key words agroecological zones, economic, GIS, maintenance research, priority setting, production system, research efficiency, systems approach

Abstract

This paper describes a quantitative, systems approach to agricultural research evaluation and priority setting The approach is designed to support (ex ante) resource allocation decisions at the research-program level To achieve this support, it links technical research parameters derived on an agroecological basis to a multi-market economic model to estimate the social benefits of research-induced technological change The direct effects of research may be exhibited in the form of yield increases, cost reductions, and natural-resource impacts

The economic-ecologic approach can be extended to allow explicit modelling of research effects at the production level This will enhance our ability to deal with research effects that exhibit a large degree of spatial variability It will also be possible to make more effective use of conventional biophysical models (e g, crop growth, land evaluation, and soil-erosion models) to estimate the likely consequences of research

Introduction

The purpose of this paper is to review some of ISNAR's on-going efforts to improve and extend existing quantitative procedures for research evaluation and priority setting. These topics are receiving greater attention from both the international and national research communities as budgets continue to tighten and calls for increased accountability grow. At the same time, however, research is being asked to address a wider range of technical and economic issues. We focus here on two areas of this larger body of work: the systems aspects of the economic-ecologic analytical framework, and the relevance of that framework to addressing resource allocation concerns. The broader conceptual and methodological aspects are described most fully in Alston, Norton and Pardey (1994), while the evolution of the agroecological dimensions of the work can be traced through Davis, Oram and Ryan (1987), Davis (1991), Wood and Pardey (1993), and Pardey and Wood (1994).

The paper not only highlights recent progress in the concepts and methods of the economic-ecologic approach, but also considers implementation lessons learned through an on-going series of collaborative pilot studies with developing-country national agricultural research systems (NARS). Interaction with NARS policymakers, managers, and scientists has brought about significant improvements in operational procedures, and in estimating and utilizing research-related parameters. It has

P Goldsworthy and F W T Penning de Vries (eds), Opportunities, use, and transfer of systems research methods in agriculture to developing countries, 45 - 66

also helped ensure that continued system developments are driven by real-world priorities.

We begin by describing briefly the economic-ecologic approach, and discussing some current developmental and implementation issues, first from a NARS perspective, and then from a systems-analysis perspective. Finally, we review possible extensions to the analytical framework with particular reference to natural resource conservation issues.

Concepts, methods, and implementation tools

The economic-ecologic approach can be viewed as a hierarchical framework of concepts, methods, and implementation tools for evaluating research that supports research policy analysis, priority setting, and resource allocation. The framework seeks to improve the information base for, and to provide a structured way of thinking through, a range of research resource-allocation decisions. In addition to providing indicators of the more direct effects of research (e.g., increased yields and productivity gains) the approach also enables the principles of welfare economics to be used to generate money measures of the size and distribution of research benefits. In extending this evaluation framework to encompass the natural resource effects (both on site and off site) of new technology, the principles of welfare analysis are particularly appropriate since they constitute an established means of addressing the various external effects and intertemporal issues involved.

Within this (still-evolving) analytical framework are a number of methodologically integrated models of the various processes involved. Technology generation, transfer, and use are defined in terms of research-related parameters with values that can be estimated or reasonably approximated by quantitative means, in conjunction with the expert judgement of experienced research managers and subject-matter specialists. These parameters include likely research-induced increases in yields or reductions in the unit cost of production, research and development time lags, probabilities of research success, and adoption rates and ceilings. The nature and complexity of individual models largely reflect perceived priorities and the (still-significant) boundaries of our knowledge. The highest priority has been to translate the yield-increasing or cost-reducing effects of research into economic measures of benefit. However, increasing attention is now being given to improving the estimation of the likely yield and cost effects themselves, as well as broadening the scope of analysis to include other aspects of agricultural production. Specifically, much is being done and remains to be done to estimate the likely natural-resource consequences of new technology—with a potentially major contribution being made by agricultural systems analysis.

At the operational level, the focus has been on improving the quantitative estimation of research-related parameters, managing information, designing computer-based analytical tools,[1] and developing meaningful ways of presenting results. The framework already makes extensive use of geographic information system (GIS) technologies for the spatial analysis of agroecological zones, adoption clusters, and market regions, but the role of GIS will grow further if, as proposed in this paper, more formal models of the productivity and natural resource consequences of research are gradually taken on-board.

The economic-ecologic approach

The economic-ecologic approach to research evaluation was originally designed to transform the yield-increasing or cost-reducing effects of research (for a given commodity or group of commodities) into measures of economic benefit. The current evaluation process consists of four steps:
1. establishing the scope and objectives of the study;
2. defining spatial units;
3. estimating the local effects of research and then aggregating them to correspond to the spatial units used for the market analysis;
4. valuing the benefits of research.

Since research evaluation is best undertaken as an interdisciplinary exercise, the first three steps depend on close collaboration between research managers, scientists, and research analysts, while the first and final steps require the additional skills of an economist.

Step 1: establishing scope and objectives
The critical first step is to identify the commodity and geopolitical dimensions of the study and the overall market context. If resource degradation is of concern, it is also necessary to identify the specific degradation processes for which research effects are to be analyzed.

Step 2: defining spatial units
Usually supported by GIS technologies and digitized thematic maps, this step defines and delineates the appropriate spatial subdivisions of the overall geographical boundaries of the study. The level of disaggregation most appropriate for these spatial units is dictated by the study objectives, the geographical and commodity scope, as well as the anticipated level of spatial variation of research-related parameters, and, often, by the limitations of the data itself. The areas defined are:

- *Agroecological zones (AEZs)*. Areas with relatively homogenous agroecological characteristics from the perspective of the production consequences of new technologies.[2]
- *Adoption clusters*. Areas within which the adoption of patterns of new technologies are likely to be relatively homogeneous. These clusters are defined by a range of economic, institutional, and technical factors. In practice, mainly because of a lack of data, these areas generally correspond to administrative regions.
- *Market regions*. The spatial units that are considered the most appropriate for the market analysis. Typically, these are administrative regions such as provinces, for national studies, or countries, for regional and global studies.

The spatial overlay of AEZs and adoption clusters define the agroecosystem areas that we propose to use to explore, where appropriate, production effects at a more disaggregated level. Where data is available, it is also valuable to delineate existing production areas.[3] Using overlay techniques, the GIS can combine these various themes and, in the process, produce area (or, preferably, production) weights. These weights are used extensively in the aggregation of research effects and the disaggregation of research benefits.

While there are several approaches to delineating AEZs, each with its own strengths and weaknesses (Wood and Pardey 1993), all benefit from the use of GIS technology. Our approach has been to work with scientists to quantify the criteria that define the AEZs most appropriate for their current research programs. A GIS is then used to map those criteria—a process almost inevitably followed by several iterations of adjustment as scientists are presented with the spatial representation of their tabular definitions. In this phase, the shortcomings of the underlying agroecological data, the lack of detailed knowledge of the environmental characterization of plant and animal performance (and their various pests and diseases), and the recognition of arbitrary (physical or mental) limits to the geographical scope of research, are often revealed. This process is illuminating for all concerned and it results in the delineation of relatively few (6-8) aggregate commodity-specific zones, primarily on the basis of thermal and moisture regimes but also according to soil and physiographic characteristics, e.g., acid sulphate soils and swamps. Once the criteria are agreed upon, we prepare a set of revised maps before discussing with scientists the specific details of the likely effects of research—which are then elicited on a zone-by-zone, technology-by-technology basis. In cases where detailed agroecological zoning studies have already been made, this operation is often reduced to identifying the aggregations of existing zones appropriate to each commodity-research program. However, it is not uncommon for these, usually generic, zoning studies to lack key-agroecological criteria from the commodity-research perspective. In such cases, a joint process of both aggregation (of unimportant zones) and disaggregation (of important but overly broad zones) is required.

Using this GIS-centred activity at the very start of the research-evaluation process has proved useful on several counts:

- it establishes communication between scientists and research analysts at an early stage and produces a jointly-developed output;
- it appears to reassure scientists that the approach has a "practical" basis and will pay due regard to the natural as well as the economic factors that affect production responses to new technologies;
- it is, surprisingly, often a "new" way for some scientists to look at the potential ramifications of their work.

Moves, such as proposed here, toward using quantitative models to help assess productivity and natural-resource effects of new technologies will further accelerate the integration of GIS into the research-evaluation framework. The ability to model both on- and off-site research effects will be of significance to the possibility of performing this type of research evaluation at a more disaggregated level (e.g., for sub-programs and major projects). This ability could be supported by surface modelling and other terrain and hydrological analysis functions built into the current generation of GIS systems, in addition to the formulation of a spatial framework with links between conventional models of crop growth, soil erosion, sediment transport, and so on.

Step 3: estimating the effect of research (the K-factor)[4]

The economic approach to estimating the benefits from research requires an estimate of the shift in the industry supply curve induced by research (see appendix A1, p. 64 of this section, for a description of the basic model).[5] This in turn requires information on the variables that quantify the research production function; a function that relates the amount and composition of research resources to actual or anticipated yield increases or per-unit cost reductions, to research lags, and to probabilities of research success.

The size of the research-induced shift in the supply curve (the so called K-factor) is a crucial determinant of the total benefits from research. While measuring K is difficult enough in an ex post study of the effects of past research, there are obviously added difficulties in developing meaningful measures of K for research that is yet to be done. Instead of estimating changes in the unit-cost of production directly, analysts typically estimate the yield-enhancing effects of research and then translate them into measures of K.[6] There are three primary sources of data for estimating yield effects:

- historical experimental and industry data;
- elicited data;
- simulated data from a variety of crop simulation, land evaluation, or empirical yield models.

In most ex ante studies, the main measurement problem is to estimate the *maximum* shift of the industry supply curve corresponding to full adoption when research is successful (i.e., K^{MAX}), and then adjust for probabilities of research success and the time path of adoption to convert the potential maximum shift into a most likely *realized* shift in time period t (i.e., K_t). The fact that changes in *industry* yields or production costs, and not simply *experimental* yield changes, are required for evaluating the economic effects of research, underscores the important point that it is the economically optimum use of new technologies by farmers and others, and not the biologically optimum response to new technologies, that is relevant here.

Historical yield-trend data (disaggregated spatially if necessary) and experimental data on past and present research effects are invaluable for benchmarking estimates about the likely outcomes of new research. But knowledge of past research effects is seldom the only reliable indicator of future research outcomes—the more so if research is being directed into new areas (e.g., biotechnology and natural-resource degradation issues) for which past experience offers little relevant information. In this case, structured elicitation (by knowledgeable scientists) of these future effects, combined with information from various sources, including formal simulation models, can provide sufficiently reliable information on potential research-induced yield gains. Research across commodities, technology areas (e.g., plant breeding, agronomy, and pest and disease control), and locations, differs in supply-shifting effects. A structured disaggregation of the elicitation process into these various commodity, technology, and spatial dimensions can help improve estimates of the research effects, and provide insights to help identify and analyze resource-allocation options.

Once information on the potential yield changes, adoption rates, and so on have been generated, it still requires some work to convert that yield effect into an estimate of the reduction in the unit cost of research induced production. In the simplest case, the *absolute* reduction in costs per unit of output i in time period t is given by

$$k_{i,t} = (1-\delta_i)^t A_{i,t} \, p_i \, PP_{i,0} \, E(y)/\varepsilon_i \tag{1}$$

where:

$E(y)$ = the proportionate increase in industry yields (after allowing for farm-level optimization of the input mix and for any differences between changes in industry and experimental yields) that would occur if the new technology were fully adopted,

$A_{i,t}$ = the proportion the total industry (area or output) adopting the new technology,

δ_i = the geometric rate of depreciation of the effectiveness of the new technology,

p_i = the probability that research will successfully lead to a new technology that will increase yields by $100E(y)$ percent,

ε_i = the elasticity of supply, and

$PP_{i,0}$ = the current producer price per unit of output i.

If the new technology involves *changes* in purchased inputs (e.g., fertilizers, fuel, or pesticides) or a change in the use of allocatable inputs (e.g., land, or operator labor and managerial inputs) then allowances will also need to be made for them when estimating $k_{i,t}$ (see Alston, Norton and Pardey 1994 for details).

Step 4: valuing research
A multi-market, equilibrium displacement model is used to convert these annual supply shifts into price and quantity changes in each of the market regions. These price and quantity changes are then used to obtain a stream of gross annual research benefits (or losses) accruing to producers, consumers, and government.[7] By specifying appropriate planning periods and discount rates, standard capital budgeting methods can then be used to calculate summary statistics such as the net present value (NPV) or the internal rate of return (IRR) to research.

The client perspective

Types of decisions

Because these evaluation approaches explicitly model the impacts of changes in government policies (e.g., taxes or subsidies on inputs or outputs) on the size and distribution of the benefits from research, they also have a valuable role to play in making more informed agricultural-research policy decisions. Moreover, they can be used to evaluate the economic consequences of pursuing alternative research programs that differ with respect to various technical parameters. Commonly there are tradeoffs to be made between the time taken to complete research and the size of the yield gains or cost reductions coming from that research. Decisions about whether to do more short-term versus long-term research, or more adaptive versus basic research, can be systematically explored in terms of the economic consequences implicit in each of these choices.

In a priority-setting context, the types of decisions to be made include:
- how much resources to allocate to research (and extension) in total;
- how to allocate the total among different programs of research;
- whether to accelerate, slow down, or even discontinue existing programs; and
- whether to introduce new programs.

There are clearly a large number of alternatives that can be considered, and not all of these alternatives justify formal analysis. The typical approach is to benchmark the analysis by first evaluating the likely economic effects of the *current* research program, and then to explore the implications of changing the research resource base (e.g., increasing or decreasing the expenditures on a program by 10 percent) or of changing the broad orientation of a program (e.g., the emphasis given to breeding versus agronomic versus crop protection work). A structured use of this type of data will enable decision makers to extrapolate the data in an informed way to consider the alternatives that were not explicitly evaluated, as well as foster an economic way of thinking about this resource-allocation problem.

While many have a direct interest in these types of decisions and provide inputs to them, only a few individuals or small group of people are ultimately responsible for making them. Usually, decisions about the overall allocation of public funds to research are made at the ministerial level, while decisions about the broad commodity, locational, and problem emphasis are delegated to senior research managers. These managers, and their analytical support staff, are the direct clients for the type of research evaluation and priority-setting tools being discussed in this paper. Decisions about the allocation of resources within a program taken by project managers, scientists, and the like are not directly dealt with here.

Degree of detail

After the various clients have been identified, the appropriate level of details poses a separate question. Our immediate emphasis is on informing strategic decisions that affect the broad orientation of a research system over a number of years. We believe that formal procedures are less useful at a detailed, disaggregated, project level for at least three reasons. First, in terms of improved research resource allocation at the program level, the costs of fine-tuning might not justify the benefits. Second, measurement problems become increasingly important as the degree of disaggregation increases.[8] Third, "micro" managing creative endeavors such as research can be counterproductive; more detailed allocation decisions are probably best guided by well-structured incentive systems instead of interventionist, "hands-on" allocative mechanisms. The last is perhaps the most important consideration. Formal evaluation and priority-setting procedures should not be used as a basis for replacing ingenuity, serendipity, and scientific entrepreneurship with bureaucratic procedures. There is a wealth of informal evidence that a successful research program rests heavily on the spirit, imagination, judgement, and integrity of agricultural scientists who are allowed freedom of enquiry. The role of research evaluation and priority setting is to help determine the boundaries within which free scientific enquiry occurs.

To evaluate research in ways that inform decision making at the national, regional, and even international level, it is necessary that scientific judgement about the likely effects of new technologies be incorporated into the evaluation exercise at compatible spatial scales. GIS-based approaches are proving useful for delineating the likely boundaries of spatial variation in research-induced shifts, and for providing a consistent basis for aggregating those shifts to the appropriate market level. To make decisions about the relative merits of one research program versus another, it is neither sufficient nor (in most instances) necessary to have overly detailed, site-specific information on the productivity or natural-resource consequences of new technologies. As a practical matter, the scale at which the spatial data are assembled and analyzed should be compatible with (i.e., at least as disaggregated as) the type of allocation decisions being addressed.

While agricultural research policy and priority setting asks many interesting questions, only some more socially important ones warrant public support. The evaluation and priority-setting approaches discussed here seek to disaggregate the formal analysis to the level that is required to make informed strategic decisions. Of course we recognize that many important resource-allocation decisions take place at lower levels in an agricultural-research agency or system. Although for the reasons noted above we advise against pushing these formal evaluation and priority-setting methods too far in this direction, the economic way of thinking about the allocation problem is just as relevant at the micro level as it is at the macro level. Therefore, structuring the research- and market-related data for use in these formal evaluation exercises is likely to also be of direct benefit to scientists and others for planning and prioritizing more narrowly defined areas of research.

Implementation aIssues

Some significant conceptual and measurement challenges remain, not least in developing tractable procedures for estimating the benefits of non-production oriented research and non-commodity specific research, e.g., social-science research. There are also a number of difficulties in dealing with cross-commodity effects. Cross-commodity effects are encountered both on-farm (e.g., the effect on crops that are planted in rotation after new, more water/nutrient efficient varieties have been grown) and off-farm, where changes in equilibrium prices and quantities affect markets for other commodities that may well serve as substitutes or complements in production or consumption. Although there are "work-around" approximations and prototype algorithms to tackle many of these issues, periodic reappraisals of the conceptual framework itself and a continuous commitment to extending and improving the methodological components of the framework are clearly needed. This should ensure that the technology generation, adoption, application, and benefit continuum that is being modelled can become an increasingly more complete, yet still tractable, representation of the real-world problems that research decision makers face.

The question of resource commitment is also relevant. The minimum dataset for the type of analysis described here is presented in table 1. However, even this list masks the underlying analysis that often needs to be performed in, say, preparing commodity-specific AEZs (practically feasible only with access to GIS technologies), and preparing consumption data, which are seldom in a format comparable with production data. Furthermore, the table does not include the potentially large amount of data that could be required in the future to support a more quantitative assessment of the productivity and natural-resource effects of research using techniques other than informed-expert elicitation. While human-resource needs may be limited in

Table 1 Minimum dataset for ex ante economic surplus models with multiple markets and AEZs

Research evaluation variables	Variable disaggregated by					
	(Sub-) commodity	Technology type	Market region	Adoption cluster	Agroecological zone	Time dependent
Market-related						
Quantities produced and consumed	✔		✔			✔
Prices received and paid	✔		✔			✔
Supply elasticities	✔		✔			
Demand elasticities	✔		✔			
Research-related						
Research-induced supply shifts	✔	✔	✔[a]	✔[b]	✔	✔[c]
R&D lags	✔	✔	✔			
Adoption profiles	✔			✔		✔
Probabilities of success	✔	✔	✔			

Note The table excludes data requirements for explicitly modelling farm-level productivity and resource effects

[a] Calculated using production/area aggregation weights

[b] Calculated using potential research effects by zone and cluster-specific adoption profiles

[c] Calculated by the evaluation model after initial conditions are specified

quantitative terms, they would be substantial in terms of skills. A minimum core team to support this type of work would comprise an agricultural economist, a natural-resources/GIS specialist, and at least one, but preferably two or three, scientists whose areas of speciality most closely match the general focus of research within the agency, e.g., a breeder, an agricultural engineer, and an agronomist.

Skills are required not only in assembling and standardizing data sets in an intelligent manner, but also in formulating appropriate research and market scenarios. These activities must be carried out in ways that are consistent with the basic assumptions of the approach, while minimizing the constraints of its imperfect representation of reality. The final—and vital—type of skill is interpreting and presenting the results of the analysis in a way that is most revealing to the clients for whom it is intended. Analysts also need good access to managers and scientists across the whole spectrum of an agricultural research system.

The analyst perspective

The systems and their components

The processes described above can be grouped into four interrelated systems:
- research and development (R&D);
- technology diffusion and adoption;
- production;
- markets (figure 1).

The conceptual model views the R&D system as investing in financial and human resources to add to the current stock of knowledge (and new technologies) that in turn have productivity (plus distributional and security) consequences. The whole process is risky—the exact costs of research are unknown when the research begins, and the likely outcomes in terms of R&D lags, potential yield-enhancing effects, and so on, are also less than certain. The technology-diffusion and -adoption system includes the processes by which new technologies are transferred from the "experiment station gate" to the "farm gate" and the conditions that govern the eventual acceptance or rejection of those technologies. The production system focuses on the actual use of adopted technologies and the effects of this use. Until now we have concerned ourselves with only more narrowly defined productivity effects (based on, for example, changes in yields and the production costs incurred by farmers). However, the research community is under mounting pressure to develop technologies that mitigate the negative effects of agricultural production for example on the environment and on human health. This calls for an expansion of the existing conceptual framework, as well as the integration of additional analytical tools (as represented by the shaded area in figure 1). Finally, the market system translates the supply-shifting effects of research into measures of economic benefits (or losses) accruing to producers, consumers, and governments, under specified market conditions.

55

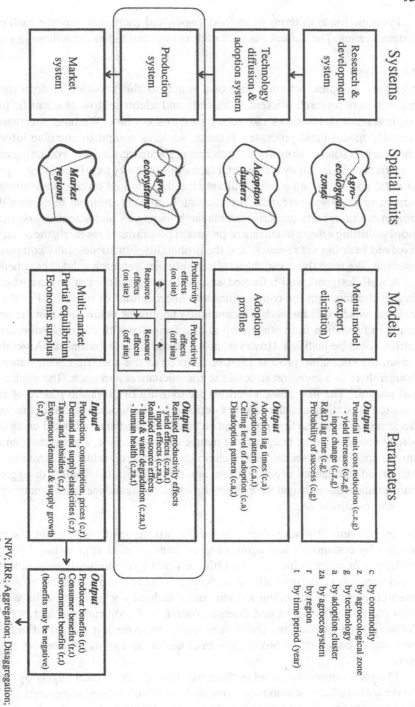

Figure 1. Systems overview of the economic-ecologic approach to the evaluation of research benefits

a Inputs additional to outputs from the three previous systems

Note: See table 2 for more details of production-related effects. The production system models are proposed extensions to the existing framework.

Progress made in terms of methodologies and models to analyze each of these systems varies. The current status of each is summarized in the following sections.

R&D system

R&D are complex activities that owe much to the knowledge, dedication, and inspiration of research workers, to a steady and adequate flow of scientific information and research resources and, sometimes, to sheer luck. We have not attempted to formally model these processes. Instead, we have sought to combine information from various sources to estimate the technical parameters used in evaluating research. Historical yield and, if available, total factor productivity trends, relevant experimental data, and so on, are used to calibrate the data elicited on the likely outcomes of current or proposed research. The elicitation process attempts to capture the best judgement of experts (experienced research managers and scientists) on the likely supply-shifting effects of current or proposed programs of research, the research lags involved in doing the research, and the probabilities of successfully completing the research. We term these variables collectively the "research-related parameters".

A well-designed, well-informed, and well-conducted elicitation is a cost-effective way of dealing with the complexities and uncertainties of research. For the spatial scales at which GIS methods are commonly applied, it is unlikely that the improvements in formation from attempting to deal formally with these measurement difficulties could be justified. However, as we have found in Indonesia, Argentina, and China, the elicitation process must be tailored to the specific circumstances, even though there is a common structure to the elicitation approach. The elicitation has two phases.[9] The first phase identifies the existing or proposed pattern of research resource allocation, and the second forecasts the likely outcomes from this research. Since results from the first phase can be cross-checked and validated using contemporary research plans, budgets, and manpower allocations, a far greater amount of time is spent on the second phase, eliciting the research-related parameters. Within this broad elicitation structure there are three additional issues: disaggregation, maintenance research, and the exploration of the consequences of alternative patterns of resource allocation.

Disaggregation. We have already noted the process of disaggregating the research domain by commodity and agroecological zone (and, if appropriate, by sub-commodities). In order to improve the estimates of the research-related parameters and to aid subsequent resource-allocation decisions, it is often also useful to disaggregate research programs according to their major technology foci. These have typically been plant breeding, pest and disease control, and crop management, although in China soil and fertilizer research was separated from the last category. For livestock, the parallel groupings of breeding, animal health, and animal husbandry have been used.

The most appropriate level of disaggregation is determined largely by balancing the desire to reduce measurement error in the research-related parameters against the need to keep the elicitation manageable, especially in its demand on key managers' and scientists' time. Typically, three to four technology groupings and six to nine AEZs for each commodity have been used to elicit disaggregated technical parameters.

Maintenance research. To measure research-induced economic benefits, it is necessary to estimate the yield or cost of production effects of research relative to what the yield or cost of production would have been otherwise. Not only does this provide the proper comparison for estimating economic benefits, but it also helps address the issue of maintenance research—the research required simply to mitigate the decline of productivity that might otherwise occur. The reasons for productivity decline are not fully understood, but some well-researched causes include the breakdown of resistance due to new pest and disease biotypes, soil erosion, and overly intensive (nutrient depleting and soil degrading) cultivation practices. Clearly, if the current on-going research program is expected to produce a 10-percent increase in yields compared with the current situation, and if without the new technology the yield of existing varieties would be expected to fall by 10 percent (over the same period), then the total yield effect of the research program is 20 percent, of which half can be considered maintenance research. Recent evidence suggests that around 30-70 percent of the agricultural research in the US is maintenance research (Adusei and Norton 1990).

Technology diffusion and adoption system
To model the technology diffusion and adoption system, a stylized adoption profile is usually generated for each commodity, and for each adoption cluster, and this profile serves to translate the elicited potential effects into a time series of realized effects of research. Whatever *S*-shaped diffusion curve is used—be it a logistic, a trapezoidal, or a Gaussian profile—the parameters required to estimate the adoption profile are similar: time to first adoption (adoption lag), adoption rate, and a ceiling level of adoption.

There are various methods by which the adoption profiles themselves can be derived (see Alston, Norton and Pardey 1994 and CIMMYT 1993). Historical data on the adoption of new technologies is patchy and sometimes of dubious quality. Nevertheless, they do provide a useful benchmark for estimating the likely adoption patterns of technologies currently being developed. Although the conceptual framework provides for spatial variation in adoption clusters independent from agroecological and market areas, the shortage of data usually dictates that adoption patterns are treated as uniform within a market region (which almost always corresponds with an administrative region).

Production system
Studies undertaken so far have not explicitly modelled the production stage. Instead, the potential effects of current or proposed research investment (given by the maximum shift in the industry supply curve, i.e., K^{MAX}), have been obtained by informed expert elicitation. This approach to estimating an industry supply shift

amounts to eliciting the joint effects of the R&D stage and the production stage (assuming complete adoption). In turn, the adoption profile is simply used to convert this maximum industry supply shift into a realized shift in each time period (reflecting incomplete adoption). This does not, however, capture the probable variation in K^{MAX} between adoption clusters within the same agroecological zone as a result of residual variation in production conditions. To take better account of such additional sources of spatial variation in the response to new technologies it may be useful to overlay relevant economic characteristics (e.g., as defined by adoption clusters) on the AEZs to form a set of *agroecosystems* that can be used to refine the spatial variability of K^{MAX}.

A corollary of recognizing an explicit production stage is that the elicitation processes may need to define only the experimental effects of research within an AEZ. A new model or group of models could then be applied at the production-system level to transform this benchmark experimental K^{MAX} into a series of K_A^{MAX}, representing the maximum supply-shifting effects for each agroecosystem. It is likely that such models already exist or, by simplifying or aggregating existing production models, could be readily developed. The ability to transform likely experimental effects into likely on-farm effects for each agroecosystem is not the only reason for explicit recognition of production. The reliability of eliciting effects only at the experimental level is probably higher. In the case of on-going research programs for which experimental results are already available, these elicited effects can also be validated. Yet another, and potentially important, reason is to provide an analytical point of entry within the overall framework to begin an exploration of the natural-resource effects of technology both on and off site.

It should be emphasized that we see the production-level analysis as a potentially valuable but optional extension to the current approach. The ability to analyze at greater levels of disaggregation requires an overhead of data, resources, and time, which may not be feasible or appropriate in all circumstances. Similarly, we do not necessarily see the type of productivity or natural-resource models used in this phase of the evaluation exercise as especially complex or sophisticated. Rather, we are attempting to develop a pragmatic capability to represent the ways in which research can have economically significant effects. Those effects, summarized in table 2, can encompass, for example, traditional yield-enhancing and cost-reducing effects, and effects on natural resources on and off site that may in turn affect productivity.

Market system

As the origin of the work (e.g., Schultz 1956, Griliches 1958, and Edwards and Freebairn 1981 and 1984) reflects, the market-related aspects of the framework have received the most methodological attention. The initial, single-commodity, closed-economy models (see appendix A1, p. 64 of this section) were extended to a horizontally disaggregated, multi-market (open-economy) framework that allowed for international and inter- regional trade in both goods and technologies. To address concerns of price- and technology-spillover effects across various levels of the production and marketing chain, vertically disaggregated models were then devel-

Table 2. Summary of the potential production-related effects of the application of new technology

Primary effects	On or off site	Commodity/resource scope	Affected attribute	Breed-ing	IPM	Irriga-tion	Soil fertility	Cropping systems	Remarks
Yields and production costs	On site	Base commodity — Main product (By-products, residues)	Growth-cycle length	1	2	2			Macagno (1990) for quality issues
			Market timing	2	1	1			
			Quality				2	2	
			Yields	1	2	2	3	2	
			Production costs	2	1	1	2	2	
		Other commodities	Yields		2	2	2	2	Multiple cropping effects in space and time
			Production costs			3	3	2	
			Quality			2	3	2	
	Off site	Base commodity / Other commodities							Off-site production effects depend largely on the off-site resource effects (this table) and factor market effects
Natural resources and human health	On site	Land and water degradation	Water demand			1	3	3	Affects on-site productivity
			Soil erosion			2	2	2	
			Soil fertility			2	1	1	
			Salinity and alkalinity			3	2		
		Human health	Human health		1	3	2		
	Off site	Land and water degradation	Water yields			1	✓	✓	
			Flooding			2	✓	✓	
			Sediment load & siltation			2	✓	✓	
			Water quality			3	2	✓	
		Human health	Human health			2	2	2	Affects off-site productivity
		Global warming	Greenhouse gases			3	3	✓	

* Intensity scale: 1 = primary effect; 2 = secondary effect; 3 = tertiary effect; blank = probably no significant effect

Note: Shaded area indicates those effects currently modelled within the overall conceptual framework

oped. Most agricultural markets face prices that are distorted by government tax and subsidy interventions, and these price distortions can directly influence the size and, in particular, the distribution of the benefits from research. Therefore, a natural extension of the earlier models was to explicitly include these aspects in the evaluation framework. More recently, economists have begun to move away from partial-equilibrium, single commodity models to consider the effects of research in a "general equilibrium" context in which the cross-commodity effects of research-induced price changes are explicitly considered.

Further extensions to the research evaluation framework
The non-market, or more specifically the external, effects of agricultural production and, related, the effects on future generations, are becoming increasingly important—at least as a policy issue in the more-developed countries. These concerns include all the "Green" issues such as environmental pollution, global climate change, sustainability, biodiversity, and human health risks. Concerns over environmental degradation, including deforestation, soil degradation, desertification, flooding, and pesticide pollution, have grown to the extent that programs of natural-resource management and conservation are gaining in support. However, there is no single concern that is all-important in all situations. While in the humid tropics, the relationship between production and soil erosion by water may be one of the greatest concerns, in arid regions problems of salinity and alkalinity build-up in irrigated areas may be more serious.

Natural resource conservation
Beyond the underlying issue of the extent to which resources directed to natural resource conservation (NRC) issues complement or substitute more traditional productivity-enhancing research, two specific questions typically arise:
- What proportion of available resources should be devoted to NRC research?
- Which aspects of NRC should be given more attention and which less?
Procedures for setting research priorities clearly must be capable of considering the effects of alternative research programs on the sustainability of the agricultural-resource base. The analyst now faces the problem of accounting for both the externalities arising from agricultural production and for the interventions (e.g., new technologies and/or institutional arrangements) designed to ameliorate these external effects. This is difficult conceptually and poses significant challenges at the empirical level. The issue of measuring research benefits in the presence of externalities has been largely ignored in the literature.[10] The problem of evaluating the benefits from research designed specifically to ameliorate such problems has also received little attention.

As shown in figure 1, our proposal to allow for an explicit modelling layer at the production stage will provide an entry point for the analysis of these natural-resource effects. However, the problems of what effects to model, and how to incorporate the information generated by such models into the overall framework, remain to be addressed. As to the question of what to model, this can only be reasonably established by the scope and objectives of the problem to be analyzed. There needs to be a prior consensus that, for example, soil erosion is, or is likely to be, significantly affected by new technologies. The corresponding measures of this potential effect must then be assembled for inclusion in the framework. The definition of "significant" in this context is also an issue; in the economic-ecologic framework, "significant soil erosion", for example, implies significance in economic terms. Although the off-site effects of soil erosion (e.g., water pollution) are often perceived as the most economically important (Swanson 1979 and McConnell 1981), an assessment of the economic costs of soil erosion in Java revealed that 80-93 percent of the costs arise from on-site (productivity-loss) effects and only 7-20 percent from off-site effects, such as the siltation of reservoirs, irrigation systems, and harbors (McGrath and Arens 1989).

There are three broad phases in quantifying the economic consequences of the natural-resource effects of research:

1. estimating the physical-resource effects of new technology, e.g., changes in the rate of soil erosion;
2. converting the physical-resource effects into economically significant consequences, e.g., the loss of soil productivity or the loss of reservoir capacity; and
3. valuing and aggregating the social costs and benefits of these consequences.

In translating this chain of analysis into operational procedures within the economic-ecologic framework, three needs arise:

■ the need to allow for additional levels of thematic spatial information that delineate areas homogenous with respect to specific NRC effects of new technologies (i.e., AEZs defined to include both production and NRC effects);
■ the need to elicit expected effects of research on parameters that govern the natural-resource effects of agricultural production; and
■ the need to broaden the basis of calculating unit-cost reductions, perhaps by creating a composite K-factor that includes all relevant market and non-market costs of production.

As an illustration, consider the case of soil erosion, in which we assume the use of the universal soil loss equation (Wischmeier and Smith 1978)[11];

$$A = R \cdot K \cdot LS \cdot C \cdot P \tag{2}$$

where:

A = the soil loss by water erosion (ton/hectare/year)
R = rainfall erosivity index
K = a soil erodibility index
LS = a topographic index that combines the influence of slope (S) and slope length (L)
C = a crop management factor
P = a factor that reflects soil management or conservation practices.

We can partition this formula into two parts: the first part ($R \cdot K \cdot LS$) can be viewed as a potential erosion hazard index (i.e., loss from bare soil), which will probably remain unaffected by research, while the second part $(C \cdot P)$ provides an opportunity to represent effects (intentional or otherwise) arising from the application of new production technologies. In this case, we could map the potential erosion hazard and then elicit from scientists the likely effect (if any) of their research on the C and P factors. From these elicited effects we could compute corresponding *changes* in soil erosion as a consequence of research, which could be translated into an anticipated change in productivity (e.g., Stocking 1984, FAO 1989). The resulting annualized impacts on productivity could, in the final stage, be merged into the overall calculation of the unit-cost effects of research for determining a corresponding stream of benefits (or losses).[12]

The extent to which the AEZs delineated for the purposes of eliciting K^{MAX} may also serve to delineate areas of relatively homogeneous natural-resource effects is open to question. If available in sufficiently disaggregated form, agroecological data can be reaggregated to provide different sets of zones to investigate productivity or natural-resource effects. In this context, the most generic agroecological database currently under development that may be appropriate for studying both productivity and natural-resource effects at a national and regional scale, is that of the Soils and Terrain Digital Database (SOTER) project (Van Engelen and Wen 1993). Indeed, proposals have recently been made to utilize these SOTER databases on a systematic basis for soil-erosion analysis in addition to their intended support of land-evaluation (land-productivity) applications (Van den Berg 1992).

Concluding comments

One consequence of the tightening of research budgets and the growing need to demonstrate an effective use of public funds is the surge of interest in more formal methods of research-impact analysis and research-priority setting. These challenging tasks are further complicated by the need to interpret past performance and to plan new initiatives in the context of a range of research objectives that transcend simple growth and stability into issues of equity and natural-resource conservation. Putting aside the fundamental question of whether agricultural research is the most efficient means at society's disposal for tackling many equity and environmental issues, this paper describes an analytical framework that can be used to do the types of research evaluation required. Some argue that the political and bureaucratic factors that help shape research-investment decisions, when added to the underlying technical complexities, negate the need for this type of structured, integrated, and essentially quantitative approach. We do not subscribe to that view. Any approach that fails to attempt to understand the likely range and intensity of research effects at both the physical and economic level cannot, in our view, contribute meaningfully to the process of allocating increasingly scarce research resources in an efficient, socially desirable manner.

The paper has focused on the environmental and economic components of the overall framework and the interface between them. It has also outlined the development path we are adopting—a path that includes strengthening the capacity to estimate the likely physical effects of technical change from both a productivity and a natural-resource conservation perspective. The first major source of strengthening this capacity is in better defining the spatial determinants of technology generation, adoption, and impact, i.e., better agroecological and/or agroecosystem characterization and zoning. The second source is the capability to interface with, or embed, suitably generalized productivity and natural-resource process models in order to make zonal estimates of research effects. Such estimates would augment, and allow extrapolation of, those estimates obtained from expert elicitation and experimental sites. The combination of the improved estimation of the magnitude of research effects with the improved understanding of their spatial variation, leads to a significantly better estimation of the subsequent price and quantity effects of technological change at the market level. This range of analytical outputs provides a solid basis for research policy and planning and strategic research-investment decisions.

We have also raised the issue of resource commitment. Our field experience to date suggests that implementation is challenging but feasible—the biggest single determinant being the degree of institutional commitment to finding better ways to inform decision making. A permanent, professional capacity to perform research evaluation requires a modest resource commitment. But we believe that the type of research evaluation described in this paper will need to become an essential task of many research agencies if they are indeed to succeed in doing more with less.

Notes

[1] ISNAR is developing a software package to support the ex ante economic surplus component of the overall framework. The "Dynamic Research Evaluation for Management" (DREAM) package, a combined database management and analytical system, is scheduled for release in 1994 to coincide with the publication of the Alston, Norton and Pardey volume.

[2] (a) "Homogenous" can often be interpreted as "known pattern of heterogeneity". For example, at one level, a soil mapping unit may be considered as homogenous, but at a more disaggregated level it may contain a number of sub-units, each with its own specific characteristics.

(b) The AEZ concept can also be applied to the identification of homogenous areas with respect to the natural-resource consequences of new technology.

[3] This is because new technologies may well change the spatial production pattern. For example, the development of more drought-resistant or saline-resistant crop varieties could promote production to be expanded into areas that were previously unsuitable for agriculture. Such spatial changes in cost structures brought about by research can have significant impact on inter-regional trade and, hence, can further affect planted area within each region.

[4] See Alston, Norton and Pardey (1994) for a more complete discussion of the conceptual and practical issues described briefly here.

[5] An industry supply curve represents the aggregation of the underlying farm-specific supply curves which, in turn, reflect farmers' economically optimal use of their resources.

[6] Although yield alone does not reflect all the research effects, it is nevertheless an extremely useful *summary statistic* that does reflect a diversity of research objectives. For instance, plant breeders have many objectives in breeding new cultivars, including improved yield potential, pest and disease resistance, tolerance of adverse environmental conditions (e.g., cold and drought), and a number of different grain characteristics that interact to determine "quality". Apart from the determinants of quality, virtually all of the other genotypic characteristics have their main impact through yields.

[7] The net revenues or losses accruing to government arise from the pattern of producer and consumer taxes and subsidies. The economic model allows these to be taken into account.

64

[8] A major difficulty of assessing research impacts at this (project) level involves apportioning observed (or predicted) changes in yields or unit-cost reductions to research-induced changes, in particular components of a technology package *while leaving other components of that technology unchanged*. Given the interrelationships inherent in many new technological packages, it is likely that spurious attribution, double counting, or both will result.

[9] Subsequent to the agreement of a set of spatial boundaries as described in the section on defining the scope and objectives of the study (p. 47).

[10] New technologies may exacerbate, diminish, or have no effect on these negative external effects of agricultural production.

[11] Anderson and Thampapillai (1990) and the references therein review some of the limitations in using this model.

[12] For example, as part of the δ_t term in equation 1.

13 Extracted (with minor changes) from Alston, Norton and Pardey (1994).

Appendix 1

The basic supply and demand (economic surplus) model of research benefits

In figure A1 the supply curve for a commodity before a research-induced technical change is denoted by S0 and the demand curve by D. The original price is P0 and the quantity supplied and demanded is Q_0. The total consumer surplus from consumption of the commodity is equal to the triangular area FaP_0 (the area beneath the demand curve less the cost of consumption). Similarly, the total producer surplus is equal to the triangular area P_0aI_0 (total revenue less total costs of production as measured by the area under the supply function). Total surplus is equal to the sum of producer and consumer surplus, as shown by the triangular area FaI_0, which is equal to the total value of consumption (the area under the demand curve) minus the total cost of production (the area under the supply curve). Changes in producer, consumer, and total economic surplus are measured as changes in these areas.

Cost-reducing or yield-enhancing research and adoption of the resulting new technologies shift the supply curve to S_1, resulting in an increase in production and consumption (from Q_0 to Q_1) and a reduction in price (from P_0 to P_1). The *change* in consumer welfare (surplus) from the supply shift is represented by the area P_0abP_1 and the *change* in producer welfare (surplus) is represented by the area $P_1bI_1 - P_0aI_0$.

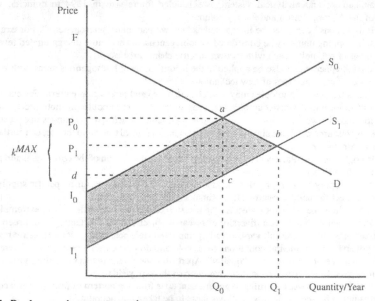

Figure A1. Producer and consumer surplus measures

Consumers necessarily gain because they receive more goods at a lower price In general, the net welfare effect on producers may be positive or negative depending on the supply and demand elasticities and the nature of the research-induced supply shift This is so because there are two effects working in opposite directions The producers sell more goods, but they must sell at lower prices Both costs and revenues are affected Producer benefits are assured if costs fall and revenues rise But under conditions that are plausible, in some cases (i e , an inelastic demand) revenue falls when supply increases In addition, when supply shifts in a pivotal fashion against an inelastic demand, revenue falls faster than costs, and producer losses are assured The nature of the supply shift can clearly have important implications for the distribution of benefits In the case drawn in figure A1, with a linear supply curve shifting in parallel, producers necessarily benefit (by an amount equal to area $P_1bcd = P_1bI_1 - P_0aI_0$) The total (or net) benefits from research is equal to the sum of the changes in producer and consumer surplus, I_0abI_1 (which, in this case of a parallel supply shift, is also equal to area P_0abcd)

Acronyms

AEZ	agroecological zone
GIS	geographic information system
ISNAR	International Service on National Agricultural Research
NARS	national agricultural research system
NRC	natural resource conservation
R&D	research and development
SOTER	soils and terrain digital database

References

Adusei E O and Norton G W (1990) The Magnitude of Agricultural Maintenance Research in the United States Journal of Production Agriculture Vol 3, No 1 (January/March) 1-6

Alston J M and Pardey P G (1993) Market Distortions and Technological Progress in Agriculture Technological Forecasting and Social Change Vol 43, No 3/4 (May/June) 301-19

Alston J M, Norton G W, Pardey P G (1994) Science Under Scarcity Principles and Practice for Research Evaluation and Priority Setting Cornell University Press, Ithaca, USA (forthcoming)

Anderson J R, Thamapapillai J (1990) Soil Conservation in Developing Countries Project and Policy Intervention Policy in Research Series No 8 World Bank, Washington DC, USA

Van den Berg M (1992) SWEAP, a computer program for water erosion assessment applied to SOTER SOTER Report No 7 ISSS-UNEP-ISRIC, Wageningen, The Netherlands

CIMMYT Economics Program (1993) The Adoption of Agricultural Technology A Guide for Survey Design D F, CIMMYT, Mexico

Davis J S, Oram P A, Ryan J G (1987) Assessment of Agricultural Research Priorities An International Perspective ACIAR Monograph No 4 ACIAR, Canberra, Australia

Davis J S (1991) Spillover Effects of Agricultural Research Importance for Research Policy and Incorporation in Research Evaluation Models ACIAR/ISNAR Project Paper No 32 (February), ACIAR and ISNAR, Canberra, Australia, and the Hague, The Netherlands

Edwards G W, Freebairn J W (1981) Measuring a Country's Gains From Research Theory and Application to Rural Research in Australia A Report to the Commonwealth Council for Rural Research and Extension Aust Govt Publ Service, Canberra, Australia

Edwards G W, Freebairn T W (1984) The Gains from Research into Tradeable Commodities American Journal of Agricultural Economics Vol 66, No 1 (February) 41-49

Van Engelen V W P, Wen Ting-Tiang (1993) Global and Natural Soil and Terrain Digital Databases (SOTER) Procedures Manual FAO, ISRIC, ISSS, UNEP, Wageningen, The Netherlands

FAO (1989) Assessment of Population Supporting Capacity for Development Planning in Kenya Soil Erosion and Productivity Working Paper No 4 Final Draft Land and Water Division, FAO, Rome

66

Griliches Z (1958) Research Costs and Social Returns: Hybrid Corn and Related Innovations. Journal of Political Economy Vol. 66:419-431.

Maredia M K (1993) The Economics of the International Transfer of Wheat Varieties. PhD Dissertation, Michigan State University, East Lansing, USA.

McGrath W and Arens P (1989) The Costs of Soil Erosion on Java: A Natural Resource Accounting Approach. Environment Department Working Paper No. 18, World Bank, Washington, D.C.

McConnell K E (1983) An Economic Model of Soil Conservation. American Journal of Agricultural Economics. (February):83-89.

Pardey P G and Wood S R (1994) Targeting Research by Agricultural Environments. Chapter 31 in Anderson J R (ed) Agricultural Technology Policy Issues for the International Community. CAB International, Wallingford, UK. (in press).

Schultz T W (1956) Reflections on Agricultural Production Output and Supply. Journal of Farm Economics, Vol. 38, No. 3:748-762.

Stocking M (1984) Erosion and Soil Productivity: A Review. Consultants Working Paper No. 1. Soil Conservation Programme, Land and Water Division, FAO, Rome, Italy.

Swanson E R (1979) Economic Evaluation of Soil Erosion: Productivity Losses and Off-Site Damages. Dep. Agr. Econ. Staff Pap. No. 79 E-77, March, University of Illinois, USA.

Wischmeier W H, Smith D D (1978) Predicting Rainfall Erosion Losses——A Guide to Conservation Planning. USDA Agricultural Handbook No. 537. USDA, Washington DC, USA.

Wood S R, Pardey P G (1993) Agroecological Dimensions of Evaluating and Prioritizing Research from a Regional Perspective: Latin America and the Caribbean. ISNAR Discussion Paper No. 93-15. ISNAR, The Hague, The Netherlands.

The role of systems analysis as an instrument in policy making and resource management

R. RABBINGE[1,2], P.A. LEFFELAAR[2] and H.C. VAN LATESTEIJN[1]
[1] Netherlands Scientific Council for Government Policy, P.O Box 20004, 2500 EA, The Hague,
  The Netherlands
[2] Dept. of Theoretical Production Ecology, P.O. Box 430, 6700 AK Wageningen, The Netherlands

Key words: aggregation level, decision making, interdisciplinarity, land evaluation, natural resource management, policy making, priority setting, simulation models, systems approach

Introduction

The use of systems approaches in agricultural research has increased considerably during the last decades. Within the field of systems approaches, scientists around the world have developed several different types of models. Conceptual, comprehensive, and summary models function to integrate multidisciplinary research at various aggregation levels. Some of these models merely describe biological and agricultural systems to generate an insight into the system itself. Summary models, based on these explanatory, comprehensive models, may help in priority setting in research. Summary models are also used in quantitative land evaluation, and may structure resource management research at high aggregation levels. Other models may help policy makers in their strategic and tactical decision making at various levels of integration and aggregation.

This paper will illustrate each of these applications of systems approaches and their application for the various purposes. We will describe some of the prerequisites and limitations of the various models and show the usefulness of systems approaches by giving some examples of appropriate use.

Systems approaches in agroecosystems

Agricultural research as well as biological research aim at understanding living production systems. Biological research aims to describe and understand basic processes, while agricultural research tries to gain insight into the ways various characteristics at the crop, cropping-system, or farming-systems level may be manipulated to improve production, both quantitatively and qualitatively. Agricultural systems are too complex to be investigated as a whole. Research therefore distinguishes between aggregation levels (e.g., region, farm, cropping system, crop level), subsystems (e.g., soil, plant, pathosystems), and subprocesses (e.g., transport processes, photosynthesis, energy balance, growth and decrease of populations) (figure 1). Processes and systems can be studied by analyzing and formulating their interactions and relationships with the environment in mathematical formulas (systems

67

P. Goldsworthy and F.W.T. Penning de Vries (eds.), Opportunities, use, and transfer of systems research methods in agriculture to developing countries, 67 - 79.
© 1994 Kluwer Academic Publishers.

68

analysis). Processes may be integrated by simulation models (systems synthesis), and the consequences of changes in the external conditions by human activity may be evaluated with integrative tools (systems evaluation). Subsystems, processes, and subprocesses usually operate on much smaller time and spatial scales than the system as a whole. The levels of aggregation at which subsystems operate may be distinguished with time coefficients as a yardstick. As a rule, the time coefficient rises with the aggregation level. Subcellular processes have time coefficients smaller than milliseconds, cells are recorded in the order of seconds, plants react in hours, crops in days or weeks, agroecosystems in months, and ecosystems have reaction times in the order of years (figure 2). The levels of aggregation with respect to their characteristic times may be distinguished with steps of 10 up to 100. Each aggregation level has underlying subsystems or processes that may be combined or even integrated to better understand the functioning of the higher aggregation level. The same hierarchical relation holds for spatial scales. Figures 1 and 2 illustrate the various biological processes that operate in plants and crops and the various disciplines involved in the study of these systems.

Constructing simulation models for systems research

Ten steps, divided over three phases, may be distinguished in simulation for scientific and applied purposes (table 1) (Rabbinge and De Wit 1989).

The first phase, the conceptual phase, comprises a clear description of the system, its various elements, and its relationship with the environment in qualitative terms. The second phase concerns the comprehensive modelling phase. Next, the evaluation and application phase, the third phase, comprises various steps and results in decision rules (when necessary) or forecasting models.

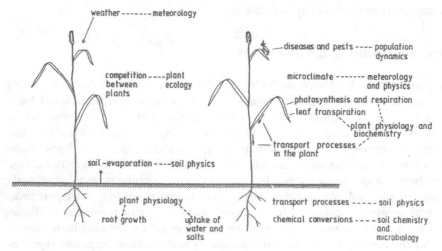

Figure 1. A crop system and some subsystems and processes that may be distinguished for its study (*Source:* De Wit 1982)

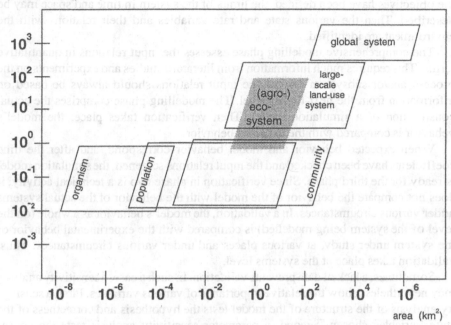

Figure 2. Spatial and temporal scales that may be distinguished in production ecological studies (*Source:* Rabbinge et al. 1993)

Table 1. Steps in model building

CONCEPTUAL PHASE	
1.	Formulation of objectives
2.	Definition of the limits of the system
3.	Conceptualization of the system

COMPREHENSIVE MODELLING PHASE	
4.	Quantification of input relations
5.	Model construction
6.	Verification of the model

EVALUATION AND APPLICATION PHASE	
7.	Validation
8.	Sensitivity analysis, feasibility studies
9.	Simplification, summary models
10.	Formulation decision rules or forecasting models

(*Source:* Rabbinge and De Wit 1989)

The conceptual phase starts with defining clearly the objectives of the study. After the objectives have been defined, the limits of the system in time and space may be described. Then the various state and rate variables and their relations with the environment are identified.

The comprehensive modelling phase assesses the input relations in quantitative terms. This requires much information from literature studies and experiments on the process and/or subsystem levels. The input relations should always be based on information from the next-higher level. The modelling phase comprises the actual construction of a simulation model. Then verification takes place; the model's behavior is compared with the expected behavior.

When expected behavior and model behavior correspond, and after the time coefficients have been checked and the input relations screened, the simulation model is ready for the third phase. Since verification in phase two is a technical activity, it does not compare the behavior of the model with the behavior of the actual systems under various circumstances. In a validation, the model's behavior as a whole (on the level of the system being modelled) is compared with the experimental behavior of the system under study, at various places and under various circumstances. Thus, validation takes place at the systems level.

Sometimes, a lack of data prevents validation. In such a case, a sensitivity analysis may nevertheless show the relative importance of various variables. First, a sensitivity analysis of the structure of the model tests the hypothesis and correctness of the state variables chosen. Second, a parameter sensitivity analysis tests the consequences of changes of (some of) the input relations. If the model validation and sensitivity analysis produce a reliable model, a feasibility study may follow. Next, the possibilities of the system are explored under circumstances that differ considerably from those used during model building. However, it is important that these circumstances stay within the limits of the structure, the parameters, and the time and spatial scales of the model; if they do not, results may be nonsensical.

The results of the sensitivity analysis may produce a simplified model, which may result in summary models that grasp the main features of the system. These summary models may be used in the tenth and last step to formulate decision rules or forecasting models. Examples of this may be found in population dynamical studies (Van Roermund and Van Lenteren 1994) and crop protection (Rabbinge and Rijsdijk 1983).

Many studies, however, end after step six. Steps are also sometimes taken in a different order. An iterative heuristic approach is often followed.

Systems research aimed at gaining understanding

Simulation models are used to bridge or connect two or at the most three aggregation levels (De Wit 1968). Quantitative knowledge at the underlying level is combined and used to understand the behavior of a system as an entity. By comparing model behavior with the behavior of the real world, e.g. by experimental research, hypotheses may be tested, thus enhancing the understanding of these systems.

The level of detail depends on the objectives of the study. For example, studies on soil aeration and denitrification (Leffelaar 1986, 1988) tried to understand the causes and results of denitrification in partially water-saturated soil. This objective requires an explanatory simulation model, interconnecting spatial and temporal scales of the organisms and transport processes involved. The spatial scale is in the order of one to ten centimeters, which is the level of soil aggregates, whereas the oxygen concentration varies from 21 to 0 percent within a few millimeters when the soil is water saturated. The time coefficients for the biological processes involved, such as respiration and denitrification, and for the physical processes, such as water flow and gas diffusion, ranged from 24 hours to 60 seconds. Therefore, a better quantitative understanding of this soil ecological system requires simulation models that use time steps of seconds and spatial units of millimeters. The information gathered through these detailed studies can then be summarized in new relationships and used in models at higher integration levels.

Another example of a model study aiming at explanation is the population dynamics of the larch bud moth (*Zeiraphera diniana*) in the Ober Engadin (a mountain valley in Southeast Switzerland) (Van den Bos and Rabbinge 1976). For more than 120 years, the larch bud moth has shown a regular population density cycle with a frequency of nine years. There is a 3000-fold difference between the maximum and minimum densities. The time coefficient in this system is in the order of days, while the spatial unit, the Ober Engadin, measures 40,000 ha. The appropriate time and spatial scales in this simulation model are therefore days, while the Ober Engadin is considered as an entity. Data for the densities of the larch bud moth were monitored by direct sampling during the last three decades and by observing tree growing rings during the period 1850-1950.

Although we could give more examples of different time and spatial scales, these two examples show the need to specify the objectives and the time and spatial boundaries of systems in explanatory studies that use systems approaches.

Systems approaches aimed at building a research agenda

Asking the right questions and doing the appropriate experiments are the most difficult parts of research. Experiments are expensive, time consuming, and pain staking, and should therefore be limited. Some pointers may help, however, such as "experiment only if you can't obtain your information in any other way", and "use an experiment only to answer an explicit question".

Suggestions such as these may generally help limit the number of experiments, but they do not help set priorities. Through feasibility and sensitivity studies, model calculations may help highlight elements of the process that need further elaboration. This is illustrated in a study on the effect of growth-reducing factors by Bastiaans (1993). Using comprehensive simulation models and summary models, Bastiaans showed the relative contribution of various growth-reducing factors on the growth and production of rice. Through this analysis, research priorities can be set in two steps (table 2). First, it appears that the relative contribution of the leaf folders is

Table 2 Simulated reduction in grain yield (Mg ha ^{1}) for rice crops infected by leaf blast (*Pyricularia oryzae*) and contribution of the various damage mechanisms Primary effects on radiation interception (RI) are caused by lesion coverage and leaf senescence The contribution of the various primary effects on radiation use efficiency (RUE) were determined separately Simulations were made for leaf blast epidemics with a maximum disease severity of 0 10 and various onset times (15, 29, and 43 days after transplanting (DAT) Bracketed numbers represent relative contributions

	Onset time of the epidemic		
	15 DAT	29 DAT	43 DAT
Total reduction due to leaf blast	1 8 (1 00)	2 8 (1 00)	3 3 (1 00)
Reduction due to primary effects on RI	0 9 (0 50)	1 7 (0 63)	2 3 (0 69)
Reduction due to primary effects on RUE	0 9 (0 50)	1 1 (0 37)	1 0 (0 31)
Contribution of the various primary effects on RUE			
- Reduced leaf photosynthetic rate	53%	48%	44%
- Increased leaf maintenance respiration	3%	4%	3%
- Assimilate uptake by the pathogen	44%	48%	53%

(*Source* Bastiaans 1993)

almost zero and that of the leaf blast is extremely high. Research should therefore be directed to the latter, while secondary pests such as rice borers should receive much lower emphasis. Second, the most important damage mechanism can be assessed. The relative contribution of maintenance respiration is negligible, but assimilate uptake and reduced leaf photosynthetic rate is equally important. Thus, research on leaf photosynthesis may be necessary.

Systems approaches aimed at quantitative and qualitative land evaluation

Explorative studies on land use often use crop growth models to investigate the potential of a specific land unit. Models for this purpose do not aim for a high accuracy for the individual fields. They should, however, produce reasonably accurate estimates for the specific crop at the specific location under the prevailing weather and management conditions, including water and nutrient management and pest, disease and weed control. Models with time steps of days and with summarizing modules for CO_2-assimilation, respiration, water balance, and nutrient uptake are appropriate. The models are applied as part of a procedure used for qualitative and quantitative land evaluation or agroecological zoning (figure 3) (Van Latesteijn and Rabbinge 1994). This procedure was applied in evaluating the possibilities for various forms of land use in the European Community (EC).

In this procedure, the first step was a qualitative land evaluation for units of the soil map that are almost comparable and exposed to the same climate. About 22,000

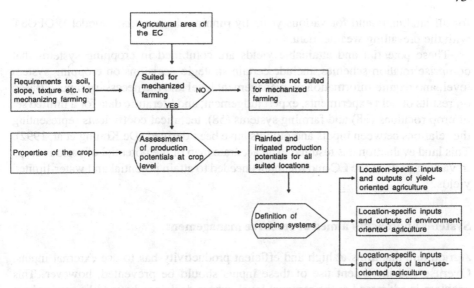

Figure 3. Procedure followed in a land evaluation study for the European Community (*Source:* Van Latesteijn and Rabbinge 1994)

units are necessary to cover the total area of the EC. The information was derived from a geographic information system (GIS) (Van Latesteijn 1993).

Characteristic factors for these units such as steepness/slopes, salinity, acidity, fractions lutum, and stoniness of the soil are used to make decisions about their suitability for the mechanized farming of grass, cereals, and root crops and their suitability for rough grazings and perennial crops (Van Lanen 1992). The qualitative land evaluation produces an estimate of the percentage of the land that is suitable for certain agricultural purposes. In Greece, for example, only eight percent of the land area is suitable for root crops, 10 percent for cereals, and about 40 percent for grassland. At the other end of the spectrum, Denmark's land area shows a suitability of 85 percent for root crops, 90 percent for cereals and nearly 100 percent for grass production.

Qualitative land evaluation is followed by a quantitative assessment of the growth potentials of crops. This step is carried out with a summary model of crop growth, WOFOST (Van Keulen and Wolf 1986). This simulation model uses as inputs soil data such as water-holding capacity, texture and fertility, climatic characteristics such as temperature and rainfall, and the most relevant crop characteristics such as the phenological, optical, geometrical, and physiological characteristics as assimilation, respiration and partitioning. With the WOFOST model, the rainfed and irrigated yields of winterwheat, maize, sugarbeet, potato, and grass are assessed. In the rainfed situation, the attainable yield is lower than the potential yield, due to water limitation. The limitation due to nutrient supply in combination with water limitation operates in a similar way. Thus, potential and attainable yields may be assessed quantitatively for all land units and for various years by running the summary model WOFOST with the prevailing weather data.

These potential and attainable yields are combined in cropping systems that comprise rotation schemes, include certain strategic decisions on cropping system level, and require information on management level and the necessary inputs. Based on results of field experiments, expert judgement, and literature data for a limited set of crop rotations (38) and farming systems (58), technical coefficients representing the relations between inputs and outputs have been derived (De Koning et al. 1992). This land evaluation has resulted in an accurate map of the potential of various crops in various parts of the EC and the inputs needed to attain potential and water-limited yields.

Systems approaches aimed at resource management

Agriculture that aims at high and efficient productivity has to use external inputs. Overuse and inefficient use of these inputs should be prevented, however. This problem is addressed at the regional level, where decisions have to be made about what land units external inputs are best used, and in what quantities or at which production levels. It has also to be addressed at the lowest aggregation level, where crops can be managed in various production situations and at various production levels, using different production orientations and technologies.

At the crop level, decades of agricultural research have resulted in many so-called dose-effect relations. Dose-effect relations show a "law of diminishing returns" for an individual external input in relation to yield. However, agricultural practice involves a combination of external inputs that, if properly mixed, may result in the law of the optimum (Liebscher 1895). In a sophisticated analysis of agricultural systems using both these laws, De Wit (1992 and 1993) shows the theoretical and empirical basis of his updated and upgraded law of the optimum (figure 4).

From this analysis, De Wit draws the important conclusion that agricultural systems are characterized more by the balanced mix of external inputs than by the simple notion of a law of diminishing returns. Agricultural systems are more complicated than the simple input-output relations characterized by production functions. The implications for resource management on micro and macro levels may be considerable. For example, more extensive, low input systems of production as a way to decrease environmental problems may have an opposite effect, because a drop in productivity is often combined with a loss in efficacy and efficiency of external inputs.

Systems approaches aimed at building a policy agenda

Explorative studies for long-term options for land use at the supra-regional level may help policy makers make strategic decisions. This is possible by using a methodology developed and illustrated in a study by the Netherlands Scientific Council for

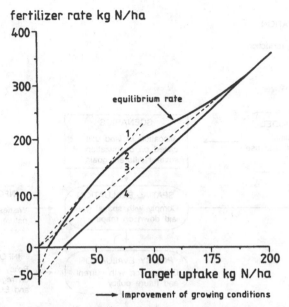

Figure 4. Resource use efficiency, an example of systems approaches. The figure shows the relation between the target nitrogen uptake and the nitrogen fertilization needed in the equilibrium situation to sustain this uptake, in case of concurrent improvement of other growing conditions. Lines 1-4 represent alternative hypotheses (*Source:* De Wit 1993).

Government Policy on options for future land use in the EC (Netherlands Scientific Council for Government Policy 1992). The methodology is explained in figure 5. The core of the methodology is an interactive multiple-goal model using linear programming techniques.

Land use exemplifies and integrates various objectives considered in this approach. Socioeconomic objectives, ecological, agronomic, and natural objectives may be distinguished. In the GOAL (General Optimal Allocation of Land use) model, eight objective functions that cover these fields have been formulated:

1. maximization of yield per ha;
2. maximization of total labor;
3. maximization of regional labor;
4. minimization of total pesticide use;
5. minimization of pesticide use per ha;
6. minimization of total N-surplus/emission;
7. minimization of N-emission surpluses per ha;
8. minimization of total costs.

These aims reflect the classification into environmental, economic, and social sustainability. To attain environmental sustainability, minimization of pesticide and fertilizer use is essential. Economic sustainability is almost guaranteed if total costs are minimized and if soil productivity continues to rise. Social sustainability can be

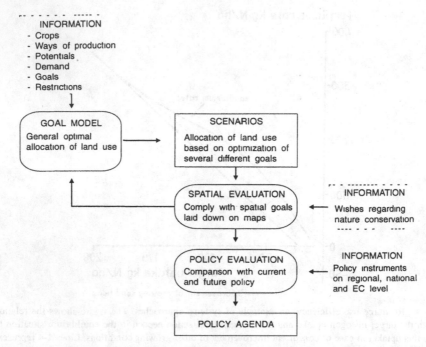

Figure 5 Procedure in *Ground for choices*, a strategic policy study for rural areas in Europe (*Source* Netherlands Scientific Council for Government Policy 1992)

achieved only if labor in the agricultural sector is ensured, or if the available labor is distributed evenly over the regions.

Four scenarios have been developed to represent four contrasting political philosophies about desired policy on land-dependent agriculture and forestry in the EC. A political philosophy here means a coherent set of preferences regarding several aims. All philosophies are based on the assumption that the ultimate aim should be to develop sustainable agriculture and protect the agricultural environment in the rural areas. The various philosophies differ considerably in their views on what must be sustained, however, which clearly illustrates the subjective nature of the concept of sustainability. The four scenarios all represent views on sustainable land use, though from different standpoints.

The following options are distinguished:
■ free market,
■ regional development;
■ nature and landscape;
■ environmental protection.

The scenarios are represented in the GOAL model by setting different preconditions to the objective functions and by varying the demand. Two examples can illustrate this.

In the free-market scenario, the costs of agricultural production are minimized, while there are no other preconditions to the objectives. Moreover, since free trade permits the import and export of products, the demand for agricultural produce within the EC is modified according to the expectations about new market balances. The model will now choose the most cost-efficient types of land use and allocate them in the most productive regions.

In the environmental-protection scenario, the costs of agricultural production are minimized, but the objective functions include strict limitations as to the use of fertilizers and pesticides. In addition, the demand for agricultural produce is fitted to self-sufficiency. Now the model will choose types of land use and allocations that agree with the imposed preconditions.

With these different options, the model calculates four different scenarios for land use. Policy makers can now see how their priorities will influence land use and how the effects are distributed over the EC. However, concerns about nature and landscape cannot be expressed in figures in ways that the model can interpret. To remedy this, a spatial evaluation is built into the procedure. One map represents the best division of land from the point of view of wildlife protection (Bischoff and Jongman 1993). The map is matched with the regional allocation of types of land use generated by the GOAL model to identify areas that are potentially problematic in terms of competing land use. The results produced by the model may have to be amended as new spatial requirements arise.

Finally, in a policy evaluation, the outcomes are used to decide to what extent current and future policy can cope with the developments in the scenarios. The effort required to achieve the aims can be estimated, depending on whether policy will have to 'go against the tide' or simply go with it. If the outcomes all point in the same direction, there is clearly a conflict between the technical possibilities and the policy, which seeks to achieve something else. In such a case, policy 'goes against the tide'. If the outcomes of the scenarios differ substantially, there is clearly greater scope for policy.

The conflict between technical possibilities and political preferences allows us to identify the extremes of the 'playing field'. Within these boundaries, choices should be made. In the study on the policy options for the rural areas of the EC, the outcomes show much variation among the scenarios. There is ample scope for choice (figure 6). All options differ considerably from the present situation, however. This holds for land use, labor, pesticide use, nutrient surpluses, costs, and regional distribution. The present policy for European rural areas aims at maintaining as much land in agricultural use as possible. This leads to the conclusion that present policies are going against the tide through indirect policies (market and price policies) as well as direct policies (such as structural changes and subsidies related to land use). The conclusion is that current policies are too expensive, environmentally unsound, and agriculturally suboptimal. This implies a clear need for more comprehensive policy reforms.

Such strategic explorations of policy options may help policy makers in their choices for the long term. Short-term policies should be geared to those long-term

78

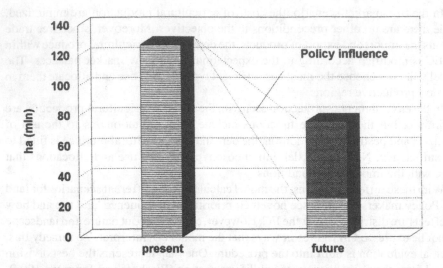

Figure 6. Results of a study on options for land use in the European Community. The two sections in the future land-use bar indicate the minimal area required for sutainable agriculture in the EC and the maximum area needed in alternative scenarios. The large differnce points at ample space for policy influences (*Source:* Netherlands Scientific Council for Government Policy 1992)

perspectives. This is often not immediately possible due to an institutional and/or cultural indisposition to action. An awareness of perspectives may refute counterproductive policies.

Concluding remarks

The examples discussed above show how various systems approaches are used to achieve various goals. A scientific goal such as gaining insight requires other tools than evaluating land or exploring policy options. The tools and objectives may be different, but the approach is usually similar.

By clearly stating aims and using a blend of systems analyses, systems synthesis and development of options, systems approaches serve many aims. They may form the backbone of new scientific integrative approaches in agricultural research and land-use studies.

Acronyms

GIS   geographic information system
GOAL   general optimal allocation of land use

References

Bastiaans L (1993) Understanding yield reduction in rice due to leaf blast PhD thesis, Wageningen Agricultural University, The Netherlands ISBN 90-5485-166-x, 127 p

Bischoff N T, Jongman R H G (1993) Development of rural areas in Europe The claim for nature Netherlands Scientific Council for Government Policy preliminary and background studies V79 Sdu Publishers, The Hague, The Netherlands 206 p

Van den Bos J, Rabbinge R (1976) Simulation of the fluctuation of the grey larch bud moth Simulation Monographs PUDOC, Wageningen, The Netherlands 91 p

Van Keulen H, Wolf J (Eds) (1986) Modelling of agricultural production weather, soils and crops Simulation Monographs PUDOC, Wageningen, The Netherlands 479 p

De Koning G H J, Jansen H, Van Keulen H (1992) Input and output coefficients of various cropping and livestock systems in the European Communities Working Documents W62, Netherlands Scientific Council for Government Policy The Hague, The Netherlands 71 p

Van Latesteijn H C (1993) A methodological framework to explore long-term options for land use Pages 445-455 in Penning de Vries F W T, Teng P and Metselaar K (Eds) Systems approaches for agricultural development Kluwer Academic Publishers, Dordrecht, The Netherlands

Van Latesteijn H C, Rabbinge R (1994) Sustainable land use in the EC An index of possibilities In Van Lier H N, Taylor P D (Eds) Sustainable land use planning (in press)

Van Lanen H A J (1992) Qualitative and quantitative physical land evaluation An operational approach PhD thesis Wageningen Agricultural University, The Netherlands 195 p

Leffelaar P A (1986) Dynamics of partial anaerobiosis, denitrification, and water in a soil aggregate Experimental Soil Science 142 325-366

Leffelaar P A (1988) Dynamics of partial anaerobiosis, denitrification, and water in a soil aggregate Simulation Soil Science 146 427-444

Liebscher G (1895) Untersuchungen uber die Bestimmung des Dungerbedurfnisses der Ackerboden und Kulturpflanzen Journal fur Landwirtschaft 43 49-125

Netherlands Scientific Council for Government Policy (1992) Ground for choices, four perspectives for the rural areas in the European community, Sdu publishers, The Hague, The Netherlands 144 p

Rabbinge R, Rijsdijk F H (1983) EPIPRE A disease and pest management system for winter wheat, taking account of micrometeorological factors EPPO Bulletin 13(2) 297-305

Rabbinge R, De Wit C T (1989) Theory of modelling and systems management Pages 3-15 in Rabbinge R, Ward S A, Van Laar H H (Eds) Simulation and systems management in crop protection Simulation Monographs 32 PUDOC, Wageningen, The Netherlands

Van Roermund H J W, Van Lenteren J C (1994) Simulation of the population dynamics of the greenhouse whitefly and the parasitoid *Encarsia formosa* on tomato (in prep)

De Wit C T (1968) Theorie en model H Veenman & Zonen NV, Wageningen 13 pp

De Wit C T (1982) Coordination of models Pages 26-31 in Van Laar H H, Penning de Vries F W T (eds) Simulation of plant growth and crop production Simulation Monographs PUDOC, Wageningen, The Netherlands

De Wit C T (1992) Resource use efficiency in agriculture Agricultural Systems 40 125-151

De Wit C T (1993) Resources use analysis in agriculture A struggle for interdisciplinarity Keynote address prepared for 'The future of the land', Symposium, 22-25 August 1993, Wageningen, The Netherlands

The development of strategies for improved agricultural systems and land-use management

R L MCCOWN, P.G. COX, B.A. KEATING, G.L. HAMMER,
P S. CARBERRY, M E. PROBERT and D.M. FREEBAIRN
Agricultural Production Systems Research Unit,
P O Box 102, Toowoomba Queensland, Australia 4350

"The structure of a farm system at any time depends on all technical economic, social, cultural, and political influences that impinge on the farmer "

Hans Ruthenberg

"The capacity for control made possible by the empirical sciences is not to be confused with the capacity for enlightened action "

Jurgen Habermas

Key words decision models, farming-systems research, Kenya, land use, land-use management, simulation models, systems research

Abstract

There is a pressing need for better management of agricultural lands in much of the developing world Understanding what changes are needed and how they might be stimulated is the central task of agricultural systems research But past experience in systems research in non-agricultural fields, where systems research methods developed, shows that a scientific approach to management has had limited effect on what managers actually do In agriculture, farming-systems research methodology has demonstrated the importance of farmer involvement if research is to change the practice of agriculture Yet, although participation appears necessary, it has proved to be insufficient in systems where there are strong resource constraints

Where such constraints exist, e g , in the semi-arid tropics, progress requires a capability to compare options that are not currently available or feasible for farmers An approach which uses improved agricultural production simulation models and economic decision models makes this possible This approach is demonstrated using experiences from work in semi-arid Kenya where investment in soil enrichment in fertility-depleted croplands is deterred by the unreliability of rainfall Although few farmers currently purchase fertilizer for maize production, augmentation of manure with modest amounts of nitrogen fertilizer appeared (a) to be needed for sustainable cropping, (b) to be profitable in the long term, and (c) to have variation in returns consistent with local farmers' attitudes concerning risk This approach also was used to compare the risks and returns of two policy strategies, i e migration of farmers to drier areas vs investment in increasing the soil fertility of existing croplands

The paper concludes that the best prospects for developing better policy and management strategies lie with skilful use of "hard" systems tools within a "soft" systems philosophical framework

Introduction

The above quotation from Ruthenberg (1980) reminds us that a way of farming is a complex adaptation, often displaying ingenuity deserving of high admiration. Yet today, there is widespread concern about the prevalence of apparently ill-adapted

P Goldsworthy and F W T Penning de Vries (eds), Opportunities, use and transfer of systems research methods in agriculture to developing countries, 81 - 96
© 1994 *Kluwer Academic Publishers*

practices of many farmers around the world, as agricultural lands come under increased pressure. In industrialized societies, economic pressure often results in management strategies that are ecologically damaging. In many less-developed countries, population pressure is causing reductions in farm size, production and returns, and forcing damaging over-exploitation of land resources. In both situations, conflicts arise when adaptations that serve farmer objectives adversely affect the interests of the farmers of the future and/or of the wider community now. Adaptations expected of farmers are increasingly complex, costly and/or risky. Increasingly, it is recognized that better government policies and research and development (R&D) planning are necessary to create environments that result in different strategy choices by farmers, and ones that can be seen as adaptive in the broader context.

This paper is about agricultural systems research methodology that facilitates the exploration of alternative production strategies and their economic and ecological consequences. We begin with a glance at the history of systems research, and it becomes clear that any new venture must be aware of perils as well as prospects. Research methods using models to optimize strategic management in complex non-agricultural production systems have been in place for over half a century. In developing-country agriculture, farming-systems research (FSR) methodology has an experience spanning nearly two decades. Both of these disparate activities have resulted in considerable disappointment and disillusionment. However, what emerges from their combined experience provides the basis for what may be a more effective systems research approach for the development of improved farming strategies. We outline such an approach and describe applications for the development of improved farming and policy strategies in Africa.

The systems movement and its "crisis"

There is a new prominence in developed-country agricultural R&D of computer-based activities, such as simulation modelling, decision-support systems, and management information systems. The establishment of International Consortium for Application of System Approaches to Agriculture (ICASA) and the holding of this workshop is indicative of the high expectations for such technologies for benefiting developing-country agricultural R&D. Although seldom acknowledged, all these information technologies have been borrowed from a mainstream "systems movement" in non-agricultural fields, which began during World War II and later differentiated into the fields of systems engineering, operational (operations) research, RAND systems analysis, and management science. These all share a rational approach to solving complex problems that features an iterative process of problem definition, identification of alternatives, and evaluation of alternatives—the latter almost always using a model (Robertshaw et al. 1978). (In this paper we will mainly use the term "operational research" to refer to the use of models to develop management strategies.)

Some of the possibilities for agriculture were recognized a generation ago by perceptive agricultural scientists and economists, e.g., Van Bavel (1953), De Wit

(1965), Morley and Spedding (1968), and Dent and Anderson (1971). But to the disappointment of many, significant use of models for such research has had to wait nearly two decades, while models capable of adequately simulating the natural processes that are central to agricultural production systems were developed. To a considerable extent, the original concept of using models in this way has had to be rediscovered. During this period of discovery of ways to successfully model agricultural systems, an enormous experience in using models in non-agricultural production and distribution systems has accrued. What can we in agriculture learn from this experience, as we now enter a phase of model use?

The overwhelming single lesson is that information systems and technology developed by professionals to aid decision makers almost never "fits" the needs of decision makers as human actors in social systems (Ackoff 1967; Churchman 1974). Failure to recognize this has resulted in a crisis in systems research due to a general failure to influence what practitioners do (Checkland 1983). In agriculture, our experience of this failure has not been sustained for a period long enough to firmly reach this conclusion, but evidence is accruing (e.g., Cox 1993). Further impediments to making generally useful contributions are due to variations in the preferences, beliefs, and abilities of individual farmers, and conflict between the objectives of the farmer and the objectives of other stakeholders in the performance of agricultural production systems. The failure of traditional agricultural science to address this human factor, together with the influence of the new systems thinking stemming from mainstream systems research experience, have resulted in the emergence of different concepts and methodologies for dealing with the "softer", people-related, agricultural system issues. One of the perversities of contemporary agricultural systems research is that after 20 years developing biophysical models, just when they are ready for credible research on management, champions of a new, rapidly-growing, "soft" systems movement are sharply critical of this approach (e.g. Wilson and Morren 1990; Bawden and Packham 1992). The effect of this tends to reinforce the influence that FSR has had on agricultural R&D.

Clients, strategies, and FSR

FSR has long since demonstrated the inadequacy of new technology development alone as a basis for improvement of farming systems in developing countries. Although under intense scrutiny as the natural resource aspects of farming increase in priority, FSR has had a major role in changing the performance of R&D in appreciating farm problems and research needs and in including the farmer-client in the research process. As the focus shifts from strategies for improved production efficiencies to include strategies for the better care of natural resources, R&D planners and government policymakers are viewed increasingly as primary clients for research on such strategies. But ultimately, the key actors are farmers: new management strategies will be their new strategies. How then do researchers contribute to improved strategies?

While many would readily sympathize with the flippant answer, "not easily", FSR methodology provides a basic framework for a meaningful answer. Professionals work with farmers in gaining knowledge of the farm system and in identifying problems and opportunities (figure 1, upper left). Testing of potential innovations is done on farms with and by farmers (figure 1, upper right). As shown in figure 1, in concept, FSR provides the linkages between farmers and professionals that are needed. Yet, the achievements of this research approach have fallen far short of expectations, and, with clear hindsight, good reasons are not hard to find.

Norman and Collinson (1985) pointed out two strategies for dealing with a constraint in the farming system: relieve it, or avoid it by exploiting flexibility in the system. They observed that it is the presence of under-utilized resources that enables

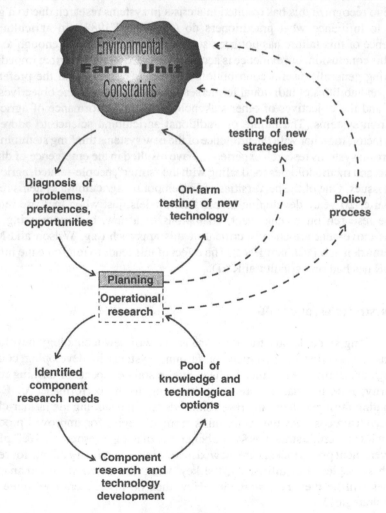

Figure 1 A schematic representation of farming systems research (adapted from Dillon and Vermani 1985)

flexibility in management, and that the success of FSR can be attributed to exploitation of such flexibility, rather than relieving constraints. They recognized that major long-term increases in productivity have to come through relieving constraints and argued that increase in productivity can provide the means to relieve a constraint. Seven years later, Waddington (1992) reported examples of successful on-farm experimentation by FSR teams. He observed that most examples could be regarded as "fine tuning" of existing technology in environments with some slack in resources. Where technologies did not already exist, and in regions with great pressure on resources, little success was experienced.

FSR workers progressively realized the probable need for government policy intervention if many farmers were to be able to adopt technology and strategies for sustainable farming (Norman and Collinson 1985; Biggs and Farrington 1992). FSR offered no methodical way to contribute to the policy process comparable to that for farm-level intervention, as indicated in figure 1 (Fox et al. 1990).

In the FSR procedure of figure 1, the key step for innovative change is that of planning (center). The potential for creative contribution by professionals is in response to questions such as the following: "Given this situation, what might be possible/feasible among the available options?"; "What new technology development initiatives are warranted?"; "What is the least unattractive option for relieving a soil-nitrogen constraint?"; "How much would the price of fertilizer need to fall before greater use could be expected?"; or "How should nitrogen fertilizer be marketed?" Traditionally, the most important ingredient for success in the planning step has been good judgement by an experienced and clever professional. While it is difficult to conceive that this ingredient will become less valuable in the future, it will increasingly need to be supported by quantitative analyses. The challenge is to introduce into this activity a methodology that enables quantitative comparison of possible management strategies in terms of their economic and ecological consequences. Not the least important contribution to be made here is demonstration of the poor chance of success of certain proffered options so that they can be dismissed, rather than forming the basis of expensive research programs.

A more analytical planning stage is not a new idea, but it is an idea whose time has come because of a new urgency and new capabilities that increase the chances of success. Dillon and Virmani (1985), in modifying the original FSR schema of Collinson (1982) substituted the term operational research (OR) for planning (see figure 1, center). While a term rarely used in agriculture, except by economists, this embodies the traditional "hard" systems approach to identifying optimal management strategies for large and complex systems, aided by models of the system or subsystems. Other FSR economists, while appreciating the value of the thinking process of OR, had previously judged that the benefits of "high-powered" economic modelling did not sufficiently outweigh the high costs of model development, testing, and application (Anderson et al. 1985). But by the late 1980s, simulation models had developed sufficiently so that marriage of this type of model (in conjunction with economic analyses) with FSR appeared to offer a new, more flexible and powerful systems approach (Thornton and McGregor 1988; McCown 1991; Thornton 1991).

FSR had a very beneficial effect in re-orienting researchers from disciplinary goals to serving clients, and the approach has been successful in identifying ways to more efficiently exploit slack in farm production resources. But in very resource-poor, climatically-variable situations, comparison of farmer options is problematic. Not only are options often apparently not economically feasible, but superiority of options will vary so much between years that comparisons in a short-term experiment cannot give a clear outcome. The quantitative methods in what is developing as an agricultural operational research approach provides a means of reducing this limitation. Simulated yields for a long period of years may indicate that an input such as fertilizer, while not profitable every season, is a good financial investment. There is also the possibility that the results could demonstrate a policy action that may make an ecologically sound farming strategy that is the "least unattractive" to the farm firm sufficiently attractive to result in its adoption as the best way to meet the farm's objectives.

Developments in using simulation modelling in an FSR perspective

The Agricultural Production Systems Research Unit (APSRU) was established in 1991 by the Queensland Department of Primary Industries and the Division of Tropical Pastures of CSIRO, Australia. Its aim is to provide benefits to a range of clients through agricultural systems research, leading to improvements in production efficiency, risk management, and sustainability.

The decision to establish a systems group of over 20 people in a period of shrinking research resources was taken on the strength of the achievements and the promise of simulation modelling as a tool in research for improved management. In spite of the harsh critique of the poor use or misuse of such "hard" approaches in other fields, they have been, and continue to be, unsurpassed in dealing with many well-structured problems (Checkland 1981; Jackson and Keys 1984). We are using our new hard systems tools in the applications that they clearly suit, and are trying to be good students, both of their limits and of complementary methodologies. In doing this, we are beginning to design a new, more effective systems research methodology. The shape this is taking, in the context of strategy development, is depicted in figure 3 and clearly has its origins in figure 2.

We view farmers as our primary clients, but we are also concerned with the decision problems faced by a range of other decision makers (figure 2, top) who have a stake in the performance of agricultural production systems and whose decisions influence, and are influenced by, farmers. Objectives vary among classes of clients, and it is to be expected that priorities for farm or farming-system performance will vary. On the other hand, some of these clients experience many of the same uncertainties as farmers, especially those concerning rainfall and prices.

Our aim is to contribute to better management and planning decisions (figure 2, right). To avoid the trap experienced in management research, which we described earlier, we invest in learning from and with clients the context and structure of decisions (figure 2, left). With farmer clients, we need to know their "rules" for

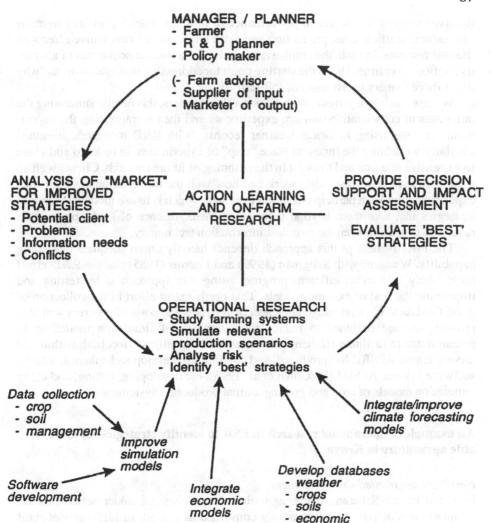

MANAGER / PLANNER
- Farmer
- R & D planner
- Policy maker

(- Farm advisor
- Supplier of input
- Marketer of output)

ANALYSIS OF "MARKET"
FOR IMPROVED
STRATEGIES
- Potential client
- Problems
- Information needs
- Conflicts

ACTION LEARNING
AND ON-FARM
RESEARCH

PROVIDE DECISION
SUPPORT AND IMPACT
ASSESSMENT

EVALUATE 'BEST'
STRATEGIES

OPERATIONAL RESEARCH
- Study farming systems
- Simulate relevant
  production scenarios
- Analyse risk
- Identify 'best' strategies

Data collection
- crop
- soil
- management

Integrate/improve
climate forecasting
models

Improve
simulation
models

Software
development

Integrate
economic
models

Develop databases
- weather
- crops
- soils
- economic

Figure 2. A systems research framework that is evolving in Australia

decisions. In many cases, farmers' rules are likely to be effective because they reflect the interdependence between different components, and the open character of agricultural production systems (Cox 1993). Change may be needed, but professional initiatives for change need to be carefully considered and tailored in the light of such knowledge. Similarly, to contribute to policy we must be knowledgeable about the policy process, the specific issues, and the niches for professional contribution.

We draw heavily from operational research principles in our systems approach (figure 2, center), but not on the stereotypic methods and algorithms and the rigid mathematical precision that have come to characterize that field. Our approach involves study of the simulated performance of farming systems primarily in terms of production efficiencies, production and price risks, and the cumulative effects on the soil resource. In each, the emphasis is on the economic consequences of alternative actions over time. This is the starting point for addressing such questions as "why don't more farmers invest more in soil fertility".

We use models in projects with clients. With farmers, this entails simulating the outcomes of collaborative on-farm experiments and then extrapolating the experiments in time, using historical weather records. With R&D managers, it entails similarly weakening the time and place "trap" of experiments, in order to add value to expensive research and to assist in the planning of future research. Clients with an appreciation of how the models work and how well the model simulates their own experiments are keen participants in the next step, which is to use models to explore strategies that take them beyond their experience. Absence of such appreciation results in distrust of even the glossiest information technology.

The effectiveness of this approach depends heavily on an adequate modelling capability. We agree with Seligman (1990) and Loomis (1985) that the R&D effort most likely to provide efficient progress using this approach is by testing and improving the best of existing models. This needs major effort in the collection of good field data. We have judged that acceptable rates and costs of progress will also require software designed to reduce the overheads of simulation modelling in research and to facilitate efficient convergence of modelling effort both within and among teams. APSRU has produced, and continues to develop and enhance, a novel software system, APSIM (McCown et al. 1993), for developing, testing, and using simulation models of crop and grazing animal production systems.

An example of operational research in FSR to identify strategies for sustainable agriculture in Kenya

Fertilizer-augmented soil enrichment

In much of sub-Saharan Africa, agricultural resources are under serious threat. Cropland productivity is being rapidly consumed as a result of high rates of rural population growth, shortage of suitable land for further expansion of cropping, and poverty, which precludes replacement of soil nutrients at rates that will sustain productivity (Broekhuyse and Allen 1988; Lynam 1978). Grazing lands are suffering similarly. The process, which impoverishes both land and people, is depicted in figure 3.

We encountered the "poverty trap" (figure 3) in a project in Ukambani in the Machakos-Kitui Districts of Eastern Kenya. Even in good seasons crop yields were low (due most conspicuously to nitrogen deficiencies), incomes from farming were low, and purchased inputs to production were correspondingly low. It seemed clear that the key to breaking this cycle was soil enrichment. But this view was not shared by our research colleagues. In the wake of a major FSR project in the region, which had focused the attention of the research establishment on farmer problems, our

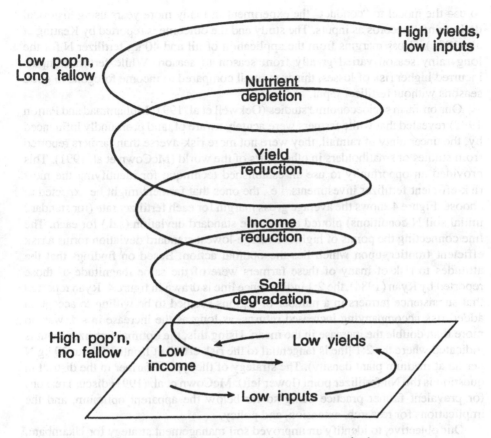

Figure 3 The poverty trap that results from human population pressure on land

proposed research focus was challenged because our proposal included investigation of the costs and benefits of commercial fertilizer (among other sources of nutrients). It was argued that this was not appropriate because "farmers here don't use fertilizer".

An important element of our research strategy was our study of a few farms where managers had come to recognize the value of fertilizer in augmenting manure applications, and were prospering from it. This sharpened the focus of the project on whether the many farmers who did not use fertilizer should be doing so in the farm's economic interests and at prices which are consistent with the manager's perception of risk. In order to quantify what the yield improvements might be in response to fertilizer inputs, we conducted a number of experiments on farmers' fields. But this could be in only a few places, and most importantly, in only a very few seasons. Since, in any given season, there is a high probability of water being more limiting than nitrogen, this would have been an inadequate approach were it not for the opportunity to use the model to "conduct" the experiment in many more years using historical daily rainfall records as inputs. The study and the outcome is reported by Keating et al. (1991). Gross margins from the application of nil and 40 kg fertilizer N for the long rainy season varied greatly from season to season. While fertilizer inputs incurred higher risk of losses, this was small compared to income foregone in good seasons without fertilizer input.

Our on-farm socioeconomic studies (Ockwell et al. 1991; Muhammad and Parton 1992) revealed that while farmers were acutely aware of, and profoundly influenced by, the uncertainty of rainfall, they were not more risk-averse than farmers reported from studies of smallholders in other parts of the world (McCown et al. 1991). This provided an opportunity to use an established technique for identifying the most risk-efficient fertilizer investments, i.e., the ones that farmers might be expected to choose. Figure 4 shows the average gross margin for each fertilizer rate (for standard initial soil N conditions) plotted against the standard deviation (s.d.) for each. The line connecting the points of highest average-lowest standard deviation forms a risk efficient frontier, upon which lies the optimal action. Based on findings that the attitudes to risk of many of these farmers were of the same magnitude of those reported by Ryan (1984), the 2:1 indifference line is drawn in figure 4. Ryan reported that subsistence farmers in a number of regions seemed to be willing to accept an added risk accompanying increased returns, as long as the increase in s.d. was no more than double the increase in the mean. Using this, the optimal fertilizer input is indicated where the 2:1 line is tangential to the risk efficient frontier (about 30 kg N per ha at medium plant density). The strategy of the typical farmer in the district in question is the Nil fertilizer point (lower left). McCown et al. (1991) discuss reasons for prevalent farmer practice being so far below the apparent optimum, and the implications for research, extension, and policy.

Our objective, to identify an improved soil management strategy for Ukambani, has been only partially achieved. The simulation study, as part of the operational research approach (figures 1, 2, center), is largely completed and indicates that a fertilizer-augmented soil enrichment (FASE) strategy has promise. But there are some ecological and economic complications (Probert et al. 1992; McCown et al. 1992). For the next steps, two activities are needed: testing of strategy with farmers, and exploring the regional fertilizer policy issue with the appropriate government bodies.

Resettlement of farmers to margins of crop production regions

Kenya's cropping regions are, and will continue to be, some of the most heavily populated rural lands in Africa (Binswanger and Pingali 1988). There is a continual out-migration of farmers from over-populated areas of higher production potential to less-populated, lower-potential areas. There is in-migration to the Midlands of Ukambani from the Highlands, and out-migration from the Midlands to Lowland pastoral areas. While this is in line with Kenyan government policy, the net consequences of this movement may be more negative than is presently appreciated.

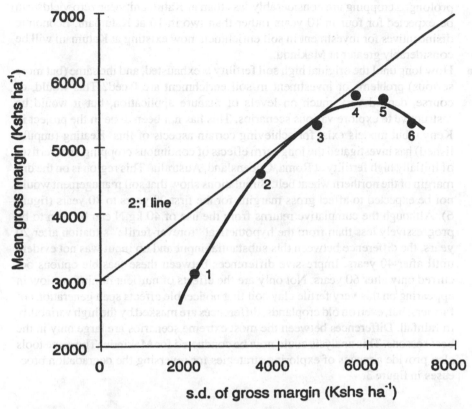

Figure 4 The values of various maize production strategies depicted in mean-standard deviation space
Strategies combined population density (plants/ha) and fertilizer nitrogen (kg/ha) Points 1-6 represent
22K, 0 kg, 27K, 15 kg, 33K, 30 kg, 38K, 45 kg, 44K, 60 kg, 55K, 80 kg

Operational research using crop models and historical rainfall records from these
regions provides a means of exploring some important questions about the implications of changing population pressure for the management of productive resources:

■ The risks of crop failure, and hence the frequency of need for food relief from
outside the region, is sufficiently great in the present croplands for its reduction
to be the major objective of local agricultural research. How much worse will this
problem be in the lands now being settled? It is possible to make a comparison
using the maize model CMKEN. Figure 5 shows, first, the decline in expected
yields in the old cropping areas, represented by Katumani, as soil fertility was
mined. Migration from a depleted shamba at Katumani to the drier Makindu
district, where newly-cultivated land has high fertility, results in large increases
in yield in the best 30 percent of years with little difference in yield distribution
in the other 70 percent of years, apart from a modest increase in the frequency of
near-zero yields. Expected yields at Makindu after the soil fertility is depleted by

prolonged cropping are considerably less than at Katumani; near zero yields can be expected for four in 10 years rather than two in 10 at Katumani. Economic disincentives for investment in soil enrichment now existing at Katumani will be considerably greater at Makindu.

■ How long until the original high soil fertility is exhausted, and the same (but more serious) problems of investment in soil enrichment are faced? This would, of course, depend very much on levels of manure application, but it would be instructive to explore various scenarios. This has not been done in the project in Kenya, but models exist for achieving certain aspects of this. Keating (unpublished) has investigated the long-term effects of continuous cropping on a vertisol of initially high fertility at Roma, Queensland, Australia. This region is on the dry margin of the northern wheat belt. Simulations show that soil management would not be expected to affect gross margins for the first 20 years to 40 years (figure 6). Although the cumulative returns from the use of 40 kg N can be seen to be progressively less than from the hypothetical "forever-fertile" situation after 20 years, the difference between this substantial input and no input was not evident until after 40 years. Impressive differences between these feasible options occurred only after 60 years. Not only are the effects of nutrient mining so slow in appearing on this very fertile clay soil that noticeable effects span generations of farmers, but, even on old croplands, differences are masked by the high variability in rainfall. Differences between the most extreme scenarios are large only in the best seasons. This analysis could now be conducted for Makindu. The same tools also provide a means of exploring strategies for reversing the degradation processes in figure 3.

The future of systems research for strategy development in agriculture

We can expect that the technologies that enable the capture, derivation, handling and integration of information will continue to improve. Improved models should contribute to improved insights into agricultural system function and new bases for enlightened management and policies. Contributing to improved simulation will be:

1. improved representation of certain aspects of cropping systems, enabling important phenomena to be better simulated, e.g., the long-term effects of crop sequences on soil N, and the competition between intercrops;
2. development of models of additional crops;
3. greater flexibility in re-combining good routines in different models to provide a superior configuration for a given task; and
4. improved software design and quality that facilitate the efficient evolution of models in response to testing and adaptation (McCown et al. 1993).

There is no doubt that pressures in both developing and developed countries will result in greater intervention by governments in the way that land is used. As the pressure on land grows, so do political pressures and conflict. Almost all agricultural professionals concerned with land resources were trained in agricultural science. They are familiar with soil processes, the effects of mismanagement on the soil, and

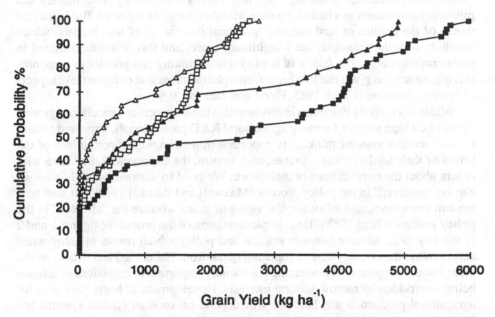

Figure 5. Cumulative probability distributions of maize production for pristine and seriously depleted nitrogen fertility states at a climatically Medium Potential site (Katumani) and a Low Potential site (Makindu). Pristine (medium) fertility represented by bold line; low fertility by light line.

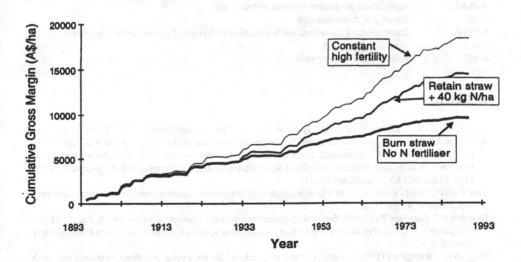

Figure 6. Cumulative gross margins of simulated wheat production at Roma, Australia for three scenarios. (The "Constant high fertility" is hypothetical.)

94

indicators of degradation. This natural science knowledge is the basis of sound "theoretical (scientific) reasoning", and is generally accepted by policymakers and public administrators as a basis for policy, all other things being equal. But in the real world of the politics of land use, the "practical reasoning" of land holders, which conflicts with the scientific, has a legitimate place, and this is institutionalized in democratic systems. The failure of hard systems thinking and practice to recognize this legitimacy has given rise to other systems philosophies that embrace these aspects of systems function (Ulrich 1983; Flood and Jackson 1991).

While it is unlikely that direct involvement in these aspects of agricultural systems should be a high priority for many agricultural R&D professionals, improved awareness of modern systems thinking is important in the shaping of perceptions of the limits of their hard systems activities, and, in turn, the expectations they raise with others about the ramifications of their work. We need to discover not only how we can be "involved" in the policy process (Maxwell and Randall 1989), but also how we can learn from, and influence, the views of those who are the "affected" in the policy process (Oliga 1988). Thus, implementation of the process of figures 1 and 2 is the key to an alliance between science and policy which results in enlightened action rather than merely control (opening quote from Habermas) and which, on the other hand, is important in preventing the socially important contributions of science being overridden by narrow sectoral interests. Development of better strategies for agricultural production and land use will depend on such an eclectic systems approach.

Acronyms

APSRU	agricultural production systems research unit
FSR	farming-systems research
ICASA	International Consortium for Application of System Approaches to Agriculture
OR	operational research
R&D	research and development

References

Ackoff R L (1967) Management misinformation systems. Management Science 14:147-156.

Anderson J R, Dillon J L, Hardaker J B (1985) Socioeconomic modeling of farming systems. Pages 77-88 in Remini J V (ed.) Agricultural systems research for developing countries: proceedings of an international workshop held at Hawkesbury Agricultural College, Richmond, N.S.W., Australia, 12-15 May 1985. ACIAR Proceedings No.11.

Van Bavel C H M (1953) A drought criterion and its application in evaluating drought incidence and hazard. Agronomy J. 45:167-172.

Bawden R J, Packham R G (1991) Improving agriculture through systemic action research. Pages 261-270 in Squires V and Tow P (eds.) Dryland farming—a systems approach. Sydney University Press, Sydney, Australia.

Biggs S D, Farrington J (1992) Farming systems research and the rural poor: A political economy approach. J. for Farming Systems Research and Extension 3:59-82.

Binswanger H, Pingali P (1988) Technological priorities for farming in sub-Saharan Africa. World Bank Research Observer 3(1):81-98.

Broekhuyse J Th, Allen A M (1988) Farming systems research on the northern Mossi Plateau. Human Organization 47 (4):330-342.

Checkland P B (1981) Rethinking a systems approach. J. of Applied Systems Analysis 8:3-14.

Checkland P B (1983) O.R. and the systems movement: Mappings and conflicts. J. Operational Research Society 34:661-675.

Churchman C W (1975) Towards a theory of application in systems science. Proc. IEEE 63:351-354.

Collinson M P (1982) Farming systems research in Eastern Africa: The experience of CIMMYT and some national agricultural research services, 1976-81. MSU International Development Paper No. 3.

Cox P G (1993) Uses and abuses of complexity in an uncertain world: Indicators for design of decision support. Pages 233-237 in Proceedings Australia Pacific Extension Conference, Gold Coast.

Dent J B, Anderson J R (1971) Systems analysis in agricultural management. John Wiley, Sydney, Australia. 394 p.

Dillon J L, Virmani S M (1985) The farming systems approach. In Muchow R C (ed.) Agro-research for the Semi-arid Tropics: North-west Australia. Univ. Queensland Press, Brisbane, Australia.

Flood R L, Jackson M C (eds.) (1991) Critical systems thinking: Directed readings. John Wiley & Sons, New York, USA.

Fox R, Finan T, Pearson S, Monke E (1990) Expanding the policy dimension of farming systems research. Agric. Systems 33:271-287.

Habermas J (1971) Toward a rational society; Student protest, science, and politics. Beacon Press, Boston, Mass, USA.

Jackson M C, Keys P (1984) Towards a system of systems methodologies. Pages 139-158 in Flood R L, Jackson M C (eds.) Critical systems thinking: Directed readings. John Wiley & Sons, New York.

Keating B A, Wafula B M, Watiki J M (1992) Development of a modeling capability for maize in semi-arid eastern Kenya. Nairobi Workshop.

Keating B A, Godwin D C, Watiki J M (1991) Optimising nitrogen inputs in response to climatic risk. Pages 329-358 in Muchow R C, Bellamy J A (eds.) Climatic risk in crop production: Models and management for the semi-arid tropics and subtropics. Ch.16. CAB International, Wallingford, UK.

Loomis R S (1985) Systems approaches for crop and pasture research. Pages 1-8 in Proceedings of the 3rd Australian Society of Agronomy Conference, 1985, Hobart, Tasmania, Australia.

Lynam J K (1978) An analysis of population growth, technical change, and risk in peasant, semi-arid farming systems: A case study of Machakos District, Kenya. PhD Thesis. Stanford University, California, USA.

Maxwell J A, Randall A R (1989) Ecological economic modeling in a pluralistic, participatory society. Ecological Economics 1:23-249.

McCown R L (1991) In Research in a farming systems framework. Pages 241-249 in Squires V, Tow P (eds.) Dryland farming—a systems approach. Sydney University Press, Sydney, Australia.

McCown R L, Hammer G L, Hargreaves J N G, Holzworth D, Huth N I (1993) APSIM: An agricultural production system simulation model for operational research. Proceedings of the International Congress on Modeling and Simulation, Modeling Change in Environmental and Socioeconomic Systems, 6-10 December 1993, Univ. Western Australia, Perth, Australia.

McCown R L, Keating B A, Probert M E, Jones R K (1992) Strategies for sustainable crop production in semi-arid Africa. Outlook on Agriculture 21:21-31.

Morley R H W, Spedding C R W (1968) Agricultural systems and grazing experiments. Herbage Abstracts 38:279-287.

Muhammad L W, Parton K A (1992) Smallholder farming in semi-arid eastern Kenya—basic issues relating to the modeling of adoption. Pages 119-123 in Probert M E (ed.) A search for strategies for sustainable dryland cropping in semi-arid Eastern Kenya. Proceedings of a symposium held in Nairobi, Kenya, 10-11 December 1990. ACIAR Proceedings No.41.

Norman D, Collinson M (1985) Farming systems research in theory and practice. Pages 16-30 in Remini J V (ed.) Agricultural systems research for developing countries. Proceedings of an international workshop held at Hawkesbury Agricultural College, Richmond, NSW, Australia, 12-15 May 1985. ACIAR Proceedings No.11.

Ockwell A P, Muhammad L, Nguluu S, Parton K A, Jones R K, McCown R L (1991) Characteristics of improved technologies that affect their adoption in the semi-arid tropics of Eastern Kenya. J. of Farming Systems Research-Extension 2:133-147.

96

Oliga J C (1988) Methodological foundations of systems methodologies Pages 159-184 in Flood R L, Jackson M C (eds) Critical systems thinking Directed readings John Wiley & Sons, New York, USA

Probert M E, Okalebo J R, Simpson J R, Jones R K (1992) The role of boma manure for improving soil fertility In Probert M E (ed) A search for strategies for sustainable dryland cropping in semi-arid Eastern Kenya Proceedings of a symposium held in Nairobi, Kenya, 10-11 December 1990 ACIAR Proceedings No 41, 63-70

Robertshaw J E, Mecca S J, Renck M N (1978) Problem solving A systems approach Petrocelli Books Inc, New York, USA 272 p

Ruthenberg H (1980) Farming systems in the tropics 3rd edition Clarendon Press, Oxford, UK

Ryan J G (1984) Efficiency and equity considerations in the design of agricultural technology in developing countries Australian J of Agric Econ 28 109-35

Seligman N G (1990) The crop model record Promise or poor show? Pages 249-258 in Rabbinge R, Goudriaan J, Van Keulen H, Penning de Vries F W T, Van Laar H H (eds) Ch 14, Theoretical production ecology Reflections and prospects PUDOC, Wageningen, The Netherlands

Thornton P (1991) Application of crop simulation models in agricultural research and development in the tropics and subtropics International Fertilizer Development Center, Paper 15 23p

Thornton P K, McGregor M J (1988) The identification of optimum management regimes for agricultural crop enterprises Outlook on Agric 17 158-162

Ulrich W (1983) Critical heuristics of social planning Paul Haupt, Berne, Switzerland

Waddington S R (1992) The future for on-farm crop experimentation in southern Africa Paper presented at the CIMMYT FSR Networkshop on Impacts of On-farm Research, 23-26 June 1992, Harare, Zimbabwe

Wilson K, Morren G E B Jr (1990) Systems approaches for improvement in agriculture and resource management Collier Macmillan, London, UK

De Wit C T (1969) Dynamic concepts in biology Paper presented to the IBP/PP Technical Meeting, Productivity of Photosynthetic Systems Models and Methods Trebon, Czechoslovakia

Discussion on Section B: The role of systems approaches as an aid to policy decisions and as practical tools for resource management

The use of systems approaches in agroecological research
The additional complexity of the issues to be addressed when NRM is integrated into the agricultural research agenda (multiple objectives, increased information needs, more complex decision procedures, and the interdisciplinary nature of NRM research) makes the use of systems approaches essential. In addition, as noted in the discussion on Section A, the methods used help in defining levels and scales at which NRM objectives can be met, and they focus attention on the relationships between production and the environment, including the social, cultural, and economic components of the environment.

For agroecological purposes it is not sufficient to describe the biological, physical, and economic characteristics of the environment. It is necessary also to be able to associate variations in these factors conceptually and quantitatively, with the variations in productivity of resources, and in the way they are used. These relationships are represented by models which, as understanding improves, progress from statistical relationships to increasingly complete descriptions of the processes involved.

Over the past 30 years or more, systems simulation techniques have been developed for use in research and decision making at different levels in the hierarchy of agricultural systems; at the level of a crop, a farm unit, a farming system, and at a national and regional level, for land-use planning. Some of the earliest work was on the use of heuristic simulation models in crop science, as research tools and integrators of new and existing knowledge on the physiological processes of plant growth. Used in this manner, the models served to focus attention on areas where knowledge was lacking and where further research was required. Currently, simulation methods are used widely for the study of a great variety of issues in NRM.

The papers in this Section have provided examples of the application of systems approaches to different kinds of resource management problems. The applications include: a combination of an economic and an ecological analysis to determine research priorities (Wood and Pardey); use of a crop growth model to compare crop and fertilizer management strategies in a dryland environment (McCown et al); and examples of the application of systems approaches as aids to different levels of decision making that concern the use of natural resources (Rabbinge et al).

How compatible are economic and ecological approaches?
There are more examples of an economic analysis having provided the basis for the determination of national policy on NRM issues than there are examples of biophysical models having influenced decision making at this level. One of the few examples where biophysical models have had an influence is the application of Multiple Goal Linear Programming (MGLP) that WARDA has used in West Africa (see Dingkuhn,

97

section F). The shared interests between economists and policy makers (many of whom are economists) may account in part for the wider affinity for economic-based approaches.

At present, economists and natural scientists differ in their perceptions of NRM issues. This is partly because their vocabularies are different, and concepts rooted in one discipline may translate poorly into the other. Economic approaches take account of the possibility of a divergence between private and social values. For instance, farmers may not adopt a resource-conserving practice because of short-term "costs". If society places a value on externalities such as impacts on future productivity, public health, conservation, or other social uses of natural resources, then a shift in the policy environment may be needed to achieve socially more favorable behavior. While an economics approach takes account of these factors, the more deterministic, process-oriented approaches used by natural scientists generally do not.

On the other hand, one of the limitations of the economic-ecologic approach described by Wood and Pardey, in which an economic surplus analysis is used, is that, in its present stage of development, the analysis is applied to single commodities. A commodity approach is generally not a satisfactory way of addressing system-level sustainability issues. Although the estimated benefits can be integrated across a number of component commodities of a production system, this does not reflect the component interactions. The process models of production systems are better able to represent these interactions, but as yet only imperfectly. Economists would therefore argue that at present the models are of only limited value for their purposes.

In spite of these incompatibilities, there was a consensus among the workshop participants that economic and ecological approaches are complementary and that both are needed. The workshop noted that there are very few examples of an effective combination of economic and ecological modelling approaches in which the issues have subsequently been pursued to the point where some change in system performance has been achieved. This self-criticism by a mixed disciplinary group was seen as both a prompt for more vigorous efforts to integrate biophysical and economic modelling with a focus on some outcome to clients, and as a cautionary note against talking-up expectations too much. A much stronger link between the two approaches needs to be sought, and more interdisciplinary dialogue is needed. The focus of the discussion was on how this can be achieved so that the strengths of one overcome the limitations of the other.

Limitations of models
The impact of changes in the pattern of resource use on the natural resource base may be evident only over long time periods, and the time frame for technology development is therefore generally also long. This makes it difficult to evaluate research priorities. For current research to be relevant it must address constraints that will affect the pattern of resource use well into the future. One of the limitations of currently available models and systems approaches is their inability to deal adequately with these long-term trends; a shortcoming which is in part a reflection of the focus on crop models in the past.

There are four elements that limit the level of confidence that can be attributed to the output from systems models that refer to distant objectives: the limited availability and quality of data; the limits of our understanding of the physical, biological, and socioeconomic processes involved, especially the stochastic elements in them; the unpredictability of human behavior (even at lower, disaggregated levels); and the evolution of the context in which problems are seen, with the passing of time (e.g., changes in prices or in the availability of capital, which greatly affect the options and objectives). This last point reflects that the world outside the boundary of the system is not static. These limitations apply particularly to applications of biophysical models to high levels of system aggregation (e.g., at an ecoregional level), and at lower levels to the predictions from behavioral models (e.g., decision trees to reflect farm management responses).

Although current models sometimes fall short of what is required to predict long-term trends, there has been significant progress in recent years in modelling production systems, and also in modelling soil processes as indicators of long-term effects on productivity. Economists perceived the ability to model trends as one of the main strengths of biophysical models. A question that remains, though, is how the validity of long-run models will be tested (e.g., for predictions of sustainability).

There is an important group of models that are intended to evaluate policy options for land-use and resource management. These options may be exclusive, in which case the outcomes from the model help structure and inform the political debate that accompanies such decisions. However, the models provide very little understanding of how the actual goals indicated by the preferred option (such as pesticide reduction or erosion control) can be achieved. In other words, they do not evaluate the relevance of policy instruments. In effect, they leave the decisions on this sort of question to the intuition of the reigning policy makers. Models help define where we want to go, but often fail to help choose the vehicle or the road to get there.

Because of their limitations, the principal use of agricultural systems models is to explore what the probable outcome of different policies or planning strategies is likely to be. Though they are imperfect, they can help identify key land-use technologies and, by improving estimates of the probable physical benefits from research, they can provide decision makers (farmers, scientists, or policy makers) with a better understanding of what is at stake and how their actions influence their expected future position.

Additional comments on the limitations of models, coming from the NARS as users, are found in the discussion following the papers in Section D.

System level

In the discussion on Section A, the need to define the level and scale of the system under study was noted as one reason why systems approaches are particularly valuable when addressing environmental concerns in agriculture. Failure to define clearly the boundaries of the system causes confusion that hinders the exchange of ideas on systems issues.

In a research program where agricultural and NRM objectives are integrated, the systems levels of interest may extend from single crops to whole land-use systems or even ecological regions. Within this range there is a continuum of intermediate levels in which the nature of the problems and the solutions changes. An optimum solution at one level is not necessarily optimum at another (as the study of fertilizer use on maize in Kenya illustrates, see McCown, p. 89).

While it is important to define clearly the level and the boundaries of the system under study, economists and biologists point out the difficulties of deciding on the appropriate place to draw the boundaries; changes at one level may well move the analysis up to another, higher level (see also Lynam, Section A, p. 10). In addition, modellers have found it difficult to establish the hierarchical linkages between systems levels that would permit easy movement from one level of analysis to another. For example, when attempting to integrate information from household surveys and anthropological studies to greater spatial scales. The workshop heard that the problem arose in the global project on "Alternatives to Slash-and-Burn Agriculture", coordinated by ICRAF and involving CIAT, IFPRI, IITA, IRRI, and CIFOR (Hoekstra, Section F, p. 305).

One of the challenges of hierarchical integration is to link the different decision-making levels for purposes of planning, e.g., at strategic and tactical levels. The IARCs have found a systems approach more useful at some levels than others. It has been more useful for strategic planning (as in CIAT's major use of its land-use databases) and at the project level, than at intermediate levels (e.g., for determining priorities between the center's research programs, see Torres and Gallopin, Section F).

Determination of priorities
A key question for NARS will be how to balance short-term production objectives with the long-term objectives of NRM research. The second of the three tasks that need to be undertaken to determine what the balance should be (namely, estimating the costs and benefits of different resource-use strategies; see discussion on Section A, p. 41-42) poses many methodological problems of measurement and evaluation. Also, the intergenerational nature of NRM objectives gives rise to questions of equity and discount rates concerning how the use of resources for current production affects the future use of those same resources. Consideration will have to be given to how these environmental considerations are to be taken into account in setting agricultural research priorities. This is another area in which systems approaches are required and where systems modelling can be used to develop measures of costs and benefits, based on estimates of changes in the physical status of both natural and man-made capital used in production, and on estimates of the change in potential productivity over time.

Recommendations

Integration of economic and ecological approaches
Economic and ecological approaches are complementary in systems research. Both are needed, and a much stronger link between the two approaches needs to be sought so that the strengths of one overcome the limitations of the other. The workshop recommends more vigorous efforts on the part of the national and international organizations to promote an interdisciplinary dialogue on ways to integrate biophysical and economic modelling with particular attention to the needs of NARS for research and development planning.

System level
By specifying their needs, model users in national and international organizations should encourage modellers to devote special attention to the hierarchical linkages between systems levels that would permit easier movement from one level of analysis to another. There is a particular need for better hierarchical integration to link the different decision-making levels for purposes of agricultural research planning (e.g., at strategic and tactical levels).

Improved methods for determining priorities
The workshop recognized that because of the additional complexity of the issues to be addressed when production-oriented objectives and research to improve the efficiency and ensure the sustainability of resource use in agricultural production are combined, IARCs and the NARS urgently need improved procedures for determining priorities and making decisions about the allocation of research resources. Procedures are needed that include appropriate measures of the costs and benefits of the environmental and natural resource consequences of different resource-use options.

The workshop recommends that the development of such procedures should be one of the main objectives of the integration of economic and ecological approaches recommended above.

SECTION C

NARS needs and priorities for addressing resource issues

Needs and priorities of NARS for the management of natural resources: Bangladesh

Z. KARIM

Member Director, Bangladesh Agricultural Research Council, Farm Gate, New Airport Road, Dhaka-15, Bangladesh

Key words: agricultural growth, Bangladesh, environment, NARS

Abstract

Bangladesh encompasses a wide range of agricultural environments reflected in the patterns of agriculture and the opportunities for agricultural development. The density of population is already one of the highest in the world (less than 0.1 ha per person) and the population is growing fast. During the last two decades agricultural production has increased at an annual rate of 1.8 percent. The area under non-cereal crops has declined and the production of pulses and spices has dropped. Forest resources and floodplain fisheries have declined significantly. The country is also highly vulnerable to environmental hazards: floods, droughts, salinity and coastal tidal surges, loss of land and soil resources, global warming, and any rise in sea level.

It is in this context that the paper discusses the needs of the national agricultural research system (NARS) and its priorities for the management of natural resources and the sustainable development of agricultural production. The strategies include improved management of land and water resources to take advantage of the environmental diversity that exist: more efficient use of inputs, particularly fertilizer, and the conservation and utilization of the genetic diversity available for the improvement of crops, livestock, fisheries and forest production.

The greater emphasis on the integration of natural resource concerns into the research agenda implies a transition from a purely commodity orientation to a systems-based organization of research in which the natural resource objectives and the production-oriented objectives complement one another. Bangladesh plans to make increasing use of systems methods as an aid to strategic policy decisions, and as practical tools for resource management and the development of sustainable agricultural production systems. The paper discusses implications of these changes for the organization and management of research in Bangladesh.

Introduction

Bangladesh occupies an area of 14.4 million hectares in a unique geographic location, extending from the mighty Himalayan mountain chains to the Bay of Bengal. The country is a vast river basin complex made up of the Ganges, the Brahmaputra, and the Meghna, and their network of tributaries. The agriculture of Bangladesh is highly diverse and complex. At present, around 75 percent of the effective land area of the country is used to produce crops, among which rice is predominant. It is grown on almost 70 percent of the cropped land during three different seasons of the year.

Bangladesh is the world's eighth, and Asia's fifth, most populous country, with less than 0.1 ha of arable land per capita. About 11 percent of the rural households do not even have a homestead, and more than 50 percent of the rural households are functionally landless, owning less than 0.20 hectares of land.

P. Goldsworthy and F.W.T. Penning de Vries (eds.), Opportunities, use, and transfer of systems research methods in agriculture to developing countries, 105 - 125.
© 1994 *Kluwer Academic Publishers.*

Bangladesh is situated in the largest delta of the world, with a catchment area of 1.66 million sq. km., of which 92.5 percent is located outside Bangladesh. Frequent natural hazards, including floods, droughts, cyclones, and a combination of north-westerly winds and tidal surges, are characteristic of the area.

Foodgrain production has increased substantially in recent years and, at present, the country is almost self-sufficient. However, the population continues to grow rapidly, and it is expected that the demand for food, fibre, and housing will also continue to grow as rapidly in the immediate future. The additional production required to meet this demand will have to come from greater efficiency in the use of the existing natural resources. This paper discusses the current status of natural resource use, the future demand for agricultural produce, and the technological needs for the sustainable management of the natural resources.

The natural environment

Bangladesh includes a wide range of agricultural environments. These, in turn, represent a great range of opportunities for, and limitations to, agricultural development (Mahtab and Karim 1992). Environmental diversity is evident not only at macro (national and regional) levels, but also at micro (district and village) levels. The small-scale complexity of soils and hydrology is a characteristic of Bangladesh's environments. Together with the considerable year-to-year variability in rainfall, temperature, and flooding, this complexity makes the planning of environment-specific agricultural development, research and extension programmes difficult.

Land and soils

Thirty-four physiographic units and subunits have been recognized in Bangladesh. The descriptions are contained in 34 reconnaissance soil survey reports produced between 1964 and 1975. Floodplain and piedmont plain units together occupy almost 80 percent of the total land area. Slightly uplifted fault blocks or terraces occupy about 8 percent, and hills about 12 percent.

About 500 soil series have been identified through reconnaissance soil surveys. These soil series are grouped into 21 general soil types, which in turn can be placed in three groups according to their association with the three main physiographic units: floodplain, terraces, and hills. Floodplain soils are formed from alluvial sediments that range from a few months to several thousand years in age. Terraces soils are associated with a wide range of relatively older soils formed over the Madhupur Clay.

On most floodplain and valley land, cropping patterns are determined primarily by the seasonal flooding regime, i.e., the dates when inundation begins and ends, the depth of inundation at peak levels, and the risk of damage to crops by early, late or deep floods. The extent of the main inundation land types in the country is given in table 1. Differences in levels of only a few centimeters between neighboring fields may influence choice of crop, varieties, or management practices.

Table 1. Extent of inundation land types in Bangladesh

Land type	Area (millions ha)	Percent of total
High land	4.200	29
Medium high land	5.040	35
Medium low land	1.771	12
Low land	1.102	8
Very low land	0.193	1
Total land area	12.305	85
River, urban, homesteads, etc.	2.178	15

Climate

The climate of Bangladesh is sub-tropical and there are three crop-growing seasons in a year; the pre-*kharif*, *kharif*, and *rabi* growing periods. The duration of each can be defined using a water-balance model that relates precipitation to potential evapotranspiration. In the *kharif* growing period, the amount of water available to crops is determined primarily by rainfall, but it is modified by seasonal inundation. About 250 mm of soil moisture is required to grow a satisfactory dryland *rabi* crop such as wheat or pulses, if planted on time.

There are two main ways in which temperature influences the cropping pattern: the duration of the cool winter period and the duration of extremely high temperatures in summer. Variations in these two are the basis for differentiating five thermal zones in the country. The mean length of the cool winter period varies from 30-40 days near the coast to more than three months in the North-West and North-East. In April and May, maximum temperatures sometimes exceed 40°C for periods of from half a day to 15 days. The West and the South-West parts of the country are most affected by these extremely high temperatures.

Information on topography, soils, land level in relation to flooding, and climate were combined to define the characteristics of 30 agroecological regions and 88 subregions (UNDP/FAO 1988). The main climatic features that were used to characterize agro-ecological zones were:

- the average duration of the pre-*kharif* period, when rainfall is intermittent and uncertain;
- the average duration of the wet (*kharif*) season that depends on rainfall, and of the dry (*rabi*) growing period;
- the average number of days in a year with a minimum temperature below 40°C;
- the average number of days in a year with a maximum temperature higher than 40°C.

Genetic diversity

Bangladesh is rich in germplasm resources of some of the world's most important crops, namely: rice, jute, cotton, sugarcane, tea, mango, banana, lime, litchi (*litchi chinensis*), bean, brinjal (*solamum melongena*), amaranths, taro (*colocasia esculenta*), and yam. Five of the ten semiautonomous agricultural research institutes of Bangladesh maintain germplasm collections. Two of the institutes between them account for about 8000 accessions, of which 5000 are rice. Smaller collections include timber, bamboo, and tea species. The extent of the loss of genetic diversity, as a consequence of the widespread adoption of high-yielding varieties of rice in particular, has not been properly assessed. There needs to be greater awareness that the exploration and conservation of genetic diversity are prerequisites for sustainable agricultural development.

Many edible species of fish are found in Bangladesh, and there is no ecological equivalent anywhere else in the world. Through genetic manipulation, the productivity of local varieties could be improved to yield a sufficient quantity of fish and fish products to feed the entire population. Not only are the local species (such as tilapia, or silver carp) more palatable than exotic ones, they are better adapted to the local ecology and offer a better prospect of ensuring an adequate supply of fish protein for future generations at an economic price.

According to recent studies, 25 forest species are endangered in Bangladesh as a result of urban development and a rapidly growing population. In the past, a vast area of Madhupur forest has been cleared and replaced with crops. The clearing continues in the South-East where hill forest is being cleared to build houses.

Water resources

Although an immense quantity of surface water flows through Bangladesh in the three major rivers, the Brahmaputra (commonly called the Jamuna), the Ganges, and the Meghna, the potential for development of these water resources is limited for a number of reasons. The most important one is that the topography is such that there are very few opportunities for either gravity diversions or surfaces storage.

Recently there has been a rapid increase in the exploitation of groundwater for irrigation in response to a change in government policies to encourage private sector investment in shallow tubewells (STWs). The installation of new STWs increased to nearly 40,000 per year between in 1988 and 1990. The irrigated area in Bangladesh has increased by almost 700,000 ha since 1985 (table 2). About 95 percent of this increase was accounted for by tubewells, of which 58 percent were STWs and 36 percent deep tubewells.

Population growth

High fertility rates have caused an increase in the population of Bangladesh from 50.8 million in 1961 to 115 million in 1990. Recent studies indicate that the country currently has a crude birth rate (CBR) of 35.2, and a total fertility rate (TFR) of 4.5, compared with 12.9 in 1961 and about 6.0 in 1990. Although these achievements seem to be appreciable, they are still well short of what is needed to control the growth of the population.

Table 2. Changes in irrigated area from 1985 to 1990

Mode	'000 ha in 1985	'000 ha in 1990	'000 ha net change since 1985	Distribution of net change percent
Low-lift pumps	600	639	+39	+5.6
Shallow tubewells	481	889	+408	+58.4
Deep tubewells	288	542	+254	+36.3
Major irrigation works	123	186	+63	+9.0
Traditional irrigation systems	432	367	-65	-9.3
Total	1,924	2,623	+699	100.0

Source: Ministry of Irrigation and Flood Control (1990)

The Bangladesh Planning Commission estimates that by the year 2000, the population will have increased to 137 million and by 2010, to 154.5 million. On the other hand, the World Bank's estimate, based on the assumption that the net reproductive rate (NRR) will be halved by 2025, forecasts a population of 145 million, 177 million, and 232 million by 2000, 2010, and 2030 respectively (World Bank 1992). These projected increases in the population represent an enormous additional demand for food in the future. With little or no additional cultivable land available, the agricultural sector faces a formidable challenge if it is to satisfy this demand, even at the present low levels of nutrition.

Growth in agricultural production

Crops
From 1965 to 1985, agricultural production in Bangladesh grew at an annual rate of 1.8 percent, with higher rates of growth (2.3 percent) in cereal production. The production of foodgrain has increased from just over 9 million tons in 1960 to 18.5 million tons. But this growth had a narrow base—it has been due entirely to the increase in production of dry-season, irrigated boro rice and wheat, and it has been achieved partly at the expense of the production of oilseeds, sugarcane, pulses, and spices. The area under non-cereal crops as a whole has declined in absolute terms, and their aggregate production increased at a rate of only 0.9 percent per annum. Within this total, the production of pulses and spices declined (table 3). In recent years, the production of wheat, which increased very rapidly at first and which constitutes about 5 percent of total cereal production, has begun to stagnate.

The wide spectrum of agricultural environments in Bangladesh, which include lowland to highland areas, provide opportunities for a diversified agriculture, particularly on the floodplains (Mahtab and Karim 1992). Thirty-six percent of net cropped area (NCA) is flood-free land, which is suitable for almost all crops.

110

Table 3. Percentage change in poduction of selected crops in Bangladesh, 1972-1991

Crops	1972-1981	1981-1991	1972-1991
Paddy			
Aus	1.05	-2.92	-0.69
Aman	1.55	-1.00	0.19
Boro	0.85	1.84	2.38
Wheat	14.10	-0.77	6.03
Potato	1.30	0.18	1.15
Jute	0.36	-0.12	0.10
Sugarcane	0.71	0.18	0.45
Vegetables	1.74	0.91	2.32
Fruits	0.06	-0.53	0.37
Spices	-2.1	0.79	-0.04
Pulses	0.04	-1.34	
Oilseeds	1.72	-1.35	
Tea	5.77	0.86	2.50

Thirty-five percent of the NCA is shallow-flood land, which usually has better soils than flood-free land and it is less prone to drought. Because of this, it produces a larger share of local transplanted aman, wheat, jute, irrigated aus, potato, pulses, spices, and other minor crops, than the flood-free land. Because the modern varieties of aman rice are shorter than local varieties, they are not suited to the entire area of shallow-flooded land. The remaining 29 percent of the NCA is deep-flooded, and the predominant crop in the wet season (*kharif*) is deep water rice (DWR). The cropping on these lands is less intense than elsewhere (only 107 percent). Broadcast aman accounts for 45 percent of the crops sown, 29 percent is boro rice, and the remaining 33 percent is made up of local transplanted aman, broadcast aus, jute, wheat, potato, pulses, and oilseeds.

Forest
The forest resources of Bangladesh are limited. There are 2.24 million hectares of forest land, of which 1.36 million hectares are hill forest, 0.12 million hectares inland sal forest, and 0.68 million hectares littoral mangrove forest. Village forest accounts for 0.27 million hectares on privately owned homestead land scattered all over the country.

The productivity of the forests is extremely poor and it is declining. From the natural hill forest it is only 0.5-1.5 m^3 ha^{-1} $year^{-1}$, and from mangrove and sal forests the productivity is even lower. The production from teak plantations in the hill forests is 2.5 m^3 ha^{-1} $year^{-1}$, but in the past it was as high as 7-10 m^3 ha^{-1} $year^{-1}$. The decline in yield is occurring in all the major forests. The growing stock in the Sunderbans mangrove forests has been depleted from an estimated 20.3 million m³ in 1960 to 13.2 million m³ in 1984. In the Chittagong Hill Tracts forest reserve the growing stock has declined from 23.8 million m³ in 1964 to less than 19.8 million m³ in 1985. The Resources Information Management System (RIMS) estimates the growing stock of coastal plantations at about 5.05 million m³ in 1986, but a recent inventory of forest resources revealed that the area of village forest plantations is decreasing. Overall, there has been a 25-percent depletion of the forest resources of Bangladesh in the past 24 years, as a result of forest land clearing for other uses and the rapidly growing demand for fuel wood.

Livestock

In the past 25 years, the most rapid growth in the livestock sector has been in poultry production, which has increased by 6 percent per year. The production of goats increased by 3 percent per year and of cattle by about 0.5 percent per year. The populations of buffalos, sheep and ducks have not changed. Cattle and buffalos are an important source of draft power for agriculture. They provide more than 98 percent of the draft power for crop production, and a substantial amount of power for rural transport and the post-harvest processing of crops. With the progressive increase in cropping intensity to meet the growing demand for food, there has been a corresponding increase in the demand for draft power. There is now an acute shortage at peak periods of land preparation, and this in turn acts as a constraint on further growth in crop production. With most of the land under crops, there is little land available for production of animal feed and forage, resulting in poor animal nutrition and low output. This further compounds the shortages of draft power at critical times in the cropping calendar, when the workload is greatest.

Fisheries

In 1988 total fish production was 820,000 million ton, of which 230,000 million ton was marine fish. The production from floodplain fisheries is declining. The construction of embankments and dykes for flood control and irrigation on the floodplain is seriously degrading the environment for fish and shrimp production in inland waters.

Vulnerability of the environment

The frequent occurrences of floods, droughts, cyclones, and a combination of northwesterly winds and tidal surges are characteristic of the delta region in which Bangladesh is located. These natural hazards all cause substantial damage to crops, and the stability of the country's agricultural production is adversely affected by these natural events. In recent years, devastating floods and cyclones have become more frequent and the loss of life and the destruction of crops and other property has been heavy.

Floods

There is a very high risk of flooding of different kinds (flash floods, river floods, local rainwater floods, and tidal floods or combinations of these) on 1.32 million ha of cultivated land and a moderate risk on a further 5.05 million ha. The floods of 1987 and 1988 were the worst on record. In 1988 more than 60 percent of the country's land area was inundated. The area affected by flash floods varies greatly from year to year. Eight of the 30 agroecological regions into which the land has been divided are vulnerable to flash floods.

Droughts

Droughts of varying severity occur in different parts of the country almost every year. About 2.3 million ha of cropped land are severely affected by drought during the *rabi* growing season, and a further 1.2 million ha during the *kharif* season. The severe drought between 1967 and 1982 caused considerable losses of crops (Karim et al. 1990). Transplanted aman rice is the main crop affected during the *kharif* season. More than 45 percent of the attainable yield may be lost on about 0.57 million ha of very severely drought-prone land, equivalent to an annual loss of about 1.5 million tons of foodgrain from existing aman rice-producing areas. During the *rabi* and pre-*kharif* seasons, wheat, potato, mustard, and broadcast aus are the principal crops affected by drought. More than 60 percent of the attainable yields of these crops are lost on 0.364 million ha in very severely affected areas. The predicted changes in the global climate mean that the area that is prone to drought is likely to increase in the future (Karim 1993).

Salinity and coastal tidal surges

More than 30 percent of the cultivated area is near the coast. Of this, 0.83 million ha is affected to different degrees by soil salinity (Karim et al. 1990). Its potential for development for agriculture is very poor, though there are cropping systems recommended as suitable, depending on the degree of salinity.

In April 1991, the worst-ever cyclone, with windspeeds of up to 225 km/hour and accompanied by tidal surges of between 3 and 6 m high, hit the southern coastal region. According to a government report, the death toll exceeded 125,000 people, but unofficial sources reported that as many as 400,000 people died.

Loss of land and soil resources

High-yielding, profitable crops remove significant quantities of plant nutrients from the soil. These must be replaced if soil fertility and the long-term productivity of the soil are to be maintained. The rate of depletion of the main plant nutrients (N,P,K) and of S from the intensive cultivated soils ranges from 180 to 250 kg^{-1} ha^{-1} $year^{-1}$ (figure 1).

The intensive use of the land has already severely depleted the organic matter content of the soil and resulted in deficiencies of secondary and micronutrients such as S, Zn, B, and Mo in some areas. Current research data from the national agricultural research institutes demonstrate that the levels of organic matter are approaching

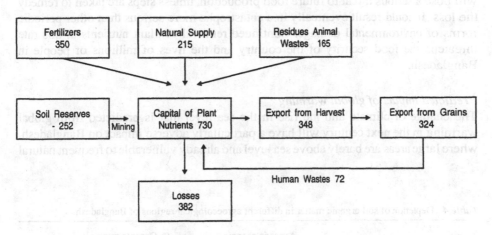

a) Nutrient Depletion in Boro (HYV) - T. Aus (HYV) - T. Aman (HYV) Pattern. Yield level : 12 - 14 t/ha/yr.

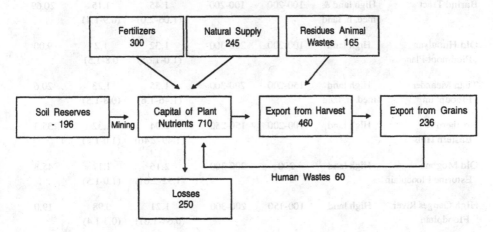

Figure 1. Depletion of nutrients under two intensive cropping systems

critically low values in more than 60 percent of the arable soils, and that these continue to decline at an alarming rate (table 4). There are more than 4 million ha of arable soils that are deficient in sulphur, and a further 2 million that show a deficiency of Zn. As a result, yields of many crops are no longer increasing. The loss of soil fertility will pose a serious threat to future food production, unless steps are taken to remedy the loss. It could result eventually in a catastrophe more serious than other periodic forms of environmental damage. Continued removal of plant nutrients at this rate threatens the food security of the country and the lives of millions of people in Bangladesh.

Predicted impact of global warming
The change of climate and the rise in the sea level that is predicted from global warming in the next century will have a particularly adverse impact on Bangladesh, where large areas are barely above sea level and already vulnerable to frequent natural

Table 4. Depletion of soil organic matter in different agroecological regions of Bangladesh

Name of AEZ	Land type	Average cropping intensity %		% Organic matter (range)		Depletion of o.m. %
		1970	1990	1969-70	1989-90	
Madhupur Track	High land	150-200	150-300	1.78 (1.3-2.4)	1.20 (0.6-1.7)	32.58
Barind Tract	High land & med. h. land	100-200	100-200	1.45 (1.06-2.0)	1.15 (0.9-1.4)	20.69
Old Himalyan Piedmont Plan	High land	100-200	200-300	1.32 (1.0-1.65)	1.2 0.8-1.5)	9.00
Tista Meander Floodplain	High land med. h. land	150-200	200-300	1.55 (1.46-1.6)	1.23 (0.8-1.5)	20.6
Northern & Eastern Hills	High land	100-200	150-250	2.04 (1.49-2.46)	1.32 (1.0-1.5)	35.3
Old Meghna Esturine Floodplain	High land	200	200-300	2.16 (1.92-2.61)	1.17 (1.0-1.5)	45.8
High Ganges River Floodplain	High land	100-150	200-300	1.21 (0.64-1.61)	0.98 (0.3-1.4)	19.0
Old Brahmaputra Floodplain	Med. h. land	150-250	200-500	1.56 (1.09-2.16)	1.23 (0.9-1.5)	21.15

Source: Reconnaissance soil survey data supplied by SRDI for the period 1970s and recent (1989-90) data obtained from NARS institutes, and computerized at BARC

calamities. According to estimates of the Intergovernmental Panel on Climate Change (IPCC), the sea level will probably have risen by 20 cm by the year 2030, and 65 cm by the end of the next century.

In low-lying areas, inundation would be increased further by water from the landward side that is held back by the rising sea level in the Bay of Bengal. The Sunderbans mangrove forests, covering more than 400,000 ha in the Southwest, would be destroyed by the increase in salinity. It would also adversely affect the supply of drinking water and the supply of water for agricultural and industrial production. Overall, the salt affected area would increase from 13 percent of the total land area at present to more than 30 percent.

The demand scenario

World Bank projections indicate that, within the next 20 years, Bangladesh will have to support an additional 45 million people. By the year 2030, the population is expected to be about 190 million, with higher per capita incomes than today. To satisfy the anticipated future demand for foods of different kinds that a population increase of this magnitude represents, food production will have to double within the next 20 years, and within the next 40 years it will need to be between 3.8 to 7.8 times the present level (table 5). To achieve this, crop production would have to increase in a sustained manner at a rate of 4.2 percent per year for the next 40 years.

These projections are based on an assumption of a moderate decline in population growth rate to 1.8 percent per annum by 2010, and an average annual growth rate of about 2.1 percent over the period. The growth in GDP over the same period is assumed to be 5 percent, or 1 percent above the growth rate in recent years, resulting in a per capita income growth of 2.9 percent per year.

Priorities for sustainable growth in agricultural production

Bangladesh has only very limited opportunities for generating export earnings with which to purchase food supplies on world markets. It therefore has little option but to find ways to produce most of what it needs from its own resources.

Since there is almost no new land to bring into cultivation, and even marginal lands are already cultivated, the only course for agriculture, if it is to produce sufficient food, fibre, and fuel in the future, is to increase the productivity of the limited land resources available, particularly in the most favorable ecological regions of Bangladesh. Traditionally, this will mean even more intensive use of those resources. Intensive cultivation in many parts of the country, however, has already resulted in stagnant or declining production. If higher yields and productivity are to be achieved and sustained, new technologies and strategies will be required, and natural resource management (NRM) concerns will have to be integrated into the production-oriented research agenda. Research will have to focus more on efficiency in the use of resources: on ways to reduce dependence on nutrient inputs, more

Table 5 Projected requirement for different kinds of food items to satisfy dietary needs and economic demand

Year	1990			2010					2030			
Popula-tion	108 3 million			153 3 million					189 9 million			
		Con-				Demand*					Demand*	
Food item	Physio need	sump-tion	Physio need	2%	3%	4%			Physio need	2%	3%	4%
Plant food												
Cereals	18 44	19 99	26 13	31 37	33 86	36 05			32 37	43 82	50 62	57 77
Pulses	2 96	0 91	4 20	1 99	2 28	2 48			5 20	3 22	4 19	4 88
Sugar/Gur	1 38	0 38	1 96	0 81	1 01	1 18			2 43	1 64	2 68	3 54
Tubers	5 34	1 56	7 55	3 08	3 59	4 06			9 36	5 14	6 86	8 89
Vegetables	8 50	5 28	112 03	10 17	11 88	12 72			14 90	16 95	22 62	26 74
Fruits	2 57	0 53	3 64	1 36	1 78	2 15			4 51	3 05	5 23	8 18
Edible oil	0 67	0 49	0 95	0 70	0 78	0 86			1 18	1 10	1 35	1 63
Animal food												
Milk	3 52	0 91	4 98	1 99	2 45	2 81			6 17	3 84	5 73	7 78
Meat	0 95	0 24	1 34	0 52	0 60	0 67			1 66	0 92	1 23	1 52
Eggs	0 36	0 23	0 50	0 45	0 53	0 59			0 62	0 77	1 07	1 35
Fish	1 56	1 55	2 24	3 11	3 80	4 34			2 77	5 90	8 65	11 52

* percent values in Demand columns indicate per capita income-growth rates

effective means to recycle nutrients, and ways to reduce nutrient losses. There is scope also for more research into developing crops tolerant to adverse conditions (e.g., acid and saline soils). There is little attempt at present to put a true valuation on the consequences for the natural resource base or the environment of different agricultural activities or technologies. The need for resource pricing and resource-use modelling is likely to become more pressing in the future, and to address these issues the NARS will need the help of resource economists. It follows that agricultural research has a central role to perform in the development of sustainable agriculture in Bangladesh.

Technological achievements will not be enough on their own. The NARS will need to understand the policy environment and develop a capacity to identify points of intervention in a policy framework. While increased food production is a necessary condition for alleviating hunger, it also requires supporting policies to increase employment and incomes, and to ensure not only equitable access to food, but also to the resources required for its production. A lack of employment opportunities, a skewed distribution of incomes, and insecure land tenure, are all obstacles to sustainable agricultural development in Bangladesh.

Strategy for sustainable management of natural resources

Reference has already been made to the diversity of the agricultural environments in Bangladesh. The agricultural production systems associated with these environments are similarly diverse, but the close integration of crop, livestock, fish, and tree production at the farm level is a characteristic of almost all of them. A systems approach is required for research on ways to improve the productivity of the use of resources in these integrated systems. Research on farming systems is already established at locations that represent nine of the main agroecological environments, but much more of this kind of research will be needed in the future.

The very comprehensive database of the natural resources of Bangladesh that has been developed is a key instrument for future land-use planning. It is being used to identify areas where, because of current production needs, resources are over-utilized, or where degradation of the natural resource base will threaten future production, unless new technologies are introduced. The identification of areas of different potentials also helps to reveal where resources are under-utilized, and where there are opportunities for improving the productivity of the land and water resources available.

With the help of information derived from this natural resource database, the 30 agroecological regions to which reference was made earlier (see p. 107), have been grouped for practical purposes into 13 agricultural production, or development units of differing characteristics and potential (figure 2).

Some of these zones represent areas of comparatively high agricultural potential in which irrigation development, the extension of irrigation to crops other than rice, measures to improve and sustain higher standards of management of soil fertility and plant nutrients, greater diversification of cropping, and new tillage technologies to permit more timely land preparation in intensive cropping systems, are some of the options available for increasing productivity.

Other zones represent areas of lower potential that include acid and saline soils, sandy floodplain soils that require special management, sloping and eroded soils, drought-prone regions or soils, and poorly drained land. In these areas, research has to focus on the development and introduction of varieties and production systems that are specificly adapted to local conditions. These are usually the more impoverished regions, and they need employment-generating technologies, integrated watershed management, measures to strengthen agricultural supply and marketing services, and systems research to assess risk and develop risk-avoidance technologies.

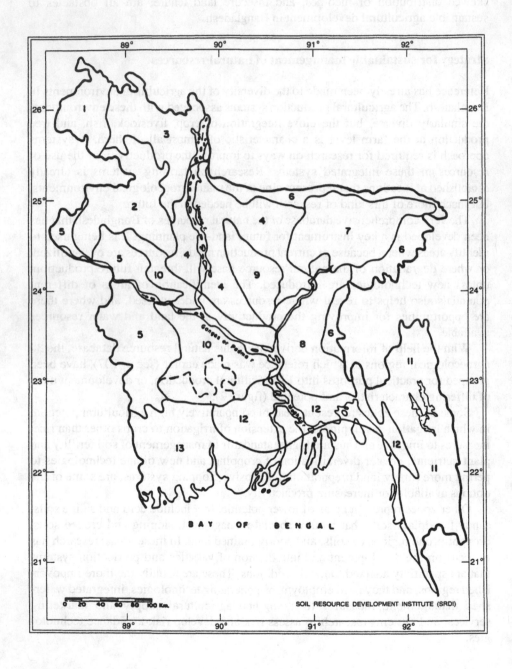

Figure 2. Agricultural production unit/development units of Bangladesh

Land allocation

Characterization of the agricultural environments in this way allows research scientists and planners to examine different options for the production of priority food crops, and to determine where each commodity can be produced best to meet the nation's needs and to make the most effective use of the land and water resources available.

The natural resource database was used, together with information derived from past research and field trials, to characterize these different agricultural production or development units, and assess their suitability for the production of particular crops. Estimates were then obtained of the maximum production that could be achieved by examining the data for different combinations of land units and commodities (table 6). From this examination, a near optimum combination was selected. Some of the main findings from this examination were as follows:

Table 6 Maximum achievable production

Crop	1990			2010			2030		
	Area	Yield	Prod	Area	Yield	Prod	Area	Yield	Prod
Rice	10440	1 71	17 85	10290	13 25	33 44	7600	6 00	45 60
Wheat	599	1 68	1 00	486	3 30	1 60	400	6 00	2 40
Maize etc	11 2	0 72	0 68	350	3 85	1 35	1300	9 00	11 70
Pulses	728	0 72	0 52	1400	1 33	1 86	2800	2 30	6 44
Sugarcane	187	40 20	7 68	215	76 00	16 34	250	150 00	37 50
Potato	169	9 98	1 78	225	20 00	4 50	300	40 00	12 00
Winter vegetable	105	7 21	0 76	215	16 80	3 61	550	40 00	2 00
Summer vegetable	65	5 05	0 33	150	8 75	1 31	200	16 00	3 20
Fruit per	100	5 21	0 52	145	8 00	1 16	200	12 00	2 40
Fruit annu	69	13 47	0 93	175	18 50	3 24	300	25 00	7 50
Oilseed	570	0 79	0 45	870	1 45	1 26	1500	3 00	4 50
Jute	584	1 65	0 96	460	2 70	1 24	400	4 40	1 76
Cotton	19	0 86	0 02	170	1 75	0 30	550	4 00	2 20
Tobacco	38	0 89	0 03	67	0 92	0 22	100	1 00	0 10
Spices	147	2 17	-	445	3 92	0 06	1350	7 00	9 45
Forage	N/A	-		200	20 00	1 74	1000	35 00	35 00
GM	N/A			350		5 60	600		
Others	N/A			550			810		
TCA	15184			17913			22110		
Cropp int	166 1%			196%			241%		

Note Prod = production, TCA = total cropped area, Cropp int = cropping intensity
Area in 1000 ha, Yield in ton/ha, Production in million tons

Rice: During the dry season, primarily irrigated rice is grown, whatever the type of land. The total area under rice has to be reduced gradually from 10.4 million ha to 7.6 million ha by the year 2030. Land released from irrigated rice cultivation would then be made available for the production of other crops in different seasons, depending on the suitability of the particular land. However, the area of transplanted aman and of transplanted aus will not be reduced, since these occupy land in the *kharif* season, on which only these crops can be grown. The major change would therefore be the release of lands from irrigated boro rice production, which is mainly on upland soils.

Wheat: No increase in the wheat area is proposed, because it is expected that wheat production will become progressively more marginal as a consequence of global warming. It is therefore more probable that there will be a slight decrease in the area of wheat by the year 2030.

Maize: This crop has the potential to become a more important crop in Bangladesh. It fits well into the agricultural environments. Its potential, both in terms of area and yield, is large. It could be grown in the *rabi* and pre-*kharif* season in different regions of Bangladesh. A much larger area could be grown on all upland soils. The area sown and the sowing date could be adjusted, depending on the date of the harvest of transplanted aman and of the gradually receding inundation water, because this crop requires well-drained soils.

Pulse crops (e.g., *Lathyrus spp.*, lentil, chickpea, mungbean, blackgram, field peas, and cowpeas): Traditionally grown in Bangladesh, these crops provide comparatively cheap and valuable sources of protein in the diet of millions of people in Bangladesh. By virtue of their ability to fix atmospheric nitrogen, they do not remove nitrogen from the soil. This feature makes them particularly valuable crops for Bangladesh, where soil fertility is declining at a rapid rate. Thus the area under pulses should be increased from 0.7 million to 2.8 million ha.

Oilseeds: Different species of oilseeds are grown. Collectively, they include adaptations to a wide range of environments. Crops like mustard need fertile soils, but groundnut requires well-drained light-textured soils. Many of these crops could be grown profitably during the *rabi* and pre-*kharif* seasons, with most of the additional area on about 0.4-0.5 million ha of the active floodplain area, as part of a double cropping system with rice. Management practices suited to such a system on the floodplain need to be developed.

Potato and sweet potatoes: These could be grown in almost all the agroecological units where there are light-textured soils. The area grown could be increased from 0.17 ha to 0.30 million ha by 2030, and most of this would be for the production of sweet potatoes.

Vegetables: Different types of vegetables could be grown in highland regions, where soils are well drained. Improved management practice for planting and for relay and mixed cropping will be required. All marginal, ridge soils (or levées) in the highlands could be used for summer vegetables. Winter vegetables could be grown in upland areas where soils drain early at the end of the summer season. The time of planting would be adjusted to suit the seasonal temperature variations of the area. Special cultivation practices need to be developed for growing vegetables along with field crops during the pre-*kharif* season.

Cotton: This is a promising crop for some well-drained ridge soils on the floodplain and for upland light-textured soils. It is now grown in a sole-crop system, but cultivation practices need to be modified to permit double cropping, with cotton as the principal crop. Short-duration varieties of cotton are to be developed for this purpose as a priority.

Green manure crops and forage: Like sesbania, legume forages, and other quick growing legume crops, these could be grown in many different situations. Areas of intensive irrigation under high yielding rice varieties could produce a green manure or forage crop at least once after every third crop. For forage production, the pre-*kharif* transition period is the most suitable, where forage would be part of the regular upland cropping pattern, before transplanted aman rice. However, there are acute seasonal shortages of forage also at other times of the year. To alleviate these shortages, the area for forage production could increase from 100,000 ha in 2000 to up to 1 million ha by the year 2030. Some of the production would come from inter-cropping and relay cropping, and most of it would have to be produced locally, where the animals to which it will be fed are kept. Some could be produced on low-quality land such as acid basin clays, peat, and coastal saline soils, which could be made permanently available for forage production. The piedmont plain and the lands in lower slopes of the hill and terrace areas could also be utilized profitably for forage production. Development of aquatic forage for the deeply-flooded land during the monsoon period should also receive research attention. Table 7 shows the rates of growth that would have to be attained to achieve the levels of production indicated in table 6. The yields that would be required are exceptionally high, but some of the best farmers in Bangladesh have shown that such yields are attainable with currently available technologies. But, clearly, to achieve anything like this as a general level of output will be an extremely difficult task. It will demand a high degree of political commitment to agricultural research and development on the part of the Government, and a readiness for significant institutional change on the part of the NARS. The needs are expressed here in terms of commodities, but to achieve these levels of production, research will have to be conducted at the level of production systems and land-use units. The focus on NRM and sustainable agricultural development means that research will no longer be seen solely as a generator of new technology, but also as a source of information for the formulation of development policy and planning.

Table 7. Annual growth rate required to achieve maximum achievable production

Food item	Production (in million tons)			Annual growth required (in %)					
				1990 -2010			2010 - 2030		
	1990	2010	2030	Prod.	Area	Yield	Prod.	Area	Yield
Cereal	18.82	35.04	48.00	3.16	-0.17	3.33	1.59	-1.48	3.07
Pulses	0.52	1.80	6.44	6.58	3.32	3.26	6.41	2.88	2.88
Sugar/Gur	0.54	1.22	3.00	4.16	0.92	3.24	4.60	1.07	3.53
Tuber	1.78	4.50	12.00	4.78	1.25	3.53	5.03	1.50	3.53
Vegetables	1.09	4.92	34.00	7.83	3.89	3.94	10.15	3.67	6.48
Fruits	1.45	4.40	12.00	5.71	3.24	2.47	5.14	2.26	2.88
Edible oil	0.15	0.49	2.97	6.10	2.77	3.33	9.42	26	6.16
Animal food	1.53	4.45	13.57	5.48			5.73		
Milk	0.37	1.87	7.55	8.44			7.23		
Meat	0.27	0.56	1.94	3.62			6.51		
Eggs	0.04	0.28	0.75	10.22			5.05		

Institutional development

The NARS of Bangladesh is a comparatively large one. It consists of an apex body, the Bangladesh Agricultural Research Council (BARC), and ten semi-autonomous research institutes. According to the Human Resources Study of the NARS (BARC 1990), 1337 scientists are working in 11 organizations of the NARS, of which about 13 percent have a PhD degree. BARC is responsible for strategic planning, coordinating research, and evaluating research programs. It is also responsible for maintaining linkages with international research organizations and donors. Two of the research institutes, the Bangladesh Agricultural Research Institute (BARI) and the Bangladesh Rice Research Institute (BRRI), are larger than the remainder and together account for about half of the total number of scientists. Besides the research institutes, there are six universities and a number of other organizations that conduct some agricultural research, but these are not usually recognized as part of the NARS.

In Bangladesh, responsibility for the 10 agricultural research institutes of the NARS is among between five different ministries. Thus, although crops, livestock, and trees commonly occur together in production systems at the farm level, they are subjects for research by quite separate organizations, often resulting in a lack of coordination. The NARS will therefore need to examine institutional and organizational options for closer interinstitutional integration. The greater complexity of research with an NRM perspective means that the organizations responsible for crops, livestock, and trees will need not only closer integration among themselves, but also closer links with natural resource and environment institutes that have not tradition-ally been considered part of the NARS.

The broader research agenda will add to the responsibilities of the NARS at a time when research budgets are declining, and it will create new institutional challenges that require a fundamental change in the way that research is organised and managed. There will be important resource and management implications for the NARS, calling for more interaction between institutions, with more sharing of responsibilities and functions.

At present, the institutes are organized into departments, and they are not accustomed to organizing research around interdisciplinary or interinstitute research teams. Very few have the kind of administrative procedures required to operate a project-based research program, with staff time allocated from different divisions or units to joint research programs and projects. This situation is currently one of the main institutional constraints on the development of a more systems-oriented ap-proach to research.

One way to accommodate the additional objectives and activities associated with the wider research agenda without shifting priorities away from production-oriented research, would be to remove areas of duplication in research, and, where necessary, merge groups or institutes and establish common priorities.

The NARS will have to set priorities for NRM issues of concern to Bangladesh and concentrate on the most urgent and important of them. To do so, it will need new and more appropriate methods of priority setting that take into account production objectives and the environmental and social consequences of different technologies. It will also have to consider the possible trade-offs between different research strategies.

The NARS does not have all the discipline skills needed for the broader research agenda. Its human resources development plans will have to reflect these future needs. It is unlikely, however, that the NARS will have the resources to acquire all it needs directly. However, by forming closer partnerships with universities and other spe-cialized institutions (e.g., for hydrology, meteorology, soil survey, and social sci-ences) the NARS could draw upon them as a source of expertise, particularly in those disciplines in which it is weak.

Conclusions

The NARS has a vital responsibility to develop the technology for the quantum increase in food production that will be required, from the same land resources, to satisfy the future needs of the people of Bangladesh, and to do this in a sustainable manner. To achieve the strategic objectives described in this paper, in the compara-tively short time that is available, will be difficult but not impossible. Greater emphasis will have to be placed on the management of natural resources. This implies a broader research agenda, and a transition from a purely commodity orientation to a systems-based organization of research in which the NRM objectives and the production-oriented objectives complement one another.

124

To make the transition, the NARS will need to re-examine its present organization. The integration of NRM concerns into the research agenda will require an intersectoral perspective, with closer integration between different NRM sectors. The resources used in agricultural production have to be weighed against the benefits of other non-agricultural uses of those resources. It will require stronger and more effective interinstitutional linkages, with procedures for cross-sectoral planning. In a comparatively large research system, as in Bangladesh, this kind of institutional change is often difficult and will take time.

It takes longer for research to have a measurable impact on the status of the natural resource base than it does on the production of individual commodities. With longer time horizons, there will be a need for more sustained support for research. This in turn implies a need for better linkages to policy makers, to increase their awareness of the value of the contribution that the broader research agenda will make to development goals.

Agricultural research will have an important role to play, not only as a generator of new technology, but also as a source of information for the formulation of development policy and planning for the sustainable use of natural resource. As this paper shows, Bangladesh already has an excellent geographically-referenced, computer database of its natural resources, and is making good use of it for strategic planning (Ministry of Irrigation and Flood Control 1990). However, the dependence to date on manual manipulation of data extracted from the database for decision making is a constraint on the ability to explore all options for the optimal combination of enterprises, because it takes time to conduct the analysis, and because the production systems and environments concerned are complex and diverse.

To overcome this constraint, the NARS plans to make increasing use of systems methods to aid strategic policy decisions, and as practical tools for resource management and the development of sustainable agricultural production systems. The use of these methods will also serve to strengthen the consensus on research and development objectives in the cross-sectoral planning process and encourage interdisciplinary research.

At present, there is a scarcity of people in the research organizations with a systems perspective, although the expertise developed as part of the farming systems studies that were a feature of research in the past, provides a valuable foundation on which to build. Meantime the NARS will draw on the expertise available in advanced research institutes to develop its own capacity in the use of these approaches.

Acronyms

BARC	Bangladesh Agricultural Research Council
DWR	deep water rice
GDP	gross domestic product
NARS	national agricultural research system
NCA	net cropped area
NRM	national resource management
STW	shallow tubewells

References

Bangladesh Agricultural Research Council (1990) Human resources study for the national agricultural research system, 4 Vols Dhaka, Bangladesh

Karim Z, Ibrahim A M, Iqbal A, Ahmed M (1990) Drought in Bangladesh agriculture and irrigation schedules for major crops Bangladesh Agricultural Research Council, Soils Pub 34 1-15

Karim Z (1993) Preliminary agricultural vulnerability assessment Drought impacts due to climate change in Bangladesh Pages 159-175 in IPCC Eastern Hemisphere Proceedings of the Workshop on Sea Level Rise and Coastal Zone Management, 3-6 August 1993, Tsukuba, Japan

Karim Z, Hussain S G, Ahmed M (1990) Salinity problems and crop intensification in the coastal regions of Bangladesh Bangladesh Agricultural Research Council, Soils Pub 33 1-63

Mahtab F U, Karim Z (1992) Population and agricultural land use Towards a sustainable food production system in Bangladesh AMBIO 21(1) 50-55

Ministry of Irrigation and Flood Control, Dhaka, Bangladesh (1990) Computerized database of the Master plan organization

UNDP/FAO (1988) Land resources appraisal of Bangladesh for agricultural development Rome, Italy

World Bank (1992) Bangladesh Selected issues in external competitiveness and economic efficiency Report No 10265 - Bangladesh

Bhutan's research needs and priorities for the management of natural resources: a small country's perspective

P.M. PRADHAN[1], K. DORJEE[1] and P.R. GOLDSWORTHY[2]
[1] Department of Agriculture, P O Box 119, Thimphu, Bhutan
[2] International Service for National Agricultural Research (ISNAR), P O Box 93375, 2509 AJ, The Hague, The Netherlands

Key words land use, natural resource management, interdisciplinary research, production systems, systems approach

Abstract
The Government of Bhutan recognizes the role of research for the development of agriculture in the uniquely diverse environments of Bhutan It has provided strong policy leadership and support for the integration of a natural resource management (NRM) perspective into the agricultural research agenda Research has been reorganized to reflect more closely the interdependencies between crops, livestock, and trees found in traditional land-use systems This account of the transformed research program illustrates some of the policy, institutional, and management changes that this implies Computerized information systems and methods of systems research are likely to be particularly useful in the future for the diverse environments in this mountainous terrain

Bhutan

Although Bhutan is a small country of 46,500 sq km, it contains a great diversity of environments, with permanently snow-capped mountains, and, at progressively lower altitudes, alpine meadows, temperate forests, and sub-tropical environments in the foothills in the south.

Six major agroecological zones are recognized (dry, humid, and wet sub-tropical; cool and moist temperate; and alpine), with substantial variations in microclimate within each zone. The rugged terrain makes communication difficult. Cultivable land occurs in small pockets, and rural communities are isolated from one another (MOA 1992). More than 75 percent of the land consists of dense forest, steep slopes, permanent snow caps, or is otherwise unsuitable for easy settlement. Land is ultimately the most limiting factor in agricultural production, with less than nine percent of the land suitable for arable cultivation. Even so, more than 90 percent of the estimated 600,000 people are engaged in subsistence agricultural systems in which crops, livestock, and use of the forest are closely integrated. The renewable natural resource (RNR) sector, which includes agriculture and forestry, is the largest and most important in the Bhutan economy, contributing about 45 percent of the GDP.

Prior to 1960, Bhutan had little close contact with the outside world. It was not until the mid 1960s that moves to develop the economy and to build a system of roads began. Since then, the opening of a traditionally self-sufficient, mainly subsistence economy, to the opportunities for trade has been one of the major forces for economic

P Goldsworthy and F W T Penning de Vries (eds), Opportunities, use, and transfer of systems research methods in agriculture to developing countries, 127 - 138

change in Bhutan. Its geographic location makes the economy increasingly dependent on its trade with India.

Patterns of agriculture and land use

In the valleys and foothills where irrigation is possible, rice is usually grown during the summer, and wheat, mustard, buckwheat, pulses, and vegetables are grown in the winter. Maize, potato, and soybean are some of the major summer crops grown on slopping drylands. Crop yields are generally low, and there is a need for better crop varieties and production practices.

There are about 420,000 head of cattle in Bhutan. Most of them have to scavenge for feed in the forest, along roadsides and in areas immediately surrounding farms. Individuals own on average five to six animals, but most of the animals are gathered into larger groups that move from the forest in the winter to the alpine pasture in the summer.

The forest has been used traditionally to satisfy the needs of rural populations, but increasingly forest products are also contributing to manufactured output and export. The present production, which accounts for about 15 percent of GDP, is only about one third of the estimated long-term sustainable production.

Six traditional land-use, or agricultural production, systems are recognized:

- *Dryland production system.* This system accounts for the largest area among the six production systems (about 54,000 ha) with about 58,000 households. Crops, livestock, and forest are all important components of the system. Productivity and incomes are low because the land is marginal for crop production, while services and transport are lacking and access to markets is limited. The main crops are maize and potatoes, with a variety of less important oilseed, pulse, vegetable, and minor cereal crops. Potatoes are becoming an important cash crop, but maize is largely for home consumption.
- *Pastoral production system.* This production system includes little or no arable agriculture, and people depend almost entirely on the yak, sheep, and cattle for their food. The herds move from the alpine regions in the summer to the lower altitudes of the warm temperate regions in the winter.
- *Plantation/orchard system.* This system extends across all but the alpine region of the country. Apples, oranges, cardamom, and areca nuts are exported to India and Bangladesh. About 16,000 households, on 10,200 ha, make their livelihood from orchard crops.
- *Wetland production system.* In this system, rice, the most important staple food in Bhutan, is the main crop. Wheat, potatoes, vegetables and pulses are becoming more important. Livestock, including cattle, pigs, and poultry, make up an important part of the rural diet. About 42,000 households depend on this system, covering about 27,000 ha.
- *Forest production system.* Forest land and plantations cover about 2.2 million ha, or about 64% of the entire land area. There is an urgent need to develop plans for

effective conservation and utilization of the forest resources, and for the protection of watershed areas.

■ *Tsheri production system.* There is 33,000 ha of Tsheri, an upland, shifting, grassland/crops system on steep slopes that supports around 32,400 households. It is an extensive system in which the traditional long rotation cycles under grass result in very little erosion. Crops, livestock, and forest are all important in the system. The main crops are maize and millet with some buckwheat. Cattle provide milk and meat.

Table 1 shows an approximate indication of the relative economic importance of the crop, livestock, and forestry subsectors and of the six production systems.

There is close integration of crops, livestock, and forestry at farm level in all these systems. Rural households depend on the forest for food, fodder, fuelwood, and timber. Farm animals provide draft power for preparing the land; they are fed crop residues, and their manure and leaf litter are used to fertilize crops. Thus there are significant resource flows between sub-sectors. A large portion of total outputs shown in table 1 are intermediate products that are inputs to other parts of the production system.

This close relation between crops, livestock and forest is characteristic of the traditional land-use systems, and in Bhutan the belief is it has to be maintained if sustainable, productive systems capable of supplying the needs of present and future generations are to be developed. Clearly, the development of one sub-sector will depend to a large extent on the development of the other sub-sectors.

This calls for a systems approach to the management of agricultural environments. Research on crops, livestock and forestry will have to be more closely integrated to address effectively the NRM issues in Bhutan. More information will be required on the agroecosystems for purposes of planning and implementation. More information will also be needed on the research process itself, for research planning and the management of the resources available for research.

Table 1. Estimate of contribution to national production by sub-sector (local currency units, millions)

Production system	Agriculture	Livestock	Forestry	Total	Percent
Pastoral	14.0	126.6	70.6	211.2	8.7
Wetland	166.1	245.8	195.0	606.9	28.1
Dryland	212.6	237.3	196.9	646.8	26.7
Tsheri	34.1	111.6	101.6	247.3	10.2
Plantation crops/orchard	217.7	65.0	34.9	317.3	13.1
Forest	1.4	190.1	199.0	390.5	16.1
Total	645.9	976.4	798.0	2420.3	
Percent	26.7	40.3	33.0		100

Policy

One of the main responsibilities of the research division of the Ministry of Agriculture (MOA) is to make available to the government, in an appropriate form, the information required for development planning. Its mandate places emphasis on finding ways to improve the productivity of the land and water resources used in agriculture in a sustainable manner.

The commitment to increasing self sufficiency in food production where there is little scope for agricultural expansion onto new land, together with an equally strong commitment to conserve the natural resource base, represents a formidable challenge for a small research system. The need to address concerns about the impact of modern agricultural production methods on the ecology is a significant additional responsibility. To meet these objectives, the research division will have to develop and maintain a cadre of well-trained and motivated scientists.

Institutional setting

To date the research system has not been able to fulfill its role adequately, mainly because there is a lack of trained personnel, funds are limited, and there are no planning procedures to focus research on priorities for development. In addition, the institutional setting has not provided effective coordination of research within the RNR sector.

Until very recently, research in the crop, livestock, and forestry sectors was being carried out in three separate departments within the MOA. This did not result in the most effective use of either human or physical resources, nor did it permit important research issues that required interdisciplinary research to be adequately addressed.

Nevertheless, within the MOA, research is considered to be one of the single most important activities, crucial for the effectiveness of other development programs. The Government therefore began a process to strengthen RNR research with the aim of creating a more effective and integrated service.

A review of research

As a first step in this direction, a diagnostic review of the national research system and of Bhutan's agricultural research needs was carried out in 1990, to provide a basis for future planning of RNR research in support of national development goals.

One purpose of the review was to examine options for an organizational structure and operational procedure that would ensure a closer integration of the research functions and strengthen the linkages between research and extension. The aim now is to create within the MOA a research organization with a greater sense of common purpose and to use the resources and facilities available to the best advantage for addressing priority research problems.

The outcome of the review process was to reaffirm the shift in thinking towards an integrated systems approach in research, and a more explicit concern for long-term productive and sustainable use of RNR. The reorganizational changes proposed as a result of the review are intended to overcome the previous shortcomings and to ensure a more interdisciplinary perspective.

Strategic planning

The next stage in the process of transforming the research system along the lines proposed in the review was the preparation of a strategic plan of research for the RNR sector in 1992. A main purpose of the plan was to provide a framework for the reorientation and restructuring of the research function into a single integrated RNR research system.

Consultation process
During the review and the strategic planning that followed, preliminary criteria for determining research priorities were established through dialogue with the stakeholders concerned. The process tried to capture the substantial knowledge demonstrated by research and extension personnel, about the current land-use systems and the constraints that rural populations face.

Five research programs
The core of the strategy that was set out is the reorganization of research into programs, each of which will focus on one of the traditional production systems that have been described. The small number of research personnel is currently the most limiting factor that determines the scope of research. In view of this, five program teams rather than six were formed, and research on the Tsheri system, which is a variant of the dryland system, has been combined into the dryland research program.

Physical facilities
The research teams will be based at four research complexes which form a network and jointly cover the six land use systems. The physical facilities at some of these research complexes will require substantial development and improvement.

Research support units
The interdisciplinary nature of the research to be carried out in the RNR sector will require close collaboration between disciplines and among scientists from different research centers. This approach calls for a range of research services and special expertise to support the research-program scientists. The support units, some of which are already established, are a biotechnology laboratory, a national plant protection-center, and a soil and plant analytical laboratory.

132

Project-based management system
Because of the nature of the terrain and the variation in the agricultural environment associated with it, each of the production systems can be found throughout the country, together forming a mosaic of different land-use patterns. Because of this pattern of distribution of agriculture, some of the scientists will be based at a center other than the one which is home to their program. It will therefore be a matrix type of structure in which individuals have a technical and professional responsibility to their program, but also a responsibility to the center where they are based for administrative affairs. This form of structure calls for a project-based management system which is more flexible than that normally associated with a departmental structure, but one which puts more demand on the managers, if it is to work successfully.

Human resources development
The small number of research staff means that the research capacity is at present quite limited in relation to the scope of the research program that is required. As one of the principal measures to develop greater research capacity, the strategy contains a 10-year human-resource development plan with provision to increase the number of scientists from around 30 at present to 61 within the next five years. The strategy document contains a 10-year projection of the financial provisions required to implement this and other parts of the plan.

Research-extension links
The plan is a comprehensive statement of the strategic priorities for research in the RNR sector of Bhutan, and of how research can be reorganized to make the most effective use of the limited numbers of research personnel and resources that are available. It also recognizes that these changes must also ensure that the links to the extension services of the MOA and other development agencies are maintained and strengthened. Accordingly, the MOA is already implementing corresponding plans for a coordinated series of institutional changes in the extension services.

A system-based planning procedure

In accordance with the strategic plan a research agenda has been developed that is based on the land-use systems. This section presents an outline of the procedure which has been adopted for formulation of the research programs. The procedure is an adaptation of that described by Collion and Kissi (1992), developed originally for a commodity-research program. The eight steps in the procedure are:
1. review the sector;
2. analyze the economic, social, and institutional constraints;
3. assess past research;
4. define research objectives;
5. identify projects;

6. determine research priorities;
7. analyze resource gaps;
8. recommend application.

Review and analysis of system constraints
A working group consisting of research and extension personnel knowledgeable about the particular system was formed for each of the five production systems. Each group reviewed the information available about their system, assessed the potential for increasing its productivity, and developed an analysis of the constraints facing it. Because these initial steps are so important, and the validity of most of what is done subsequently depends upon them, the group's analysis was repeatedly reviewed and revised as necessary, until there was agreement that it accurately represented the real situation. Then, based on the analysis of constraints, the research objectives for the production system were defined and projects were identified.

Priorities between programs
Priorities were determined at two levels; first between production systems and then between projects within systems.

The contributions to national production from the different sub-sectors and from the different production systems were shown in table 1, and it was noted that a large portion of the output was in the form of intermediate products that are used as inputs in other parts of the agricultural system.

To illustrate the extent of these resource flows, a rough estimate of the consumable outputs and the value of the recycled products within the RNR sector is shown in figure 1. Further breakdown of the origin and destination of the recycled internal products clearly demonstrates the complexity of the interaction between sub-sectors, as shown in figure 2.

Estimates such as these will provide an important basis for determining research priorities between and within production systems in the future. For the time being, though, the priority-setting procedure is hindered by a lack of quantitative information on the contributions from each of the land-use systems to national agricultural development. Because of this lack of data, and the difficulty of collating the data sets required for a complex production systems approach, the figures in table 1 and figures 1 and 2 are at best only first rough estimates of system outputs and interdependencies. Further refinement and confirmation will be required in the future.

More information is also required on research costs, development potential, importance of particular technologies, probabilities of research success and adoption rates. This information and a better understanding of the functioning of the systems has to be gained through active interdisciplinary research experience. To this end, a diagnostic survey of the production and farming systems of the country is to be carried out during the first two years of the reorganized research program.

Meantime, by using the data available for present and projected levels of national production of different commodities, subjective group estimates were made of the probable contribution to national production and of the potential for increases in

134

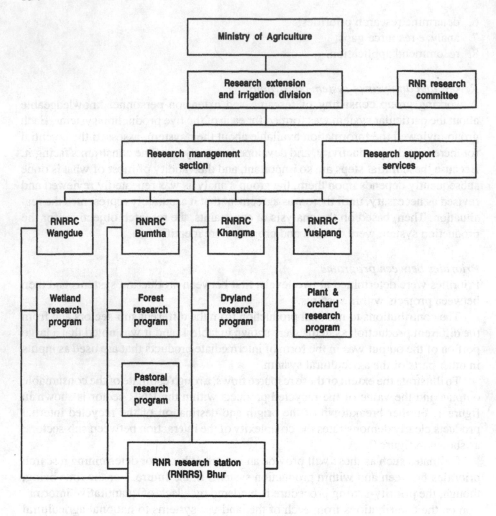

Figure 1. The organizational structure of the RNR research programs

production, for each commodity produced in a given production system. Historical yield and experimental data were used where available. By aggregating the corresponding values of these increases across commodities, for each production system, it was possible to make some comparisons between the different production systems.

First, weights were assigned by the working groups to the overall research objectives of economic efficiency, foreign-exchange earnings, and to goals of equity and sustainability. Then, on the basis of the current data and using only the gross value of the potential production increase as the criterion, the order of ranking of the

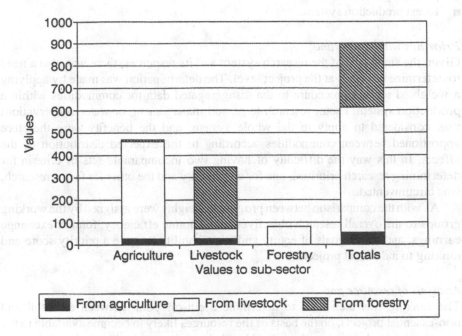

Figure 2. System interactions: distribution of internal values (local currency units)

production systems, in terms of the value of their contribution, would be dryland, forest, wetland, plantation/orchard, and pastoral system.

However, if equity considerations are also taken into account by expressing, for example, the gross value of the potential production increase in terms of the value per farming household within the system, then the dryland production system provides the greatest return per farming household, and about three times as much as the return to households in the plantation and orchard system. Similarly, if foreign-exchange earnings are an objective, then forestry provides the greatest return and the pastoral system the least.

Using the information in this way, a simple ranking of the relative importance of the production systems can be obtained. Given the relative weights assigned by the production system working groups to efficiency, equity, and foreign exchange benefits, an indicative overall ranking of the production systems in terms of the probable returns to research would be:

■ dryland production system;
■ wetland production system;
■ plantation and orchard production system;

- pastoral production system;
- forest production system.

Priorities within programs

Given the small size of the research system and its resources, there was also a need to determine priorities at the project level. The determination was made by applying a weighted scoring procedure to the disaggregated data for commodities within a production system. Factor research (e.g., soil management, or water conservation) was considered to apply to the whole system, and the benefits were therefore apportioned between commodities according to the expected distribution of the effects. In this way the difficulty of having two incomparable sets of criteria for determining research priorities, one for commodity and the other for factor research, was circumvented.

As with the comparison between programs, weights were assigned by the working groups to the overall research objectives of economic efficiency, foreign-exchange earnings, and to the goals of equity and sustainability, to give a priority score and ranking to individual projects.

Analysis of resource gap

The ranking was then used to determine a cut-off point between essential and non-essential projects, on the basis of the resources likely to become available to the program, and judgement on the critical level of activity required to produce an impact on the system (Collion and Kissi, 1992). The latter is also a check on undue dispersion of effort.

When reviewing the ranking, account was taken of the fact that some aspects of research may have to be undertaken before others, or at a particular time. High-priority activities usually deserve immediate attention.

To match program to resources, an inventory was compiled of research staff, by discipline, percentage of time allocated to the program, and location. The resource gap is then the difference between this information and the estimate of resources required to carry out the essential projects. A resource gap may arise because of limitations in the numbers of the research staff, or the mix of disciplines available. It can also occur because there are no resources for expensive equipment or facilities. From the result of this comparison, a final list of proposed projects was prepared.

Features of process

Though the reliability of the data used remains questionable because of the many assumptions made, and the ranking is at best only a guideline to be interpreted with care, there were several reasons why following a simple priority setting procedure was useful in this situation.

An analytical procedure: First, the steps of the planning and priority-setting procedure helped scientists who had little previous experience of formal research planning as a group, through a thought process that relates their work to the work of others, and to the ultimate development objectives which the research is designed to support. It demonstrated where information about the production systems is lacking, and what kinds of data need to be collected to provide an acceptable level of confidence in the analysis of research priorities in the future.

Group dynamics: The process served to elicit information based on the collective experience of the group—information that would not otherwise have been readily accessible.

An iterative process: Judgements reached at each step could subsequently be reviewed quite easily, if required.

Consensus building: This was a vital part of the transformation from a situation in which research activities depended largely on individual initiative and perceptions (with a consequent dissipation of effort), to one in which agreement has to be reached on the most effective use of limited resources. The principle of the weighted scoring method is that, if resources are limiting, the choice of activities to be implemented is based on the value of each activity's contribution to the overall objectives of efficiency, equity, and sustainability, and the estimates of these contributions are based on a group decision.

The two-stage priority-setting procedure also helped. For want of any alternative for the present, the potential increase in the production of a system was estimated from the sum of the potentials of the individual commodities produced in the system. This potential will be reached only if all the constraints are alleviated. But since a single project addresses only part of the constraint, and therefore contributes only a portion of the potential production increase, this rather crude summation procedure uses the consensus process to moderate unrealistic claims for any given project, and to reduce the number of projects to a manageable level.

Future prospects

A systems perspective
The RNR research will have a systems perspective, the focus of which will be the six traditional land-use or production systems. A key issue will be the increased analytical complexity required to balance the multiple objectives, components, and interdependencies that are features of the systems. In the difficult terrain of Bhutan computerized models to simulate the impact of new interventions may eventually provide a valuable complement to the limited testing that can be done in empirical experiments. A wider range of options might then be explored, and long-term effects could be included more readily.

RNR database

A systems perspective requires reliable data. A geographically referenced inventory of natural resources that includes biophysical and economic data on land capability and on patterns of land and water use will be a key requirement for development planning, as well as for RNR research. It can indicate how Bhutan can increase production by making optimal use of resources, while conserving the resource base to meet future needs. It could also be used to make geographical extrapolations to complement local research, with information from research done at similar sites elsewhere—an advantage of special interest to a small country with such diverse environments as Bhutan.

Establishing and maintaining such a database will require specialized skills and it will be a difficult challenge for a small research system. Even so, geographic information systems (GIS) and remote sensing are technologies that the RNR research division will need to use to fulfill its responsibilities.

Decision support systems

There will be a need for priority-setting methods that will accommodate measures of the economic, environmental, and social costs and benefits of different research strategies, and that allows different user groups to be involved.

Management information system

NRM research agenda requires an intersectoral perspective and interdisciplinary teamwork, both of which put demands on the management of national agricultural research systems (NARS) that go beyond discipline or commodity research alone. Successful management will require flexible procedures for project planning, priority setting, and allocation of resources, as well as a computerized management information system to monitor how research resources are linked to priority research projects.

Acronyms

GDP	gross domestic product
MOA	Ministry of Agriculture
NRM	natural resource management
RNR	renewable natural resources

References

Collion M H, Kissi A (1992) A research program planning method—illustration from faba bean in Morroco. ISNAR, The Hague, The Netherlands. (unpublished).

The institutional and organizational implications of integrating natural resource management and production-oriented research

P.R. GOLDSWORTHY, P. EYZAGUIRRE and L. BOERBOOM
International Service for National Agricultural Research (ISNAR), P.O. Box 93375, 2509 AJ The Hague, The Netherlands

Key words: cross-sectoral planning, decentralization, information systems, institutional development, interdisciplinary research, national agricultural research systems (NARS), natural resource management, organization of research, participatory approach, research management

Abstract
This paper discusses some of the organizational and institutional implications for national agricultural research systems (NARS) of integrating natural resource management and environmental concerns into their agricultural research agendas. Integrating these concerns adds significantly to the responsibilities of NARS at a time when there is already a heavy demand on their shrinking research resources. NARS face the difficult task of maintaining a balance between production-oriented research and environmental concerns. This paper discusses the implications for NARS of four key issues related to organizational and institutional change: cross-sectoral planning as the key to integrating environmental and development goals; information as a key requirement for the planning process; a participatory, decentralized approach in both planning and implementation as a key to success; and institution building as an essential objective in all sectors related to agriculture.

Introduction

Developing countries will experience increasing pressure on their natural resource base. The world's population is expected to double by the year 2050, with 80 percent of that increase in developing countries (UN 1992). The production increases to meet demand will increasingly depend on whether the productivity of the existing land, water, and biological resources can be improved. In addition, these resources are coming under greater pressure for competing uses or are being depleted by the production necessary to meet the current demand for food (Cunha and Sawyer 1991; Pingali 1991; Crosson and Anderson 1992).

The growing demand for food will continue to be the highest priority of governments of developing countries, while agriculture, livestock, and forestry will continue to use the majority of the land and water resources (Antle 1993). The increasing pressure on and competition for natural resources may well deplete the land, water, and biological resources crucial to the survival and well-being of future generations. It is because of this gloomy prospect that development and agricultural policies have begun to put greater emphasis on environmental concerns (Douglass 1984; Oldeman et al. 1990; Davies et al. 1991), culminating in the UNCED 1992 "Earth Summit".

Making agriculture more efficient in its use of natural resources, while at the same time conserving resources for future productivity and general well-being, requires

139

P. Goldsworthy and F.W.T. Penning de Vries (eds.), Opportunities, use, and transfer of systems research methods in agriculture to developing countries, 139 - 152.
© 1994 *Kluwer Academic Publishers.*

informed policies, new technologies, and better information about the condition of natural resources. Agricultural research, therefore, lies at the heart of NRM. Sustainability issues will increasingly have to play an integral part in any agricultural research policy. In this respect, agricultural research has to carry out various essential tasks: provide information on the state of natural resources, identify production systems that optimize the use of those resources, and produce or identify more efficient production technologies and conservation techniques.

These tasks overlap with the current scope of many national agricultural research systems (NARS) but may require a new perspective. In some cases, agricultural research may need to assume new tasks, which will have organizational and management implications for NARS. In other cases, agricultural research will have to forge stronger linkages with policymakers, local institutions, and non-government organizations (NGOs) involved in natural resource management (NRM). Thus, further improvements in agricultural productivity will depend not only on new technologies but also on institutional reform and ecosystem management (TAC/CGIAR 1989; Ruttan 1991).

This paper emphasizes the organizational and institutional implications for NARS in developing countries of integrating NRM and environmental concerns into their research agendas. Four key issues related to organizational and institutional change are discussed: cross-sectoral planning as the key to integrating environmental and development goals; information as a key requirement for the planning process; a participatory, decentralized approach in both planning and implementation as a key to success; and institution building as an essential condition for the sustainable development of agriculture and a more rational use of resources.

New criteria and broader scope in sustainability and NRM

Lynam, in an earlier paper (Lynam and Herdt 1989) as well as in this volume (see Section A), discusses the definition and implications of "sustainability" for agricultural research. As an evaluation criterion, "sustainability" is best used at the level of a farming system. There is now general agreement that the natural resource factors and the ways in which they are used in production are crucial determinants of sustainable systems. Hence the focus on total factor productivity as in the definition proposed by Crosson (1993), cited by Lynam (p. 4).

Resource economists consider the concept of total factor productivity the most appropriate measure of the sustainability of a production system (Lynam and Herdt 1989; Graham-Tomasi 1991; Crosson and Anderson 1993). Changes in total factor productivity then serve as an indicator of agricultural development and performance. Its validity depends on the correct measuring of the quantities and values of outputs and inputs over time—a highly difficult task (Crosson and Anderson 1993).

NRM is a broad concept that encompasses resource management and use, including, but not restricted to, agricultural production. It moves the level of discussion away from a focus on production systems to the level of land-use systems. These systems, in turn, have a hierarchy of levels, including the major ecoregions.

The concerns about NRM and sustainable agricultural development have introduced new criteria for designing and evaluating new technologies and production systems. However, the concerns themselves hardly guide research managers in developing sustainable production and resource management systems (Chopra and Rao 1991).

Defining the system: levels and boundaries

A major problem for research institutions and the research leaders who manage them is that much of the work focuses on systems that operate at different levels with different actors. Sustainability is a relevant criterion for evaluating agricultural technologies only if the boundaries and the inputs and outputs of a system using a certain technology can be well defined. Sorting out the various objectives, levels, and relevant factors that define the system is a difficult task that few national systems have been able to accomplish. Several conditions must be present, two of which include:

- Clear definitions: The objectives, the system to be sustained, and the relationship between sustainability objectives and other objectives need to be defined clearly. This may mean, for example, that the sustainability criterion can be applied only up to the level of a farming system. Above this level, systems are often too complex to be described adequately (Lynam and Herdt 1989; Anderson and Hardaker 1992). Considering effects outside the production system under study (such as the functioning of markets or the consequences on natural resources and the environment), can often take the problem to a higher system level (Barbier 1991; Siamwalla 1991; and Lynam, Section A in this volume). Therefore, it may be best to begin by defining sustainability at the highest level that can be adequately described, and then define it progressively downwards. The sustainability of a system does not necessarily depend on the sustainability of all of its sub-systems (Lynam and Herdt 1989; Graham-Tomasi 1991; Crosson and Anderson 1993).
- Clear policy environment: The actions and effectiveness of NARS with respect to NRM issues will be affected by the wider policy environment and the macroeconomic environment, including exchange rates, structural adjustment, and the policy bias to urban populations. The performance of NARS will also be affected by relative prices, credit, subsidies, labor policies and inputs and outputs prices. NARS will need to be able to identify points of intervention in a policy framework (ISNAR 1990). The emphasis on natural resource and environmental issues means that the research agenda has to be broadened to include, in addition to the development of commodity technologies, improvements in NRM as a source of growth in total productivity.

It follows that NRM and commodity research must be viewed as complementary; the success of the one depends to some extent on the success of the other. Therefore, when formulating their research strategies, NARS will need to balance their natural resource and commodity research programs.

The implications for NARS

Integrating natural resources and environmental concerns into the research agenda will add substantially to the responsibilities of the already overburdened NARS. Responsibilities will expand at a time when most NARS see their budgets declining (Oram 1991; Pardey and Roseboom 1989; Pardey et al 1991). The fact that NRM broadens the agenda of NARS presents clear pitfalls for many NARS institutions. First, some agricultural research institutions will be marginalized as attention and funding are increasingly channelled towards environmental issues instead of to production. However, there is little doubt that in most developing countries, agricultural research is of central concern to the sustainable use of land, water, and biological resources. Second, NARS institutions may shift their focus away from production, where they may have a comparative advantage, to environmental studies. In such fields, however, they may be insufficiently equipped to demonstrate impact. Given the complexity of the methods, the scope of the problem and the pitfalls involved, one can understand why many NARS are reluctant to integrate NRM into their research agendas.

In spite of such obstacles, there are several sound reasons for NARS to integrate NRM into agricultural research sooner rather than later (Arntzen 1993). First, the significant resource and environmental problems of many developing countries are becoming priorities for national policy. Second, greater attention to NRM will help reduce production losses that result from natural resource degradation. Third, preventing degradation is cheaper than rehabilitating a degraded environment. Fourth, NRM approaches make it possible to identify opportunities for developing under-utilized resources, or to identify the optimal use of natural resources, thus increasing the total factor productivity. Finally, an NRM approach can improve the prospects for the development of agricultural systems for marginal environments.

NRM and commodity research are not necessarily complementary, however. Environmental concerns may impede urgent productivity objectives. The time frame of NRM objectives is long and involves the concerns of future generations. There are questions as to how equity and discount rates concerning the use of resources for current production affect the future use of those resources. All these are difficult measures to include in agricultural research planning and evaluation. The long time horizon of the NRM approach also affects farmers' readiness to adopt new technologies that may not bring about immediate benefits. It also bears on policymakers: the impact of NRM is difficult to assess, while current expenditures on NRM research are hard to justify politically. The long time frame also affects researchers, because research tasks are becoming increasingly complex.

Integrating NRM into agricultural research agendas will require new efforts in research policy, organization, and management. In planning the kind of research that combines NRM and commodity research to increase the total factor productivity, NARS need to be able to compare the input and output consequences of different agricultural technologies and production systems. To achieve such a comparison, researchers need to identify and measure the natural resource consequences of new technology. They also need to carry out socioeconomic research to acquire a better understanding of why resource users follow practices that have adverse environmental consequences, or conversely, why they fail to adopt practices that are considered more "efficient" in the use of resources (Vosti et al. 1991).

Information about the environmental impacts and processes must then be taken into account in setting research priorities. The current priority-setting methods work well for conventional research programs, but the NARS will need new and more appropriate priority-setting methods that take into account environmental and social costs and benefits, and the possible trade-offs between different research strategies.

It will be essential for NARS to balance production-oriented research and concern for the environment. They will need to ensure that agricultural research is being recognized as central to achieving the new environmental agenda, and that it continues to receive political and financial support. It is therefore of critical importance for them to state the objectives of their research programs clearly.

Intersectoral and transboundary issues

NRM requires an intersectoral perspective. The benefits of using resources for agricultural production have to be weighed against the benefits of using these resources for non-agricultural uses. Faced with an expanded research agenda, with few or no additional research resources, NARS will need to examine the institutional and organizational options for cross-sectoral planning. Each country must decide for itself how it can best integrate natural resource conservation and management into production-oriented research.

Some resource management and environmental concerns that span an international or regional scale involve trans-boundary issues. At this scale, the main responsibility for managing natural resources rests with international organizations and national governments. The CGIAR's ecoregional approach (TAC/CDWG 1993) is essentially a planning tool that is likely to involve more than one country. It will also involve cross-sectoral issues that include non-agricultural uses of both land and water. An outstanding question for NARS is what institutional and organizational structures they will require to interact effectively in ecoregional activities.

Many other natural resource issues are more locally defined, and these are likely to preoccupy NARS first. These issues result from increased demands and competition for the use of the resources, from an unfavorable physical climate, or from the relative intensity of the land use of existing systems within the country.

Organization and management of the research system

For many NARS, the integration of NRM into the research agenda will require a fundamental change in the way their research is organized and managed. The structure and organization of the research system determine the way the system operates and its capacity to achieve its objectives. Structure and organization also affect the system's links with its environment. Research on sustainable production systems requires linkages between different spatial levels and sub-systems, which requires interdisciplinary work. This puts special demands on national research institutes—demands that are not made if research is carried out separately in a number of different fields of study (Arntzen 1993).

Donors have already urged most developing countries to tackle NRM and environmental problems. One question is in how far these new concerns coincide with what NARS institutions view as their own priorities. Another is how the existing institutional structures and processes will handle complex systems issues. A more integrated approach within the subsectors of agriculture (crops, livestock, forestry, and fisheries), and between agriculture and other sectors, may bring about shifts in the relative importance of institutions and the way they are structured. Such institutional changes are likely to generate at least some resistance from researchers and institutions, accustomed to more narrowly defined approaches.

Changing the organization of NARS may be the only solution to handle additional NRM objectives and activities without undercutting production-oriented research. As NRM agendas are implemented, NARS may consider new national research policies, new mechanisms for formulating priorities and coordination, interinstitutional linkages, including consortia and even mergers between NARS institutions.

Here we discuss four key issues related to organizational and management change:

- cross-sectoral planning—crucial to integrate environmental and agricultural development goals;
- information—a key requirement for the planning process and a major output of NRM research;
- a participatory, decentralized approach in both planning and implementation;
- institution building—an essential objective in all sectors related to agriculture.

Cross-sectoral planning

In many developing countries, the responsibilities for higher education, research, extension and information activities involving NRM are distributed among several ministries and departments (Lele 1989). Ministries and other bodies dealing with natural resource and environmental issues may have research agendas that are not linked to those of the NARS, often resulting from a lack of coordination.

A broader research agenda for NARS creates new institutional challenges, because it will require more interaction between institutions and more sharing of responsibilities and functions. For example, some of the most urgent problems in irrigation systems arise from agricultural activities that, although they take place within the catchment area, are outside the responsibility of the irrigation agencies. Although crops, livestock, and trees are often found together in production systems at the farm level, they are usually researched by different institutes. The organizations responsible for these research topics need to integrate more closely and forge links with natural resource and environment institutes that have not traditionally been considered part of the NARS (e.g., soil-survey, meteorological-services, and hydrology departments).

In some developing countries, such as the Philippines and India, the agricultural faculties of universities are integral parts of the NARS. In many other countries, universities are not directly engaged in development-related research in spite of their status as sources of scientific capacity (Oram 1991). However, universities ultimately determine the quality of the national institutional capacity (Bunting et al. 1986). By forming closer partnerships with universities, NARS could draw upon them as sources of expertise, particularly in disciplines where the NARS are weak. Such closer relationships would also ensure that what is being taught at university is relevant to the future agricultural development needs of the country.

Institutional and organizational options for cross-sectoral planning are likely to be considered in integrating NRM and commodity-oriented research. NARS will need to design new procedures for setting research objectives, determining priorities, and designing programs that cut across the institutional structure. If countries are to develop more coherent NARS with in cross-sectoral planning, the programs will require at least the following: effective channels for information exchange, reliable mechanisms for the transfer of resources between decision-making centers; the mechanism to allocate responsibility, and incentive mechanisms that focus the scientific effort on the countries' significant agricultural problems (Ruttan 1984; Winpenney 1991).

Appropriate information systems
Integrating NRM research into agricultural research agendas will mean that research is no longer solely the provider of new technology, but also a source of information for formulating development policy and planning. The spatial and temporal complexities in the production system will often make modelling the impact of technology a more effective research strategy than traditional empirical experiments. This research process is already used in policy research. A familiarity with the techniques used in systems research can forge links across system levels.

Farmers may not be the only or even the principal clients for research. The output of research may feed directly into policy-making levels of governments. Governments also need ways to anticipate the likely consequences of the growing demands on a country's natural resources. They will need indicators of the use of land and water resources, and these indicators should take into account environmental, economic, social, demographic, cultural, and political factors. Faced with competing intersectoral interests in the use of land and water resources, policymakers need information about the economic and environmental costs to make informed decisions. Policymakers are already asking their agricultural research services to provide information, in an accessible form, about the technical options to meet both environmental and developmental goals (Rudder 1992).

An inventory of natural resources that includes physical and economic data on land capability, land use, and management patterns is essential for planning development and integrating agricultural and NRM research properly. Information-based technologies such as crop and land-use modelling, geographic information systems (GIS), and the use of remote sensing are among the tools that NARS will be expected to use as part of their NRM research. It is a public-sector responsibility to ensure that an inventory of this kind is maintained. Collecting information about the natural resource base and the impact of agricultural activities on it and the environment, in the manner described by Crosson and Anderson (1993), will pose a considerable challenge for NARS. Whether NARS will be able to collect this kind of information will depend on whether and to what extent they can access information management systems. So far, only a few NARS have access to such systems.

Much of the information on agricultural environments has been organized and collected by international and regional research organizations, or national agencies whose mandates are neither in research nor in agriculture. UNEP, FAO, international agricultural research centers, and international scientific institutes such as the French ORSTOM, the British NRI, and the Dutch KIT are all major sources of information on global agricultural environments. Valuable efforts have been made to improve the consistency, concordance, and use of these data for research and development purposes (Bunting 1987).

Efforts to bring together the needs of the developing countries and the expertise of advanced institutes in this field are increasing. The International Consortium for Application of System Approaches to Agriculture (ICASA) is one example of how experience in developing systems-analysis methods is being introduced to the service of NARS in developing countries. These methods help make policy decisions and can serve as practical tools for resource management and developing sustainable production systems.

Guidelines for the future information requirements of NARS, including procedures for collecting data about the sustainable use of natural resources, need to be developed. To formulate informed agricultural and environmental policies, the expertise from different disciplines needs to be integrated during the collecting of data, to ensure that physical and economic data are compatible and that scientific research and policy analysis can be linked (Antle and Just 1992). Information about agricultural environments provided by a database can then be used to designate areas for production to meet the future demand (e.g., Karim, this Section, p. 117). For many NARS, an effective approach may be to look for opportunities to use research results and techniques developed for similar situations outside the country, as part of an active adaptive research program. An inventory of natural resources helps them identify these opportunities much more readily—an advantage of particular interest to small countries (Eyzaguirre 1993).

NARS will need to acquire the skills to assemble, maintain, and manage environmental databases. Two kinds of skills may be needed: one is to build and maintain local databases on research and the state of natural resources; the other is to use and apply where possible the information available in global databases. New collaborative mechanisms will be required to facilitate NARS to access sources of information held by others, and for a NARS to allow others to use its database. The CGIAR/UNEP Global Resources Information Database (GRID) is an example of an international initiative to develop an environmental database that is compatible with the needs of both international agricultural research centers (IARCs) and NARS (UNEP 1991). NARS will need to strengthen their capacity to contribute to and use the growing body of information on the state of land, water, and biological resources.

Decentralization and local institutions

At the local level, cross-sectoral planning for development or research involves all uses of land, water, forests, and other natural resources in a defined system. Since the planning unit could be an ecosystem or a watershed, the emphasis on research is usually specific to a local production system and culture (ISNAR 1993). Research and development components of NRM such as soil conservation, the use of community forest, and management of common grazing, should be community driven (Arntzen 1993). Whether new technologies will be adopted often depends on social institutions that control access and use of resources, the knowledge of how to use the resources, and on access to markets, rather than on the technologies themselves (Nelson 1985).

Government agricultural services have a poor record of working successfully with local communities on these issues. Many NGOs are nearer to the grass roots of communities and are more responsive to their needs (FAO 1991). Agenda 21 suggests that for research and development that involves the use of local knowledge and an understanding of farmers' incentives, NGOs are often better equipped than government services, while public-sector research organizations may have reservations about them as full partners (Lele 1989; Bebbington and Farrington 1993). There is also evidence that local NGOs working in resource management could benefit from the scientific and policy guidance that NARS could provide.

The greater decentralization that is likely to be associated with locale-specific research will be an opportunity for NARS to build new and stronger technology-transfer links with the help of NGOs and informal organizations (Petit and Gnaegy 1991). The change in emphasis, from simple production increase to the more complex requirements of changing the component parts of production systems, will mean that technology transfer will take place in a slower and more difficult manner. Given the intergenerational objectives of NRM technologies, whether farmers will adopt new technologies depends on whether they are convinced that future benefits will be substantial. Identifying the conditions that foster an environment receptive to NRM is an important area for research-policy linkages. Following the Earth Summit, there will be more pressure on governments and NARS to promote the active participation of non-governmental and informal organizations, and to recognize them as partners in the research and development process (Farrington et al. 1993).

Institutional development and change

Unlike the production-oriented, disciplinary research that NARS are familiar with, research with a sustainability perspective has to operate at the level of production systems. Such research is more interdisciplinary and process oriented, and it requires longer time horizons (Eyzaguirre 1993). It deals with the role of livestock and all, or parts, of the continuum from crop plants to forestry. It also includes meteorology, soils, hydrology, social, and economic disciplines. Recent examples suggest that this type of research has to be planned and organized somewhat differently than production-oriented, disciplinary research (ISNAR 1993).

Building multidisciplinary research teams that focus on NRM problems may need the participation of scientists from different institutions, whose mandates may vary widely. Some may focus on one commodity while others may have a much wider scope of research. Often organized into disciplinary departments, many lack experience in working in interdisciplinary or interinstitute research teams (Bengtsson 1987). Very few have administrative procedures with the flexibility required to operate a project-based NRM research program. It is difficult to plan and manage staff time when staff are allocated from different divisions or units to joint programs and projects. There is usually no established institutional base for research at the systems level (Oram 1991). This situation remains one of the main institutional constraints that threaten to impede both the development of natural resources management research and the emergence of sound land-use policies (National Academy of Sciences 1977).

NARS scientists usually have not been educated or trained in analyzing production systems. Educational systems will need fundamental changes to remedy the present narrow disciplinary range in NARS (Oram 1991). This narrow educational background is also reflected in the past focus on commodity-production training courses provided by the IARCs. The training needed should include elements of various basic disciplines, applied sciences, and specialized subjects, in addition to the usual plant sciences (e.g., geology, hydrology, meteorology, soil science, ecology, land-use and survey methods, natural resource economics, post-harvest technology, and research management).

In a review of the scientists engaged in agricultural research in developing countries in the 1980s, Oram (1991) found that out of a total of about 36,000 scientists, the distribution of disciplines heavily favors the natural sciences. The most conspicuous deficiency is in the social sciences. Social scientists usually work in other agencies, often with poor links to the NARS. But the NARS will need economists and scientists of other kinds, including social, to understand how farmers reach their decisions. Natural and social sciences are also becoming increasingly interdependent to balance technological progress and institutional development (Ruttan 1984).

NARS will need a clear idea not only of the types of expertise required for research on NRM, but also of the desirable weight in the overall staffing of a NARS. This will require forward-looking human-resource development plans that adequately reflect these needs, with information on training requirements and the means to match this information to worldwide training opportunities. Given the high costs and special skills required, it is important for NARS to move into the area of NRM in the most effective and efficient way possible.

Since few NARS have resources to gather all of these discipline skills, the search for new institutional arrangements is essential. These new arrangements may include:

reorganizing or merging institutes, establishing new partnerships with NGOs and Universities, collaborating with private-sector research organizations, and exploring opportunities to cooperate with nations with similar natural resource and environmental problems (National Academy of Sciences 1977).

Research management

As mentioned earlier in this paper, a major part of institutional development will consist of establishing appropriate procedures for planning projects, setting priorities, and allocating resources.

Whether a research program that reflects the broader research agenda will be managed successfully also depends on an effective management information system (MIS) to support these tasks. A computerized database will be a key element in managing interdisciplinary research projects. It will help establish a clear connection between priority research programs and the necessary staff and budget to implement them. The database should form part of a reference facility that would make all relevant planning and evaluation documents readily available to those concerned with the project. An MIS can be a powerful management tool to assist in reviewing and assessing research priorities and in allocating resources.

To help manage programs, the database should contain information on the various organizations, their budgets and staffing. It should also permit the allocation of resources to be examined by type of research (e.g., by commodities, type of natural resource, region, or disciplines involved). Attempt should also be made to standardize the information so that it can be used both at the aggregate level and at the institutional level.

The paper noted earlier that changes in total factor productivity provide a measure of the sustainability of a production system and of its effects on the surrounding environment (e.g., biodiversity). Though measuring is difficult, the increased complexity of the study of production systems makes it all the more urgent to put in place monitoring and evaluation procedures to ensure the relevance and effectiveness of the NRM research program.

Topics requiring further research

Rather than being interested in the NRM technologies themselves, ISNAR is most interested and has most expertise in the research-policy, organization, and management issues that NARS will have to deal with. Some of the issues that will require attention in ISNAR's plans for research, advisory services, and training activities include the following:
- development of procedures that developing countries can use for measuring both the economic and environmental costs and benefits of alternative NRM practices;
- development of appropriate methods for NARS to determine relative priorities between NRM research and commodity research in their research agendas;

- formulation of appropriate research policy for addressing NRM and environmental concerns;
- identification of constraints to adoption of practices that satisfy both the needs for growth in production and natural resource and environmental concerns;
- identification of the role of land tenure on NRM, and other producer-related issues such as limited access to both knowledge and choices of technology;
- devise and test methods for NARS to handle the institutional and managerial aspects of cross-sectoral planning, based on an ecological unit (e.g., an ecosystem, watershed, or ecoregion);
- assemble and disseminate knowledge on how NARS are handling NRM issues.

ISNAR's work on the policy, organizational, and management implications of NRM for national research systems is based on three premises. First, the NRM issue is central to agricultural research and development, and NRM cannot and should not be avoided. Second, national research systems in developing countries have a vital role to play in finding solutions to the environmental problems that threaten future productivity and personal well-being. Third, the methods and perspectives in an NRM approach and in conducting NRM research provide cost-effective opportunities for scientists and policymakers to apply science to solve some of the developing world's most urgent problems.

Acronyms

CGIAR	Consultative Group on International Agricultural Research
FAO	Food and Agriculture Organization
GIS	geographic information system
IARC	international agricultural research center
ICASA	International Consortium for Application of System Approaches to Agriculture
KIT	Koninklijk Instituut voor de Tropen
MIS	management information system
NARS	national agricultural research system
NGO	non-governmental organization
NRI	Natural Resources Institute
NRM	natural resource management
ORSTOM	Office de la Recherche Scientifique et Technique Outre Mer
UNCED	United Nations Conference on Environment and Development
UNEP	United Nations Environment Programme

References

Anderson J R, Hardaker J B (1992) Efficacy and efficiency in agricultural research: A systems view. Pages 105-124 in Teng P S, Penning de Vries F P V (Eds.) Systems Approaches for Agricultural Development. Elsevier Science Publishers Ltd, London-New York.

Antle J M (1993) Environment, Development, and Trade between High- and Low-income Countries. American J. of Agric. Economics 75:784-788.

Antle J M, Just R E (1992) Conceptual and Empirical Foundations for Agricultural-Environmental Policy Analysis. J. of Environmental Quality, 21:307-316.

Arntzen J (1993) Natural resource management and sustainable agriculture Policy implications for research systems in small developing countries ISNAR Discussion Paper 93-02, The Hague, The Netherlands

Barbier E B (1991) Macroeconomic and sectoral policies, natural resources, and sustainable agricultural growth Pages 167-184 in Agricultural sustainability, growth, and poverty alleviation Issues and policies Proceedings of a Deutsche Stiftung fur Internationale Entwicklung (DSE) & International Food Policy Research Institute (IFPRI) conference, 23-27 September 1991, Feldafing, Germany

Bebbington A, Farrington J (1993) Governments, NGOs and Agricultural Development J of Development Studies 29, 2 199-219, London, UK

Bengtsson B (1987) Research and development priorities and systems for increasing food production and rural income of small holding sectors in Africa Pages 294-307 in Proceedings of the FAO/SIDA seminar on increased food production through low-cost food crops technology, 2-17 March, Harare, Zimbabwe

Bunting A H (1986) Development of Irrigation in Africa needs and justification Consultants Report prepared for the FAO Consultation on Irrigation in Africa, Lomé, Togo, 21-25th April 1986 @REF-ERENCE = Bunting A H (Ed) (1987) Agricultural Environments Characterisation, classification and mapping Page 335 in Proceedings of workshop on agro-ecological characaterisation, classification and mapping, April 1986, Rome CAB International, Wallingford, UK

Chopra K, Rao C H H (1991) The links between sustainable agricultural growth and poverty Page 53 in Agricultural sustainability, growth, and poverty alleviation Issues and policies Op cit

Crosson P (1993) Sustainable agriculture A global perspective Choices (2) 38-42

Crosson P, Anderson J R (1992) Resources and Global Food Prospects Supply and Demand for Cereals to 2030 World Bank Technical Paper No 184, World Bank, Washington DC, USA

Crosson P, Anderson J R (1993) Concerns for Sustainability Integration of Natural Resource and Environmental Issues in the Research Agendas of NARS Research Report 4 ISNAR, The Hague, The Netherlands

Cunha A S, Sawyer D (1991) Agricultural growth and sustainability Conditions for their compatibility in the humid and sub-humid tropics of South America Page 311 in Agricultural sustainability, growth, and poverty alleviation Issues and policies Op cit

Davies S, Leach M, David R (1991) Food security and the environment Conflict or complementarity? IDS Discussion Paper No 285, Institute of Development Studies, UK

Douglass G K (1984) Agricultural Sustainability in a Changing World Order Westview, Boulder CO, USA

Eyzaguirre P (1993) Research on Natural Resource Management A New Beginning or Another Agenda Item? Chapter 5 in Small is Feasible Innovative Strategies for Research on Agriculture and Natural Resources from Small Countries ISNAR, The Hague, The Netherlands (to be published)

Farrington J, Bebbington A, Lewis D J, Wellard K (1993)) Reluctant Partners? Non-governmental organisations, the state and sustainable agricultural development Poutledge, London

Food and Agriculture Organisation of the United Nations (1991) Issues and perspectives in sustainable agriculture and rural development Paper for FAO/Netherlands Conference on Agriculture and the Environment Strategies and Tools for Sustainable Agriculture and Rural Development, 15-19 April 1991, 's-Hertogenbosch, The Netherlands FAO, Rome, Italy

Graham-Tomasi T (1991) Sustainability Concepts and implications for agricultural research policy Pages 81-102 in Pardey P G, Roseboom J, Anderson J R (eds) Agricultural research policy International quantitative perspectives Cambridge University Press, Cambridge, UK

International Service for National Agricultural Research (ISNAR) (1990) Highlights of Consultation Meeting on Agricultural Research Policy, Organisation, and Management for Sustainable Agriculture, 13-15 October 1990 ISNAR, The Hague, The Netherlands

International Service for National Agricultural Research (ISNAR) (1993) Agenda 21 Issues for National Agricultural Research Briefing Paper No 4 ISNAR, The Hague, The Netherlands

Lele U (1989) Agricultural growth, domestic policies, the external environment, and assistance to Africa Lessons of a quarter century MADIA Discussion Paper 1 World Bank, Washington DC, USA

Lynam J K, Herdt R (1989) Sense and Sustainability Sustainability as an Objective in National Agricultural Research Agricultural Economics 3 381-398

National Academy of Sciences (1977) World food and nutrition study; The potential contributions of research. National Academy of Sciences, Washington DC, USA.

Nelson M (1985) Institutional and economic issues in development of humid tropical lands. Pages 59-75 in Proceedings of an IBSRAM workshop on tropical land clearing for sustainable agriculture.

Oldeman L R, Hakkeling R T A, Soemroek W G (1990) World Map of the Status of Human Induced Soil Degradation. UNEP, Nairobi, Kenya.

Oram P (1991) Institutions and technological change. Page 245 in Agricultural sustainability, growth, and poverty alleviation: Issues and policies. Op. cit.

Pardey P G, Roseboom J (1989) A Global evaluation of National Agricultural Research Investments: 1960-1985. Pages 163-177 in Javier E, Renborg E Q (eds.) The changing dynamics of global agriculture: A Seminar/Workshop on Research Policy Implications for National Agricultural Research Systems, 22-28 September 1988, Feldafing, Germany. ISNAR, The Hague, The Netherlands.

Pardey P G, Roseboom J, Anderson J R (1991) Regional perspectives on national agricultural research. Pages 197-264 in Pardey P G, Roseboom J, Anderson J R (eds.) Agricultural research policy: International quantitative perspectives. Cambridge University Press, Cambridge, UK.

Petit M, Gnaegy S (1991) International development assistance for sustainable agricultural growth. Page 119 in Agricultural sustainability, growth, and poverty alleviation: Issues and policies. Op.cit.

Pingali P L (1991) Agricultural growth and the environment: Conditions for their compatibility in Asia's humid tropics. Page 295 in Agricultural sustainability, growth, and poverty alleviation: Issues and policies. Op. cit.

Ruttan V W (1984) Induced innovation and agricultural development. Pages 107-134 in Douglass G K (ed.) Agricultural sustainability in a changing world order. Westview, Boulder CO, USA.

Ruttan V W (1991) Sustainable growth in agricultural production: Poetry, policy and science. Page 13 in Agricultural sustainability, growth, and poverty alleviation: Issues and policies. Op. cit.

Rudder, Winston R (1992) How Policy Determines the Demand and Scope of Agricultural Research in Trinidad and Tobago. Paper presented at the International Workshop on Management Strategies and Policies for Agricultural Research in Small Countries, May 1992, Réduit, Mauritius. ISNAR, The Hague, The Netherlands.

Siamwalla A (1991) The relationship between trade and environment, with special reference to agriculture. Page 105 in Agricultural sustainability, growth, and poverty alleviation: Issues and policies. Proceedings of a Deutsche Stiftung für Internationale Entwicklung (DSE) & International Food Policy Research Institute (IFPRI) conference, 23-27 September 1991, Feldafing, Germany.

TAC/Centre Directors Working Group (1993) The Ecological Approach to Research in the CGIAR. Sixtieth Meeting, TAC Secretariat. FAO, Rome, Italy.

TAC/CGIAR (1989) Sustainable Agricultural Production: Implications for International Agricultural Research. FAO, Rome, Italy.

United Nations (1992) Long-range World Population Projections: Two Centuries of Population Growth 1950-2150. United Nations, New York.

United Nations Environment Programme (1991) Increased sustainable agricultural productivity in Africa through the use of intelligent geographic information systems (IGIS) within the Consultative Group for International Agricultural Research (CGIAR). Report of a Scientific Workshop held at UNEP, Nairobi, January 14th-18th 1991. UNEP, Nairobi, Kenya.

Vosti S, Reardon T, Von Urff W (1991) Overarching policy priorities and research issues. Page 513 in Agricultural sustainability, growth, and poverty alleviation: Issues and policies. Op.cit.

Winpenney J T (1991) Development Research: The environmental challenge. Overseas Development Institute, London, UK.

Data requirements for agricultural systems research and applications

J.T. RITCHIE[1] and J.B. DENT[2]

[1] Department of Crop and Soil Sciences, Michigan State University, East Lansing, MI 48824, USA
[2] Institute of Ecology and Resource Management, University of Edinburgh, Edinburgh EH3 9JG, UK

Key words biophysical data, database, decision making, farm household model, pedotransfer function, production system, simulation models, socioeconomic data, systems approach, validation

Abstract
The quantitative assessment of agricultural systems requires a balance between simulation and data to support the simulations Data for systems research is needed as inputs or as constraints for simulation models and as validation of simulations to establish the credibility of the systems approach If agricultural and land-management simulation models are to be useful for solving problems, they need to include in the detail of the simulation representation of processes in several subject-matter areas The level of detail required as output in a simulation determines the amount of data necessary as inputs and for validation This paper suggests five levels of modelling detail, ranging from local experience to very detailed mechanistic simulation models Biophysical models that require daily weather data have evolved into the most useful form of models for the transfer of description based on agrotechnology information However, some elements of the required daily weather data, especially global radiation, are not always available In such instances, substitutes such as hours of sunshine or a weather generator can be used

Data that describe the biophysical characteristics of a system are required as an input for the simulation process They include data on weather, soils, crop management, the characteristics of the cultivar, and the incidence of pest Information on the crop is needed for model validation Socioeconomic data required include input costs and selling prices, attitudes toward risk, financial and human resource constraints, and data on several social factors such as land tenure, education and wealth

Scientists within the ICASA group are committed to the development of practical and balanced systems approaches that will assist in many important aspects of decision-making processes in agriculture A systems approach can increase the efficiency of agricultural research and provide more realistic information to policymakers

Introduction

The simulation of several important components of the soil-plant-atmosphere continuum is becoming an increasingly valuable tool to assist decision making at both the farm and the regional level. One of the principal limitations in our ability to apply simulation systems more widely is the lack of good-quality data required as input. The basic data needed for realistic simulation of the performance of crop plants must be adequate to describe the solar energy, water, and plant nutrients required to produce growth in dry weight and economic yield. Once the dry-weight production and yield are simulated, there are usually also the effects of plant pests in the system that have to be accounted for, as potential competitors for the biomass produced or for the resources from the soil. There are also social, economic, and policy issues that influence the decisions regarding realistic land-use alternatives. These decisions will likely affect the biophysical system, especially in terms of external inputs and

153

P Goldsworthy and F W T Penning de Vries (eds), Opportunities, use, and transfer of systems research methods in agriculture to developing countries, 153 - 166
© 1994 Kluwer Academic Publishers

constraints to production. The important socioeconomic dimensions of agricultural systems require basic data for the application of realistic agricultural systems.

This paper focuses on data requirements for modelling the soil-plant-atmosphere system and for linking pests and socioeconomic dimensions to the biophysical models. Two types of data may be distinguished. First, basic information needed in equations that drive the biophysical system, such as the daily temperature of air, the sowing date of the crop, and wilting-point water content of the soil. Second, information derived from monitoring the crop-soil system, such as observations of the date of anthesis, crop yield, and soil water contents during the growing season for model validation.

The amount and type of data required for the realistic simulation of a particular system depends on the level of detail of a particular simulation model and the goals of the decision maker or researcher who will use the system.

Simulation models of the soil-plant-atmosphere system

Simulation models of crop production systems can be produced for different purposes and at various levels of detail. Some objectives for simulation with simple models may require little detail, while other objectives may require complex models that require more detailed data. Bouma et al (1993) listed five levels of detail for models that have been used:

Level 1: models based on farmer knowledge;

Level 2: models based on expert knowledge;

Level 3: empirical models that use statistical or capacity concepts;

Level 4: mechanistic models that emphasize rate concepts;

Level 5: complex mechanistic models for subprocesses of larger systems.

Scientific groups within ICASA have successfully used models in recent years for various problem-solving purposes throughout the world. Models that use the capacity concepts of level 3 (often termed functional models) and the mechanistic models of level 4 have evolved into the most useful models for technology transfer beyond site-specific results. Scientists affiliated to ICASA are developing practical and useful models of agricultural systems that require a minimum amount of data to assess specific goals. Most of the level-4 models of the soil-plant-atmosphere system have made simplifying assumptions that require a reduced amount of input data to make them as useful as possible. An example of this simplification is that daily-weather data is used as input for the model and within-day variations of the weather are approximated with assumed functions. The approximations are used to generate reasonable weather data within the day as required for the relatively short time increments needed in rate calculations of mechanistic models.

Mechanistic models appear to have the greatest practical potential for general use in agricultural systems applications and technology transfer. The data requirements to be discussed in the paper are therefore focused on these two levels of detail. The minimum data requirement, usually associated with level-3 models, will be distinguished where relevant from the additional data required for level-4 models. Data requirements will also be discussed as either input-related data or validation data.

Biophysical data for model inputs

Weather
Much of the risk involved in agricultural systems is related to the uncertainty of the weather. The analysis of agricultural systems provides a means to take advantage of information on the variability of the weather in the past. A summary of weather data requirements for simulation is presented in table 1. In many regions of the world, the main weather element of concern is the amount and distribution of rainfall. However, if models are to predict the performance of crops in most circumstances, then solar radiation and air temperature are the minimum data needed to calculate photosynthesis and evapotranspiration of the crop. The temperature is needed to calculate the rates of plant development, to modify growth rates, and to evaluate extreme ranges of temperature that may kill or adversely affect the crop. Within season, year-to-year variations in these weather data often account for the large variations in crop productivity. The optional weather data help predict the evapotranspiration and growth better, and, along with the minimum data, they help predict pest growth and crop damage. The intensity of rainfall is needed when more accurate estimates of erosion and of runoff or water ponding are important.

Ideally, the weather data should be recorded at a site near where the model is to be used. The temperature and radiation data are probably fairly similar when the weather recordings are taken some distance from the site, provided that the elevation of the weather recorder and site are about the same. However, rainfall differs spatially to such an extent that it must be measured at the site of interest for model testing.

Most weather stations have historical rainfall and temperature data, but they usually do not have global radiation data. This becomes a limitation when long weather records are required for simulations for a succession of years. When data about hours of sunshine or percent sunshine is available, solar radiation can be approximated using empirical equations (van Keulen and Wolf 1986, p. 64).

Weather data can be obtained from national meteorological services, local weather stations, the FAO (Frère 1987) and, in some instances, from CGIAR centers

Table 1. Data types for weather specification from standard weather stations

Minimum	Optional
Daily precipitation	Daily dew point temperature
Daily maximum air temperature	Daily wind run
Daily minimum air temperature	Daily net radiation
Daily total global radiation	Hourly values of each of above variables
	Rainfall intensity during a storm

(e.g., Oldeman et al. 1987). Whatever the source of the weather data, the quality of the data should be checked because instrument calibrations may be biased. Records should be checked by an experienced agricultural meteorologist.

Approximation of a long time sequence of weather variables
When long-term daily weather records are unavailable, an alternative procedure is to use stochastic time-series modelling to generate a sequence of weather data with statistical properties similar to historical sequences. To produce longer sequences, a short sequence of daily weather data is used to determine several stochastic coefficients that describe the weather variations. The coefficients are used to generate a longer sequence randomly. However, this longer stochastic sequence may not contain all the possible variations of a longer time sequence. One widely-used form of this approach is the weather generator (WGEN) of Richardson and Wright (1984). The WGEN approach has been modified and extended by Geng et al (1985a,b). In some instances, when only monthly mean weather data are available, daily sequences can be generated from known statistical characteristics of weather variables within a region. For example, Geng et al. (1986) used monthly precipitation and number of wet days per month to provide a realistic daily sequence of rainfall for a daily incrementing simulation model.

Soil
Table 2 shows soil data required by simulation models. On-site measurement of soil properties is important where the aim is to validate the model for a particular site. It is recommended that as much data as possible for simulations be taken during monitoring in the field, rather than from small samples taken from the field to the laboratory. For example, in simulations of systems where soil water is a critical constraint, it is better to measure the lowest water content in the field than to estimate it in the laboratory from samples removed from the field. A field measurement where the plants are grown until almost dead or dormant due to a lack of water will give an insight into the effective depth of rooting, another important soil property. The depth of rooting can be constrained by soil properties that prevent root penetration, or it can simply be a function of the duration and rate of downward root movement. The root depth is extremely important when simulations are made under water-limiting conditions. For soils that have little or no internal drainage limitations, it is best to measure the "drained upper limit" in the field using water content of soils that have been thoroughly wetted and allowed to drain without irrigation.

The drainage time to obtain the data related to the availability of soil water may range from one to 15 days, depending on the soil properties. For soils with poor internal drainage, measurements of the hydraulic resistance of a restricted layer in the field are likely to be more accurate than from cores removed from the field. When the water table is within one meter of the depth of rooting, more detailed hydraulic soil properties are needed to simulate the supply of water for plants that moves upward from the water table. When soils are poorly drained, or when the water-table depth

Table 2. Data types for soil specification

Minimum	Desirable for specific applications	Initial conditions
Lower limit water content at 10-20 cm depths	Hydraulic conductivity and water retention curves at 10-20 cm depths	Water content at 10-20 cm depths
Field capacity soil water content at 10-20 cm depths	Runoff curve number	Nitrate concentrations at 10-20 cm depths
Crop rooting depth	Surface albedo	Ammonium at 10-20 cm depths
Hydraulic conductivity at soil depths that restrict water flow	pH at 10-20 cm depths	Extractable phosphorous at 10-20 cm depths
	Organic carbon in upper depths	
	Textural characterization for 10-20 cm depths	Fresh plant residue or manure amounts and depth of incorporation
	Surface water ponding capacity	
	Bulk density at 10-20 cm depths	
	Groundwater depth by-pass flow fraction	

is within the reach of plant roots, the "drained upper limit" capacity concept is not useful.

If simulations are to be applied to an area larger than a field, data on soils can sometimes be obtained from traditional soils descriptions, which are available from soil scientists in national research services. Large-scale soils maps and descriptions are available from FAO.

The most practical method to obtain good-quality soil data as an input for simulation is to derive it from data that is already easily available. Although the base data is not needed for model inputs, empirical relationships can be derived to convert basic data to the required input values. These conversion relationships are termed pedotransfer functions (Bouma et al. 1993). Some of the input for both functional models (level 3) and mechanistic models (level 4) are not easily available and are difficult and costly to obtain.

Several research groups involved in systems simulations have successfully used soil characteristics such as texture, bulk density, and organic-matter content to approximate the characteristics of soil hydraulics for mechanistic water-flow models (Wagenet et al. 1991; Vereecken et al. 1990; Cosby et al. 1984; Driessen 1986). The use of pedogenetic soil horizons as "carriers" of physical information about soils such as conductivity and retention has also been successful (Bouma et al. 1993). It is often possible to describe several distinct pedogenetic soil horizons with one or two functional classification groups, thus reducing the amount of input data required to model the known variability in pedogenetic soil horizons.

Equations that use soil texture, bulk density and organic matter to estimate the lower limit of the availability of soil water, the field capacity and saturated water content for use in functional models have been used successfully (Ritchie and Crum 1989). Runoff curve numbers, soil albedo, and drainage coefficients can be reasonably estimated from the soil classification descriptions of hydrologic groups, surface color, and internal drainage.

As the systems approach gains in popularity, more pedotransfer functions are likely to be developed. These functions make existing soil data much more useful in agricultural decision making. Bouma et al. (1993) has pointed out that by using regression equations or by expressing data for horizons in statistical terms, the accuracy of pedotransfer functions can be determined. This information can be used, in turn, to express the accuracy of simulation outcomes.

To obtain data on the initial condition of soils can also be costly and time consuming. Some procedures can approximate satisfactorily the initial conditions in many circumstances. For example, in many regions of the world, the initial soil water content a few days or weeks before sowing is either uniformly high or uniformly low. The initially wet condition occurs frequently in temperate regions, and the initially dry condition in tropical regions. Soil water balance simulations can be started some time before sowing by assuming an initial value of soil water content near the drained upper limit or the lower limit water contents or from their corresponding matrix potentials.

It is also possible to use expert knowledge to approximate initial conditions. Initial residue amounts, nitrate and ammonium concentration, and soil water content can be reasonably obtained from local experts who know whether the status of the soil is low, medium, or high in some of the input values that are needed. This approach is not recommended in situations were model validation is to be done, but is usually satisfactory for strategic planning with the systems approach.

Management input data
Table 3 lists some of the management practices used in simulations to determine combinations of management strategies needed to accomplish specific goals. If the goal is maximum profitability, then various management strategies can be simulated to obtain yields. Yield output from the biophysical simulation is linked to an economic model containing costs and prices such that the output value can be determined for each strategy simulated. Other goals may be yield stability, income stability, minimizing nitrate leaching, etc. The management input combinations can be costed or constrained by resource availability or by economic constraints to evaluate the best managements strategies within the limits of those constraints. Since management options are at the heart of decision making in most agricultural systems, simulation is a powerful tool in providing decision support.

The characteristics of the crop variety or cultivar are of great importance to management decision making. Exploiting variations between cultivars to find those

Table 3. Data types needed for management

Minimum	Optional
Crop cultivar characteristics	Sowing date
Planting or transplanting date	Row spacing
Plant population density	Row direction
Irrigation inputs (dates and amounts)	Pesticide inputs (dates, types, amounts)
Fertilizer inputs (dates, type, amounts)	Harvest date
Crop residue or manure inputs (dates, quality, amounts)	

most suited to a region has been a primary focus of many research programs. The main variations between cultivars that have been useful in simulations have been associated with differences in the duration of successive phases in the development of plants and the partitioning of dry weight to the reproductive organs. To simulate the development phases, there are two levels of detail of cultivar characteristics to choose from as inputs. The simplest level is to input the number of days or degree days the plant is expected to grow within each development stage. Practically all functional and mechanistic crop models use thermal time (the time-temperature combination in degree days) to express the duration of growth stages. The more detailed level of cultivar information is to use sensitivity to photoperiod, and, for winter cereals, to vernalization, because these factors influence the thermal time of the vegetative stage. There are major differences between cultivars in the thermal time required to complete the juvenile phase, and minor differences in the thermal time required for other stages of development such as grain filling. Quantitative, genetic coefficients can be used to express these cultivar differences (Ritchie 1993). The duration of growth of cultivars can be predicted for any crop season-weather sequence by using the correct genetic coefficients. Variations in cultivar characteristics that determine the components of yield have not been successfully used in simulation, especially in situations where there are small differences between yields of cultivars. Attempts are usually made to characterize specific components of yields such as grain size and number of spikelets, which are known to differ between cultivars. Some models express differences between cultivars by variations in the coefficients used to represent assimilate partitioning.

If crop models are linked to pest models, then the degree of resistance to the pest of concern becomes another important cultivar characteristic. When the pest-crop linkages are known, it is possible to use a genetic coefficient to describe the degree of pest resistance. These coefficients could be used in simulations of alternative agricultural systems in which the use of pesticides is unacceptable or where the quality of the product is a concern.

Crop validation data

A balanced set of data on crop performance is essential to establish the credibility of simulation models and to assist with interpretations and comparisons with other studies. Simulation of the duration of crop growth stages is essential to the successful modelling of plant growth and yield. Practically every successful systems simulation must start by using good-quality data on field crops to establish the accuracy and sensitivity of the simulation system to the variables of interest in a particular situation. Data required for validation are summarized in table 4.

Data needed for linking pest damage to crop simulation

There are various types of pest models for practical and for research purposes. As with crop models, models of pest damage can be based on farmer knowledge, expert knowledge, statistically derived functions, and mechanistic principles. Because of the almost infinite number of possible crop-pest combinations, it is possible to discuss crop-pest models and their data requirements only in a general way. The data needs listed in table 5 are of primary importance for crop-pest models at a level of detail which is most useful for transfer of technology outside the region from which they were developed, ie. for the functional and mechanistic level models.

Farmers everywhere have to make critical decisions about when to start applying a pesticide, how often to apply it, and when to stop applying it. Methods to predict what is required are needed to support these decisions because of increasing economic pressures and environmental concerns. Crop models that are properly validated can be linked to pest models to provide scientists with (a) a reliable means to estimate

Table 4. Data required for validation of crop models

Minimum	Desirable
End of season biomass	Yield components
Economic yield	Leaf area index on several dates
Dates of important phenological events	Biomass on several dates
	Soil water content on several dates and depths
	Number of leaves produced on main stem
	Nutrient concentration in grain
	Nutrient concentration in vegetation
	Pest losses amount and type

Table 5. Data for pest/crop interactions

Identification of pest and its linkage point to plant
Within-season decreases in plant population density
Pest intensity within season
Spatial patterns of pests within a field

yield of a pest-free crop and of a pest-infested crop (and hence economic losses) under different conditions, and (b) a means to develop least-loss strategies due to specific pests. Models of crop pests also need to be validated, and they need to provide reliable information on pest populations and how environmental factors affect them. Pests often alter the spatial patterns of plants in a crop. The resulting gaps in the crop would also have to be accounted for in the crop-pest model linkage (Hughes 1988). Because a crop model and pest models need to be run with the same set of weather data, little extra information may be needed for running both. There is a critical need to define linkage points (the known, quantifiable effects of a pest on a crop) between the two models. Pests can be classified according to their potential effects on a crop (i.e. the type of linkage points). The main linkage points include (Boote et al. 1993; Teng 1988):

1. reduction in plant population, where the number of plants and the weight of plant material in a system is diminished (e.g., damping-off fungi);
2. reduction in photosynthetic rate, in which the remaining host tissues have a reduced rate of carbon uptake (e.g., certain virus pathogens);
3. acceleration of leaf senescence, in which the senescence or abscission of leaves is affected by a known level of the pest population (e.g., some leafspotting fungi);
4. shading, in which photosynthetically active radiation is removed from plant parts that can utilize it (e.g., weeds);
5. assimilate removal, in which soluble assimilate otherwise used for plant growth and development are used by pests for their own growth and development (e.g., leafhoppers);
6. tissue consumption, in which plant tissue is physically removed from the system (e.g., defoliating insects);
7. turgor reduction, in which there is an interruption of the vascular system to affect xylem and phloem transport (e.g., wilt pathogens);
8. metabolic diversion, in which metabolites to be used by one type of host tissue or host part are diverted to another (e.g., root lesion nematodes);
9. resource competition, in which the pest competes for and utilizes resources before they enter the plant (e.g., weeds);
10. translocation disruption, in which the translocation system is rendered totally dysfunctional (e.g., the neck blast syndrome in rice);
11. tissue disruption, in which the pest causes cells to be disoriented and disrupt (e.g., root lesion nematode, rust pathogens).

The first step in coupling a pest model to a crop model is to identify which linkage points best describe the particular pest-crop combination (Heym et al. 1990).

Weed models are probably the most difficult type to link with crop models. This is partly because weed populations may contain several species and they are spatially variable. An intercrop model (of a mixed population of species) is needed for this application in which the weed species and crop species must have different growth and development patterns. Kropff (1988) has made reasonably good progress in modelling the effect of weeds on crop production, with details of model types and case studies in weed-crop modelling (Kropff 1993). Fischer et al. (1990) has described a relatively simple model of weeds in a wheat system. This model allows multiple germination dates where germination occurs after a set of given near-surface soil-water conditions exist.

Socioeconomic data for models

There are two approaches to incorporate socioeconomic data into simulation systems. The first is a bottom-up approach, which examines planting and harvesting decision making in detail and identifies the information required (Dent and McGregor 1994). The second is a top-down approach, which assumes that many social factors impinge on every decision (Brossier et al. 1990). The top-down approach is closer to the classic practice of anthropologists to collect data and formulate a hypothesis. The bottom-up approach lends itself more easily to crop simulation but has the disadvantage that it may fail to take account of social factors that, though important, remain unperceived. Conversely, the top-down approach recognizes the importance of social factors, but does not lend itself to linkage with crop simulation. It is difficult to know when to stop collecting data with the top-down approach.

Advocates of either approach usually have difficulty perceiving the problems of the other. Unfortunately, because of the differences in approaches to collecting and manipulating the data, the socioeconomic dimension of agricultural systems applications is more difficult to define. Most of the efforts to link crop simulation to socioeconomic work done in recent years have emphasized the use of rather simple economic and resource-constraint data.

The economic and sociocultural framework of the farm family substantially influences their acceptance of new ideas and technologies, and, therefore, the rate at which these ideas or technologies are adopted. Even comprehensive results from field trials or demonstrations that emphasize an improved technical management strategy for individual enterprises play a relatively small part in influencing decision making in farm households. An understanding of the biological inputs required for production, and hence an understanding of biological information, is important, but the essential driving force for change is social and cultural in nature.

Past attempts to develop simple empirical models that relate the rate of change to, for example, the age of the farmer, the size of holding, and land-tenure agreements, give no clear understanding of the decision-making process of the farm family. A much more acceptable model needs to be presented in a behavioral context. Here, one might say that a farmer will continue with the same management system and land allocation and the same distribution of resources if he found the last year or two satisfactory, and if the exogenous and endogenous conditions for the farm remain the same. In adverse conditions, the farmer may want to assess the situation. He may wish to modify the land-use pattern by varying the relative area of crops, or to modify the type of crops, including the choice of cultivar, or to change the intensity of management. Within this range of adjustments, technology (e.g., new crop cultivars) may feature as a supporting element. Together, these factors will determine the cropping system or the land-use system for the coming year.

The exogenous factors that impose on the farm decision-making process may be divided into three classifications (see table 6).

- unpredictable factors, such as the weather and prices;
- factors that relate to provision by regional or national agencies, such as credit, availability of technology, supply of information and education and health services in a region;
- factors that relate to the community in which the individual farmer lives and the culture which is established within that community, creating peer group pressures and kinship relationships.

The endogenous factors in decision making in households consist of biological elements, such as the selected farm enterprises and their management, the tenure of the land, and the factors related to the basic needs of the farm family for survival and motivation, attitudes and objectives. Some of these will be determined by other endogenous factors related to the age of the farmer and the size of his family.

To understand the process of change on the farm (and in particular the underlying decision-making processes of the farm household), we must understand the interrelationships between the endogenous and exogenous variables in any farming situation. The process of local change cannot be separated from the general environment in which the farm family and farm community exist. National policies for agriculture and for rural communities therefore have an impact on the way farm families respond to technological opportunities. It is clear that appropriate credit schemes, together with good extension back-up, can speed up the adoption of technology. Where levels of health and education are generally poor and where the male work force has to migrate to urban areas, technology is likely to be at a low level and it is likely to be adopted slowly. Research to drive the development of technology, and even the existence of technology per se, barely reflects on the decision-making process of households. It is the total mixture of conditions under which farming takes place that provides the impetus for change.

The above discussion highlights the fact that research, technology development, and the adjustment of farming systems are closely linked to the policy environment

Table 6. Socioeconomic data for modelling purposes

Economic data	Social related data	Attitudinal data
Variability & seasonality of prices	Family size	Potential for:
	Family education level	- off-farm employment
Variability & seasonality of costs	Ownership or tenancy of land	- use of non-family labor
	Place in community	- use of credit
Availability of inputs	Religion	- use of extension
Market opportunities	Peer/kinship pressures	
	Wealth	

for agriculture. Policy is effective only if individual farm families decide to use it as a basis for change. The appropriate policy instruments and their delivery mechanisms can be determined only if the decision-making processes of the farm family are understood. The farm household, then, plays a central role in decisions that policy-makers make about the development of technology, the research behind it, and the mechanisms to deliver the technology.

Understanding the process of decision making at farm households thus plays a central role in higher-level decision making about deploying research resources and developing technology. Such an understanding is also central to the process of decision making related to drawing up and delivering national and regional policy. If this is true, then decisions about deploying resources for research and technology development and decisions about establishing policies cannot be taken independently of each other. A better understanding will lead an improved ability to predict farm-family response by way of some appropriate form of model.

What steps are needed to allow such prediction?

1. From an anthropological point of view, all households, indeed all people, are unique. But in terms of understanding the decision making at farm households with regard to technology adoption and policy, families may be classified. In a similar manner in which agroecosystems have been defined as homogenous areas for the purpose of the definition, it is possible to envisage socioagroecosystems, in which a socioeconomic classification is "laid on top" of defined agroecosystems. Within a defined socioagroecosystem, all farmers are expected to respond in a similar way, although perhaps with different dynamics. Classification may be based on some or all of the information listed under social related data in table 6.

2. For each class, determine a minimum set of data required to drive the decision model of the farm household. The model needs to be specified in generic form and should be dynamic because of the nature of the process of change as described above. This is an essential requirement for the model to be applied to farms that can represent various groups without structural modification. The model requires the kind of social related data in table 6 to particularize it for a specific target group. Some elements of the data set are obvious, non-contentious and relatively easy to collect by routine rural-survey methods. Others, in particular those related to attitudes, lifestyle, and the place in the community, require careful definition and description. Indeed, some may not be able to be captured in quantitative form at all. The model will almost certainly need to be able to accommodate qualitative as well as quantitative information. Such data must also be elicited by survey methods because anthropological methods are too expensive and time consuming. Recent experience shows how attitudinal characteristics may be determined and verified with psychometric tests. Progress in this methodology is likely to be fairly rapid, although matters concerning classification processes through the analysis of the principal components will prove troublesome with respect to sample size and qualitative analysis.

Conclusions

It is essential that agricultural sciences become more integrated in the future because problem solving in the real world needs data from several subject-matter areas. The systematic incorporation of information from these areas is possible through the simulation approach. Such an organized systems approach can potentially take agricultural sciences into the information age. It also improves research that has site-specific limitations, and it allows a more rapid transfer of agrotechnology. For a systems approach to be used in solving problems that concern specific farmer or regional goals, it is obvious that pest and socioeconomic dimensions must be a part of the system. ICASA proposes to make a major attempt in this direction by combining efforts of leading systems scientists, national agricultural researchers, and researchers in international agricultural research centers. In order to accomplish this ambitious goal, there will need to be a balance between the simulation systems that help the decision making and the data necessary for inputs into the system and its validation. This paper has attempted to describe the minimum and desirable data required to make good progress in agricultural systems application.

Acronyms

CGIAR	Consultative Group on International Agricultural Research
FAO	Food and Agriculture Organization
ICASA	International Consortium for Application of System Approaches to Agriculture

References

Boote K J, Batchelor W D, Jones J W, Pinnschmidt H, Bourgeois G (1993) Pest damage relations at the field level. Pages 277-296 in Penning de Vries F W T et al. (Eds.) Systems approaches for agricultural development. Kluwer Academic Publishers, Dordrecht, The Netherlands.

Bouma J, Wopereis M C S, Wösten J H M, Stein A (1993) Soil data for crop-soil models. Pages 207-220 in Penning de Vries F W T et al. (Eds.) Systems approaches for agricultural development. Kluwer Academic Publishers, Dordrecht, The Netherlands.

Brossier J, Vissac B, Le Moigne J L (1990) Modélisation systémique et système agraire. Décision et organisation. INRA Publications, Paris, France. 365 p.

Cosby B J, Hornberger G M, Clapp R B, Ginn T R (1984) A statistical exploration of the relationship of soil moisture characteristics to the physical properties of soil. Water Resour. Res. 20:682-690.

Dent J B, McGregor J M (1994) Integrating livestock and socioeconomic systems within complex models. EC Symposium on Systems Analysis in Livestock Production Research. Zaragoza, Spain. (in press).

Driessen P M (1986) The water balance of soil. Pages 76-116 in Van Keulen H, Wolf J (Eds.) Modelling of agricultural production: Weather, soils and crops. Simulation Monographs, PUDOC, Wageningen, The Netherlands.

Fischer R A, Armstrong J S, Stapper M (1990) Simulation of soil water storage and sowing day probabilities with fallow and no-fallow in southern New South Wales: I. Model and long-term mean effects. Agric. Sys. 33:215-240.

Frère M (1987) The FAO agroclimatological data base: Its use in studying rice-weather relationships. Pages 41-46 in The impact of weather parameters on the growth and yield of rice. International Rice Research Institute, Los Baños, The Philippines.

166

Geng S, Penning de Vries F W T, Supit I (1985a) Analysis and simulation of weather variables: Rain and wind in Wageningen. Simulation Reports CABO-TT No 4. Centre for Agrobiological Research, Wageningen, The Netherlands.

Geng S, Penning de Vries F W T, Supit I (1985b) Analysis and simulation of weather variables: Temperature and solar radiation. Simulation Reports CABO-TT No 5. Centre for Agrobiological Research, Wageningen, The Netherlands.

Geng S, Penning de Vries F W T, Supit I (1986) A simple method for generating daily rainfall data. Agricultural and Forest Meteorology 36:363-376.

Heym W D, Ewing E E, Nicholson A G, Sandian K P (1990) Simulation by crop growth models of defoliation, derived from field estimates of percent defoliation. Agric. Sys. 33:257-270.

Hughes G (1988) Models of crop growth. Nature 332:16.

Van Keulen H, Wolf J (1986) (Eds.) Modelling of agricultural production: Weather, soils and crops. Simulation Monographs, PUDOC, Wageningen, The Netherlands. 464 pp.

Kropff M J (1988) Modelling the effects of weeds on crop production. Weed Res. 28:465-471.

Kropff M J, Van Laar H H (Eds.) (1993) Modelling crop-weed interactions. CAB International, Wallingford, UK. 274 pp.

Oldeman L R, Seshu D V, Cady F B (1987) Response of rice to weather variables. Pages 5-40 in The impact of weather parameters on the growth and yield of rice. International Rice Research Institute, Los Baños, Philippines.

Richardson C W, Wright D A (1984) WGEN: A model for generating daily weather variables. US Dept. of Agric., Agric. Res. Service, ARS-8. 83 p.

Ritchie J T (1993) Genetic specific data for crop modelling. Pages 77-93 in Penning de Vries F W T et al. (Eds.) Systems approaches for agricultural development. Kluwer Academic Publishers, Dordrecht, The Netherlands.

Ritchie J T, Crum J (1989) Converting soil survey characterization data into IBSNAT crop model input. Pages 155-167 in Bouma J, Bregt A K (Eds.) Land qualities in space and time. Wageningen, The Netherlands.

Teng P S (1988) Pests and pest-loss models. Agrotechnology Transfer 8:5-10.

Vereecken H, Maes J, Feyen J (1990) Estimating unsaturated hydraulic conductivity from easily measured soil properties. Soil Sci. 149(1):12-32.

Wagenet R J, Bouma J, Grossman R B (1991) Minimum datasets for use of taxonomic information in soil interpretive models. Pages 161-183 in Mausbach M J, Wilding L P (Eds.) Spatial variabilities of soils and landforms. Soil Sci. Soc. Am. Special Publication 28.

Issues of development of an international soils and terrain database

L.R. OLDEMAN

International Soil Reference and Information Centre (ISRIC), P O Box 353, 6700 AJ Wageningen, The Netherlands

Key words biophysical data, database, database management, erosion, GIS, GRID, socioeconomic data, soil degradation, SOTER

Introduction

The present intense and increased pressure on land and water resources leads to the degradation and pollution of those resources, as well as to a partial or complete loss of the productive capacity of the soil. This calls for an approach that will strengthen not only the awareness of users of these resources to the dangers of inappropriate management, but also, at the same time, the capability of national soil/land resource institutions of delivering reliable, up-to-date information on land resources in an accessible format to a wide audience.

The development of an internationally accepted geo-referenced system capable of providing accurate, useful and timely information on soil and terrain resources is a prerequisite for policymakers, decision makers, resource managers, and the scientific community at large in their assessment of the productive capacity of soils, of the status, risks and rates of soil degradation, and of global change.

The environmental links between soils, food crops, and water supplies mean that if soils are degraded or polluted, food chains and drinking water will probably also be affected. Quantification of processes of soil degradation (and pollution) requires that researchers have access to compatible, uniform databases that include data layers on climate, landform, soils, and land cover/land use and that are linked to a geographic information system (GIS).

The link between research and development can be established through a package of technology that consists of two major elements:

- a detailed database on natural and human resources established in such a way that the information it contains can be immediately and easily accessed and combined;
- the establishment of scientifically valid methods, to analyze the land, water, climate, vegetation, land- use and socioeconomic information for evaluation of land-use options, both in relation to food and other human requirements and for assessing environmental impact.

In close cooperation with ISSS, FAO, and UNEP, the International Soil Reference and Information Centre (ISRIC) in Wageningen, The Netherlands, has developed an internationally endorsed land-resource information system, called World Soil and

167

P Goldsworthy and F W T Penning de Vries (eds), Opportunities, use, and transfer of systems research methods in agriculture to developing countries, 167 - 178
© 1994 *Kluwer Academic Publishers*

Terrain Digital Database (SOTER). This paper discusses the issues related to the development of this georeferenced agroecological database.

The need for a natural resource database

Throughout history, people have collected information on natural resources. The wealth of information generated by national and international institutions all over the world is stored in different forms, such as maps, tabular data, and descriptive reports. Storage of this information is initially geared towards the needs of the individual institute or country. As a result, the material stored in one place is often not compatible with information stored elsewhere. The large amounts of information, whether on soils, climate, or land cover, has generally been assembled and then ordered, grouped, or classified, and given symbols or names so that scientists within a certain country or within a certain discipline could understand what characteristics were common to a certain soil or climate. Since no systems in the past were available for storing information on natural resources in a standardized or systematic way, scientists had no other option than to classify the natural resources according to their own system. In national soil classification systems, soils were given local names which were perhaps meaningful to local soil scientists but did not help a wider audience of non-soil scientists to better understand the characteristics of the soil. It became extremely difficult to extrapolate research results obtained at one site to larger areas—it was virtually impossible to exchange research results with other countries.

To solve the problems of international barriers to exchanging natural resource information, international classification schemes were developed. In soil science, two major schools of classification emerged. The first one was the soil taxonomy system, developed by the Soil Conservation Service in the United States. The other international system was developed by FAO's Land and Water Division within the framework of their mapping activity for a soil map of the world. Both systems focused their classification systems on attributes of the soils that could be measured in the field or analyzed in the laboratory. In contrast, the French soil classification system focused on soil processes and soil formation.

These international classification systems, now universally adopted, were—and still are—an essential and comprehensive methodology for characterizing the soils across national boundaries. They have an important consequence: only very experienced soil scientists understand them. For the non-soil science audience, the systems are incomprehensible.

Every new soil-survey report published in a country is more advanced, more complex and technically more difficult than previously published reports. However, the majority of potential users of soil and terrain information are not technically trained in soils. Consequently, these users will not or cannot extract the specific information they need from a modern soil-survey report. Purnell (1992) stated that there is a need for more reliable and more timely information about soils. This information must be more accessible and intelligible to non-soil scientists: "Clear language, free from jargon, does not come easily to soil surveyors, but soil-survey reports are of no value if not read."

The urgent need for a global inventory of the Earth's natural resources has been expressed by many international research organization and policymakers (UNEP 1992; UNSO/UNDP, UNEP 1992; WRI, Cal.#Tech 1992; ISNAR 1992). While there have been substantial advances in our understanding of environmental processes, and a consensus has emerged that all forms of development must be sustainable, there are still important gaps in our understanding of how to assess adequately the impact of environmental damage. The many international initiatives in modelling the effects of an environmental change in combination with population growth in the developing countries are in dire need of up-to-date spatial databases which have been verified from ground observations, on agroclimatic conditions, basic landform-soil relationships, water resources, agrohydrology, present-day land cover and land use, and the status and hazard of land degradation (Sombroek 1992).

There is no doubt that what is needed at the global level also applies at the national and regional level. Rural development planning in developing countries increasingly depends on reliable natural resource information. The history of agricultural development projects during the past decades is full of examples of how the results might have been more positive if accurate and timely information about land and water resources had been available to the planners (ISSS 1986).

For an improved assessment of land resources for the development of sustained use of the land, the following are necessary (Anon. 1992):

- Database development. A database should contain all available information on topography, soil attributes, climatic parameters, vegetation, land use, population (both human and animal) infrastructure, as well as socioeconomic factors such as food requirements and preferences, skills, costs of input (including labour), market availability, and stability.
- Geographic referencing. Each item of information should be linked to its precise geographic position, using tools such as GIS and remote sensing.
- Estimation of biomass. Relevant modelling approaches can be used as a tool to assess biomass from each land unit to evaluate levels of production at different levels of input.
- Estimation of the status and risk of land degradation. Risks such as loss of soils by water and wind erosion, nutrient decline, salinization, acidification, soil physical degradation, vegetation degradation, and soil, water and air pollution can be assessed for given land units with specific land-use and production systems to evaluate environmental impacts.
- Improved access to existing knowledge. This will require the development of a documentation information system on development-oriented research activities such as those reported in national and international journals.

The development of a geo-referenced natural resource information system with internationally accepted standards will provide policymakers, decision makers and resource managers with important and indispensable tools to reverse the current trends of soil degradation in developing and industrialized countries, and to implement a program for soil conservation and sustainable use of the land.

Issues of development of databases

The objective of a natural resource information system is to utilize information technology to produce an internationally accepted, standardized database, containing digitized boundaries to map units in a GIS, with their attribute database in a relational database management system (RDBMS). SOTER is an example of such an information system—one that will hold the data necessary for the improved mapping and characterization of world soil and terrain resources and for monitoring changes in it. This database is organized in such a way that it provides a comprehensive framework for the storage and retrieval of uniform soil and terrain data that can be used for a wide range of applications at different scales.

Organization of databases for agroecological characterization

Two types of database are required in every discipline engaged in the mapping of spatial phenomena: (1) geometric data, i.e., the location and extent of an object represented by a point, a line, or a surface, including its topology (shapes, neighbors, and hierarchy of delineations), and (2) attribute data, i.e., the characteristics of the object. The geometry is stored in the part of the database that is handled by GIS software, while the attribute data is stored in a separate set of attribute files, manipulated by a RDBMS. A unique label attached to both the geometric and the attribute database connect these two types of information for each mapped unit (figure 1). The overall systems (GIS plus RDBMS) stores and handles both the geometric and attribute database.

The geometric database. This database contains information on the delineations of the mapped unit. It also includes data on features of the base topographic map, such as roads and towns, the hydrological network, and administrative boundaries. In order to enhance the usefulness of the database, it will be possible to include additional overlays for boundaries outside the physiographic unit mosaic. Examples of such overlays could be socioeconomic areas (population densities), hydrological units (watersheds), or other natural resource patterns (vegetation, agroecological zones).

The attribute database. A relational database is considered to be one of the most effective and flexible tools for storing and managing nonspatial attributes (Pulles 1988). Under such a system, the data is stored in tables whose records are related to each other through specific identification codes (primary keys). These codes form the link between the various subsections of the database (e.g., the terrain table, the terrain component table, and the soil component tables in SOTER). Another characteristic of the relational database is that when two or more components are similar, their attribute data need only be entered once. Figure 2 gives a schematic representation of the organization of the attribute database. The blocks represent tables in

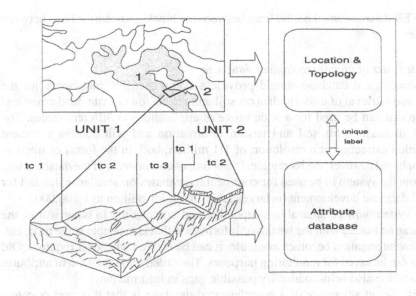

Figure 1. SOTER units, their terrain components (tc), attributes, and location

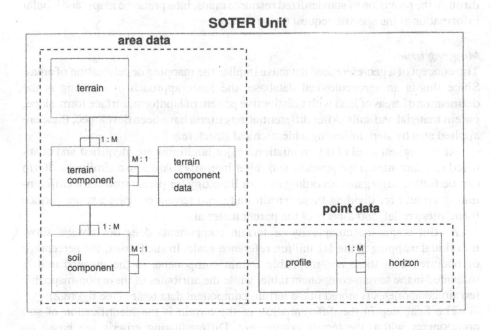

Figure 2. SOTER attribute database structure with area and point data (1:M = one to many; M:1 = many to one)

the SOTER database and the solid lines between the blocks indicate the links between the tables.

Characteristics of an agroecological database

An agroecological database should provide a comprehensive framework for the storage and retrieval of uniform data on soil and terrain, on climate, land cover and land use that can be used for a wide range of applications at different scales. The SOTER database holds soil and terrain information and will contain sufficient information extraction at a resolution of 1:1 million, both in the forms of maps as well as tables. However, the hierarchical structure of components of the database will also allow the system to be used for continental databases on smaller scales and for national database development on larger scales from 1:1 million to 1:100,000.

The system will be amenable to updating and completing. In other words, the system can be loaded with the wealth of information that is available at the time, but, when new information becomes available, it can be entered whenever required. Old data sets can be saved for monitoring purposes. The orderly arrangement of attributes in the database also helps to identify possible gaps in information.

A significant advantage of a georeferenced database is that it is user oriented. While conventional natural resource mapping is a one-off exercise, directed at a few selected users, a georeferenced database provides information to a broad array of international, regional and national environmental specialists and policymakers through the provision of standardized resource maps, interpretative maps, and tabular information at the specific request of the user.

Mapping issues

The concept of a georeferenced database implies the mapping or delineation of areas. Since this is an agroecological database, the basic approach in mapping is the delineation of areas of land with a distinctive pattern of landform, surface form, slope, parent material and soils. After differentiating criteria have been developed, these are applied step by step, following a hierarchical structure.

At the highest level of differentiation, major landforms are identified and quantified (by dominant slope gradient and relief intensity). Areas in a similar landform can be further segregated according to their lithology (or parent material). Differentiating criteria for dividing these terrain units into terrain components are surface form, meso relief, and aspects of the parent material.

At this stage, the complexity of terrain components does not always allow individual mapping at the 1:1 million reference scale. In such cases, the percentage of occurrence of these non-mappable terrain components in the mapped unit is indicated in the terrain-component table, while the attributes of these non-mappable terrain components is stored in the terrain component data tables (see figure 2).

The final step in the differentiation of the terrain is the identification of soil components within the terrain component. Differentiating criteria are based on diagnostic horizons and properties as formulated by FAO (FAO 1988). As with terrain components, these soil components can be mappable or nonmappable. Most likely, these terrain components comprise (at the reference scale of 1:1 million) a number of soil associations or soil complexes. The percentage of occurrences within the mapped unit is indicated in the soil component table, while the attributes are stored in the soil profile and soil horizon data tables.

Content of the database

The attributes of the database are elements that can be quantified, either through visual observations in the field or by measurements in the laboratory. The general approach adopted by SOTER is to screen all existing soil and terrain data in a geo-referenced area and to complement the terrain information with remote sensing data where necessary. The various attributes of the terrain, terrain component, and soil component tables are listed in table 1. Each attribute is described in detail in the SOTER Procedures Manual (Van Engelen and Wen Tin Tiang 1993). Some issues related to the content of the database are:

- Descriptive attributes (land forms, lithology, surface forms). Descriptions are based foremost on morphology and not on genetic origin or processes responsible shape.
- Class values versus numeric values. Numeric values are needed for any algorithms for which attribute values are needed. Class values are important for land evaluation purposes.
- Mandatory versus optional values. The attributes for terrain and terrain components are either directly available or can be derived from other parameters during the compilation of the database. Many of the soil parameters consist of measured values, the availability of which varies considerably. A minimum set of soil attributes is generally needed for any realistic interpretation. Their presence in the database is mandatory. If data is not available for some of these mandatory attributes, the database management allows expert estimates to be used. Measured and estimated values will be stored separately.
- Spatial variability. A major problem in any survey is the reconciliation of point data and area data. In SOTER, for every soil component, at least one, but preferably more, fully described and analyzed reference profiles should be available from existing soil information sources. Following judicious selection by the local soil-survey team, one of these reference profiles will be designated as the representative profile for the soil component. In order to give some degree of variability for soil profile data, single values taken from the representative profile are complemented with maximum values and minimum values taken from all available profiles within the soil component. These values are an indication of the range of variation that exists within the component. Although statistical parameters such as standard deviation and means could give a better idea of the variability, it is in many cases not realistic to achieve this, considering the small number of profiles generally available for the compilation of the soil component.

Table 1. Non-spatial attributes of a SOTER unit

Terrain:
1 SOTER unit-ID
2 year of data collection
3 map-ID
4 minimum elevation
5 maximum elevation
6 slope gradient
7 relief intensity
8 major landform
9 regional slope
10 hypsometry
11 dissection
12 general lithology
13 permanent water surface

Terrain component:
14 SOTER unit-ID
15 terrain component number
16 proportion of SOTER unit
17 terrain component data-ID

Terrain component data:
18 terrain component data-ID
19 dominant slope
20 length of slope
21 form of slope
22 local surface form
23 average height
24 coverage
25 surface lithology
26 texture group non-consolidated parent material
27 depth to bedrock
28 surface drainage
29 depth to groundwater
30 frequency of flooding
31 duration of flooding
32 start of flooding

Soil component:
33 SOTER unit-ID
34 terrain component number

35 soil component number
36 proportion of SOTER unit
37 profile-ID
38 number of reference profiles
39 position in terrain component
40 surface rockiness
41 surface stoniness
42 types of erosion/deposition
43 area affected
44 degree of erosion
45 sensitivity to capping
46 rootable depth
47 relation with other soil components

Profile:
48 profile-ID
49 profile database-ID
50 latitude
51 longitude
52 elevation
53 sampling date
54 lab-ID
55 drainage
56 infiltration rate
57 surface organic matter
58 classification FAO
59 classification version
60 national classification
61 Soil Taxonomy
62 phase

Horizon (* = mandatory):
63 profile-ID*
64 horizon number*
65 diagnostic horizon*
66 diagnostic property*
67 horizon designation
68 lower depth*
69 distinctness of transition

70 moist colour*
71 dry colour
72 grade of structure
73 size of structure elements
74 type of structure*
75 abundance of coarse fragments*
76 size of coarse fragments
77 very coarse sand
78 coarse sand
79 medium sand
80 fine sand
81 very fine sand
82 total sand*
83 silt*
84 clay*
85 particle size class
86 bulk density*
87 moisture content at various tensions
88 hydraulic conductivity
89 infiltration rate
90 pH H_2O*
91 pH KCl
92 electrical conductivity
93 exchangeable Ca^{++}
94 exchangeable Mg^{++}
95 exchangeable Na^+
96 exchangeable K^+
97 exchangeable Al^{+++}
98 exchangeable acidity
99 CEC soil*
100 total carbonate equivalent
101 gypsum
102 total carbon*
103 total nitrogen
104 P_2O_5
105 phosphate retention
106 Fe dithionite
107 Al dithionite
108 Fe pyrophosphate
109 Al pyrophosphate
110 clay mineralogy

Source materials

A major issue of the development of databases is the availability of information. An agroecological information system like SOTER is based on existing soil and terrain information, and specifically excludes new land-resource surveys within its program. Basic data sources for the construction of SOTER mapping units are existing topographic, morphological, geological and soil maps. All soil maps that are accompanied by sufficient analytical data for soil characterization can be used, with the data entered into the database. One of the major advantages of the orderly arrangement of information in a database is that gaps in information can be more easily identified. Some extra analytical work may then have to be carried out to complement existing information.

A database like SOTER should include a reference file that contains information on the source materials used for the compilation of the database. In the case of SOTER, this would include information on the source of the map, the laboratories that analyzed the soil samples, the laboratory methods employed and the organizations responsible for the national profile database. An ideal situation in which all soil samples are analyzed according to standardized analytical methods does not exist.

Associated data

The issues of the development and management of databases discussed so far are related to soil and terrain databases. Climatic data form an inseparable part of the basic inventory of natural resources. Nevertheless, climatic data are treated separately from soil and terrain data, since this set of information is not directly linked to the physiographic SOTER areas. Climatic data are based on point observations only. The link with the soil and terrain information is formed by the geographical location of these points.

Land-cover characteristics (vegetation and land use) are also stored in separate files. Unlike the more stable terrain and soil characteristics, land cover is considered a more dynamic entity, which can change quickly. Thus there may be a frequent need for more recent data. Moreover, other groups are working on global databases for land use and vegetation. At present, such databases are not available, but they should be included in a database for agroecological characterization. In SOTER, the land-cover information is given at the level of the mapped unit. Land-use classes are defined in a hierarchical system (Remmelzwaal 1990), while for vegetation, the generalized hierarchical description of the physiognomy of the present native vegetation developed by Unesco is followed (Unesco 1973).

Operational issues

A project proposal for a World Soil and Terrain Digital database at a scale of 1:1 million was developed on the basis of recommendations from an ISSS international workshop on "The Structure of a Digital International Soil Resources Map Annex database" (ISSS 1986). This proposal received further endorsement at the 1986 ISSS Congress in Hamburg, Germany, as well as financial support from UNEP in 1987 to develop a SOTER Procedures Manual and for a first testing of the SOTER concept in a pilot area, covering portions of Argentina, Brazil, and Uruguay. Application of the SOTER methodology was also carried out in an area along the border between the USA and Canada.

Based on the experience obtained in these pilot test areas, the Procedures Manual was revised and evaluated at an international workshop at UNEP, Nairobi (UNEP

1992b). At this workshop FAO, expressed its full support for SOTER and indicated that is was prepared to use the SOTER methodology for storing and updating its own data on world soil and terrain resources. This workshop also recommended applying the SOTER approach to continental coverage.

At a national level, many countries are quite acutely aware of the need—and some are already attempting—to establish computerized resource databases. The SOTER program will create an excellent vehicle for training a cadre of national environmental specialists in the use of modern information technologies. This will enable them to provide answers to requests from planning agencies involved in soil conservation and sustainable food productivity. A polit training course was held in South America. Participating countries were provided with the necessary hardware and software and are now implementing SOTER activities at the national level with technical advisory support from ISRIC in Wageningen.

The SOTER program has now progressed from its initial development and testing phase into an operational phase. Currently, joint cooperative SOTER projects are implemented in Uruguay, Argentina, Brazil, Venezuela, Kenya, and Hungary, while proposals for international funding have been submitted for six countries in West Africa, parts of China, and parts of the former USSR. Member countries belonging to the Asian Network of Problem Soils have also recommended the use of SOTER at their expert meeting in Bangkok (FAO 1993). SOTER is being implemented at a scale of 1:5 million for continental coverage of Latin America, 1:1 million in Uruguay, Argentina, and Kenya, 1:500,000 in Hungary, and 1:100,000 for selected areas in Uruguay and Argentina.

Copies of the resulting digital databases generated by UNEP-funded projects will be incorporated into Global Resources Information Database (GRID) archives of UNEP in Nairobi for release according to the "GRID Data Release Policy and Data Archive Access" guidelines.

Conclusion

The rapid growth of the world population leads to the key question of whether the land supply will accommodate a doubling of global agricultural demands in 40 years' time. There are estimates (Buringh and Dudal 1987) that around 1800 million hectares of land not yet used for agricultural food crops are still available and suitable for growing crops. However, this land is not being used for agriculture for good reasons: most of it has moderate to low productivity, the spatial distribution of available land is not in proportion to the population density, the infrastructure that would be needed to transport the produce to domestic and foreign markets is usually very poorly developed or absent, the hazard of soil erosion and soil degradation increases when these areas—now under forest or grasslands—are transformed into croplands, and conversion of tropical forests into cropland is a hotly debated issue.

Past and present human intervention in the use and manipulation of environmental resources has had unanticipated consequences. There is now greater public awareness about the consequences of indiscriminate destruction of forests and woodlands, leading to land degradation, decreased productivity, with dire social consequences. The Brundtland report stated: "There is a growing realization in national and international institutions that not only do many forms of economic development erode the environmental resources upon which they are based, but at the same time environmental degradation can undermine economic development" (Brundtland et al. 1987). Soil degradation implies by definition a social problem. Environmental processes such as leaching and erosion occur with or without human interference, but for these processes to be described as "degradation" involves social criteria that relate land to its actual or possible uses (Blaikie and Brookfield 1987). The "World Map of the Status of Human-Induced Soil Degradation" (Oldeman, Hakkeling and Sombroek 1991; Oldeman 1993) revealed that almost 2000 million hectares of agricultural land, permanent pastures, forest and woodland have been degraded. Worldwide, almost 40 percent of agricultural land is affected. Some 900 million hectares have a moderate degree of soil degradation, indicating greatly reduced productivity.

The need for a georeferenced socioeconomic database is evident. The issue is whether it could then be linked directly to a georeferenced agroecological database and whether it is possible to link socioeconomic attributes to the physiographic mapped units of the agroecological database. Socioeconomic information is usually assembled for administrative areas, which, generally, do not conform to physiographic delineations. A further difficulty is that socioeconomic attributes can change relatively rapidly (as land-use and land-cover attributes do).

The need for a holistic approach to agroecological characterization is obvious. Anon. (1992): "The task of all people concerned with land and water resources is to direct their interest not just to the physical, chemical and biological aspects, but also to those environmental, economic, social, legal and technical aspects that affect soil use". Life and livelihood on earth depend largely on the capacity of the soil to produce, whether by agricultural, industrial or other practices. The soil is a natural resource, non renewable, or very difficult to renew in the short term, and expensive either to reclaim or to improve following erosion, physical degradation, or chemical depletion. It is our duty to maintain it for the future, as well as to obtain the best benefit from its use today.

Acronyms

FAO	Food and Agriculture Organization
GIS	geographic information system
GRID	global resources information database
ISRIC	International Soil Reference and Information Centre
RDBMS	relational database management system
SOTER	Soil and Terrain Digital Database
UNEP	United Nations Environment Programme

References

Anon. (1992) New challenges for soil research in developing countries: A holistic approach. Proceedings of the Workshop funded by the European Community Life Sciences and Technologies for Developing Countries (STD 3 Programme), March 1992. Rennes, France.

Baumgardner M F, Oldeman L R (Eds.) (1986) Proceedings of an international workshop on the Structure of a Digital International Soil Resources Map. Annex database. SOTER report 1. International Society of Soil Science, Wageningen, The Netherlands.

Blaikie P, Brookfield H (1987) Land degradation and society. Methuen, London and New York.

Brundtland G H, Khalid M, et al. (1987) Our common future. Report of World Commission on Environment and Development presented to the Chairman of the Intergovernmental Intersessional Preparatory Committee, UNEP Governing Council. Oxford University Press, Oxford, UK.

Buringh P, Dudal R (1987) Agricultural land use in space and time. Pages 9-43 in Wolman M G, Fournier F G A (Eds.), Land transformation in agriculture. SCOPE 32. John Wiley & Sons, Chichester, New York, USA.

Van Engelen V W P, Wen Ting-Tiang (1993) Global and national soil and terrain digital databases (SOTER), procedures manual. International Soil Reference and Information Centre, Wageningen, The Netherlands.

FAO (1988) Soil map of the world, revised legend. World Soil Resources Report 60. Food and Agriculture Organization of the United Nations, Rome, Italy.

FAO (1993) Proceedings of the Third Expert Consultation of the Asian Network of Problem Soils, 25-29 October 1993, Bangkok, Food and Agriculture Organization of the United Nations, Rome, Italy. (in press).

ISNAR (1992) Summary of agricultural research policy: International Quantitative Perspectives. International Service for National Agricultural Research, The Hague, The Netherlands.

Oldeman L R (1993) Global extent of soil degradation. Pages 19-36 in Bi-annual report 1991-1992. International Soil Reference and Information Centre, Wageningen, The Netherlands.

Oldeman L R, Hakkeling R T A, Sombroek W G (1990) World map of the status of human-induced soil degradation: An explanatory note. International Soil Reference and Information Centre, Wageningen, The Netherlands.

Pulles J H M (1988) A model for a soils and terrain digital database. Working Paper and Preprint No. 88/5. International Soil Reference and Information Centre, Wageningen, The Netherlands.

Purnell M F (1992) Offer and demand of soil information: International policies and stimulation programmes. In Soil survey, perspectives and strategies for the 21st Century: Proceedings of an international workshop for heads of national soil survey organizations (keynote speeches). International Institute for Aerospace Survey and Earth Sciences, Enschede, The Netherlands.

Remmelzwaal A (1990) Classification of land and land use, first approach. Food and Agriculture Organization of the United Nations, Rome, Italy. (unpublished manuscript).

Sombroek W G (1992) Some thoughts of the new director, AGL. Land and water (36).

UNEP (1992a) Two decades of achievement and challenge: The United Nations Environment Programme 20th Anniversary. Our Planet 4(5).

UNEP (1992b) Proceedings of the ad-hoc expert group meeting to discuss global soil databases and appraisal of GLASOD/SOTER, 24-28 February 1992. United Nations Environment Programme, Nairobi, Kenya.

UNESCO (1973) International classification and mapping of vegetation. Ecology Conservation 6. United Nations Educational, Scientific and Cultural Organization, Paris, France.

UNSO/UNDP/UNEP (1992) Draft report on the expert meeting on desertification, land degradation and the global environment facility, 28-30 November 1992, Nairobi. United Nations Environment Programme, Nairobi, Kenya.

WRI/Cal Tech (1992) Global environmental monitoring: Pathways to responsible planetary management. A proposal by the World Resources Institute, Washington, DC, and the California Institute of Technology, Pasadena, CA. World Resources Institute, Washington DC, USA.

Discussion on Section C: NARS needs and priorities for addressing natural resource issues

Strategies for incorporating NRM concerns into agricultural research will vary with the size and level of development of the agricultural sector as well as the size and structure of the research and development institutions responsible for promoting sustained growth. The differences between the needs and priorities of large and small countries is one dimension to the problem. Another is the institutional and organizational options that countries have for reorienting their research efforts and mobilizing new actors to address sustainability issues in agriculture.

The first point raised was the role of policy. Policy decisions are crucial in deciding the degree to which NRM concerns can be integrated in the agricultural research agenda. Policymakers in developing countries have accepted the importance of sustainability in agriculture. It is not purely an issue imposed by the industrialized countries, although there is a significant difference between rich and poor countries in their perceptions of environmental issues. Evidence of the new concern about sustainable agriculture can be seen in the numerous high-level commissions, government councils, and new agencies to specifically address natural resource and environmental issues in, for example, Brazil, Bangladesh, the Philippines, and Kenya. However, many of the conflicting objectives noted in previous sections, combined with short-term planning, unstable funding, lack of priorities, and resistance to institutional change, impede the implementation of NRM perspectives in research.

Supporting interinstitutional research
Responsibility for policy issues concerning resource management is often fragmented among several administrative divisions and institutions. This frustrates attempts to formulate coherent policies that take into account the diverse and often conflicting interests involved. Major organizational and institutional innovations are required to facilitate the linking of agricultural and NRM research, and the introduction of systems approaches.

At the research level, NARS rarely include meteorological services; remote sensing, cartography, and soil survey are usually separate; and economic and social sciences are usually the concern of yet other agencies.

Agenda 21 provides forceful arguments for countries to re-examine the structure of their NARS, to cut out duplication, to merge separate bodies if necessary, and to enforce coordinated priorities. This may be the only way with the research resources available to handle the additional objectives and activities in NRM without shifting priorities away from research, which is directed to increasing agricultural production.

The complexity of NRM research requires an integrated cross-sectoral approach to planning, between agriculture and other sectors, between the sub-sectors of agriculture, and within agriculture as a whole. This in turn calls for more interdisciplinary research and closer inter-institutional links. While some countries have effected, or are considering, such changes, institutional constraints still loom large in most developing countries.

It was a failure in the past to give sufficient attention to the institutionalization of systems approaches in national systems that led to the isolation and failure of many farming systems programs once donor funding declined. As a minimum, many NARS are likely to need mechanisms and procedures for setting NRM research objectives and designing programs that cut across the institutional structure to prevent this from happening again.

The discussion highlighted that planning research on NRM often did not include all the actors involved in its implementation. Experience has shown that it is much easier to create mechanisms for interinstitutional planning than it is to implement and manage joint NRM programs. NRM research calls for more flexible and responsive institutional structures than those currently found within the civil-service administrations in many countries. For example, the usual practice of allocating funds to organizations rather than to programs is seen as a barrier to institutional collaboration and the development of more interdisciplinary research. It was suggested to use funding mechanisms that would encourage research organizations to contribute jointly to solving resource management problems.

Strengthening linkages between research and policy
Policy support to research remains weak and subject to rapid changes. This is particularly harmful to some of the longer-term programs that are essential for systems and NRM research. Policy support will be crucial to building interinstitutional linkages and the organizational innovation needed to implement agroecological systems approaches. The level of support will depend very much on maintaining an effective dialogue between program and policy levels of the organization or research system. Bhutan is an example where strong political support and leadership made possible a reorganization of the agricultural research system into one that will allow future research to focus on five main agroecosystems, and, in doing so, to integrate the related research on crop, livestock, and forestry production. The need to integrate research around the ecosystem may be more apparent in a small country, where research cannot deal separately with the priority crops and the environmental issues. Also, the small size of the institutions makes it possible for leadership to effect more rapid changes than it could in larger, more bureaucratic organizations.

The outputs from the methods use to study agroecological systems and the environment will often be information and recommendations to policymakers and planners to guide their decisions on the rational and sustainable use of resources. Policymakers should be made more aware of the opportunities for improving the overall productivity of land, water, and biological resources that can result from applying these systems methods and NRM approaches. Increasingly, research will be expected to participate more directly in the policy dialogue. NARS should give priority to strengthening linkages between natural resource users, scientists, planners, and policymakers.

Supporting long-term goals
Short-term planning and funding is inconsistent with the long-term commitment needed to address NRM issues. NRM research often requires long time horizons with stable funding commitments to achieve results. The impacts of changes in the resource bases resulting from new and improved resource management practices are only apparent after a lapse of years and provided that implementation is achieved at the requisite scale. In many cases the objectives, impacts, and benefits are for future generations. And yet, in many developing countries, planning horizons are short, often no more than five years. Funding commitments are often even shorter and unreliable.

Understandably, the short time span of political office in many countries makes policymakers hesitant to commit funds to projects whose benefits will be reaped by future generations of political officeholders. There is no simple answer to the instability of public funding for long-term systems research. However, it may be possible to show how systems and NRM research can make policymakers more informed and effective decision makers, thus enhancing their prestige.

Donor funding is not necessarily more stable than public national funding. Donor-driven efforts at institution building are often inappropriate, given national circumstances and capacity. Donor priorities and interests also change. In this case, changes are due to feelings and political pressures in the developed countries rather than the needs of the developing countries that are dependent upon their assistance. The vulnerability of NARS to externally funded priorities makes it imperative that they develop a capability to determine their own priorities to negotiate the allocation of investments with donors.

Decentralization, institutional diversity, and change
Applying systems approaches to NRM helps scientists and planners to define the geographic scale and temporal dimension of the problems to be solved. Depending on this scale (e.g., local, national, regional, or global) and the time frame (e.g., short, medium, and long term, or intergenerational), the users of the research product as well as the institutions that generate new information or technology will vary. Among the users of the outputs from NRM research are policy makers, development agencies, other research organizations, producers, or local communities. The benefits of NRM and systems research for governments are likely to include increased economic activity and better information on which to formulate policies on the sustainable use of natural resources and to evaluate the implications of new technology for national development.

However, it is often farmers and pastoralists who are custodians and users of a large share of a country's land resources. Ultimately, technological innovation and changes in NRM practices must pass through their hands. Even commodity research must take the local ecology and social environment into account. In Bangladesh, systematizing this knowledge in a national database is helping to improve the relevance and targeting of research results to the actual conditions that exist in the fields and villages. Adding socioeconomic data to environmental classification should enhance the usefulness of these databases.

In particular, little information exists in either the databases or systems models that take into account the socioeconomic and cultural factors affecting decisions about resource use at the farm and village level. These non-technical factors that guide decision making by resource users may become increasingly important in introducing new resource management technologies and practices. The longer time scales implicit in many of these NRM research outputs make it difficult for producers to discern visible benefits. Whatever benefits there are may be shared among many producers. Thus a better understanding of decision making and how benefits are distributed within local communities may be crucial to effecting change.

In the discussion below, the importance and difficulty of integrating socioeconomic and biophysical information in environmental databases is considered.

Natural resource databases

The papers in this section make it clear that a systems perspective requires reliable data. National and international agricultural research organizations are interested in terrain and soil, rainfall, and temperature, and their variations in space and time. Most of them are also interested in the seasonal course of water balance. A geographically referenced inventory of natural resources that includes this biophysical data, but also economic data from which to assess land capability and land- and water-use patterns, is a key requirement for development planning and for the integration of agricultural and NRM research (see Ritchie and Dent; and Oldeman). It can indicate where within a country, production to meet future demand can most productively and most safely be placed.

A distinction has to be made between data on the components of the environment and the ways in which this data is to be used. NARS and the international agricultural research centers require the same kind of primary environmental data, though they use it in different ways and with different models. NARS want rational methods to quantify environmental resources and to plan how best to use them for research and for development. Methods of geographical extrapolation using GIS may be particularly valuable to them. Such methods can help complement local research with information from research done at similar sites elsewhere; an advantage of special interest to small countries.

The whole question of the requirements for data on natural resources, including issues of data collection, data quality, and the maintenance and access to geographically referenced natural resource databases, was the subject of study and discussion at an earlier inter-center workshop, organised by ICARDA under the auspices of the CGIAR and FAO, in Rome in 1986.

Since then, a further workshop was hosted by UNEP in Nairobi in 1991, and one of the main recommendations it made was that the IARCs, UNEP, other institutions, and concerned donors work together to ensure that environmental-data-management technology is being generated and used in a coordinated manner throughout the CGIAR system. In late 1991, UNEP's GRID program commissioned a consultancy to review the information needs of the centers, with particular reference to the use of GIS technology. A lack of data was identified as the main constraint to the effective application of the technology, and presumably therefore of other systems methods. Recommendations were made for this issue to be addressed through cooperative action between the CGIAR, GRID, and other agencies, and for more exchange of expertise and experience among centers and cooperating organizations.

However, it was evident from many of the statements made at this workshop that the availability of this kind of data is still limited in developing as well as some developed countries. While the proceedings of the workshop were being prepared, ISNAR learned that UNEP/GRID has proposed a project to further strengthen the links between the CGIAR and UNEP, and to harmonize the research activities, technology, and environmental information management throughout the CGIAR.

The Rome workshop in 1986 recognized the role of national agencies as both the source and the ultimate users of much of the environmental and socioeconomic data required (Bunting 1987). The UNEP proposal focuses more on the requirements at an international level. While the IARCs will be able to give more support to the NARS once they have their own house in order, it will be important for the IARCs to remember that the NARS also need to be seen to support the national meteorological services, soil survey, and hydrological institutes that provide the primary data. These services are currently threatened in many countries as budgets for public services are squeezed. In the case of meteorological services, the demand for their services declines as their traditional clients turn elsewhere for their information. Aviation, for example, now obtains its weather information from international and global telecommunication networks. Agriculture has become the main client for the national services, and it is in the common interest of national and international agricultural research services, and of international and bilateral donors to arrest the decline in the quality of data and services that is occurring because of a lack of funds. Some of the ways in which decline could be halted were discussed during this workshop and at the earlier international workshop on agricultural environments referred to above.

Establishing and maintaining an environmental database requires specialized skills and it represents a difficult challenge to many NARS. Even so, GIS and remote sensing are technologies that NARS will need to use more to fulfill their NRM responsibilities.

It would clearly be useful if all the agricultural organizations involved used the same internationally accepted units and conventions for assembling and managing the primary environmental data. WMO provides a manual on codes for national services, and the package CLICOM (climate computing), which was designed for use by NARS, is readily available. It includes facilities for data entry, quality control, and statistical operations. As reported by Oldeman, ISRIC maintains up-to-date worldwide data on soils and terrain.

UNEP has set up two databases, GRID, based in Geneva, and the Global Environmental Monitoring System (GEMS), in Nairobi. They are intended to provide services to a network of national GIS, rather than to serve as central global databases.

Socioeconomic data for environmental databases
Adding socioeconomic data to environmental databases would enhance the usefulness of the environmental classification. It would involve nation-wide surveys to characterize land-use patterns, farming systems, social characteristics, and the economic importance of the main agricultural activities. The aim would be to establish the relation between ecological and socioeconomic characteristics, to explain the effects of environment on farmer's choice of enterprises and technical methods, and to clarify their attitudes to innovation and risk (see Wood and Pardey, Section A).

Collecting socioeconomic data of this kind is a slow job. From past experience, surveys often start off by seeking answers to too many questions. There is therefore a need to define a minimum data set for description and diagnosis, but so far it has proved difficult to find one standard data set, equivalent to the basic set needed to describe the physical characteristics of the environment, that would satisfy all needs.

Although at this workshop and two before it (Bunting 1987; CGIAR Aug. 1992) it was agreed that primary socioeconomic data are valuable, there was no agreement on what constituted a minimum dataset. There were also reservations about the inclusion of socioeconomic information in databases together with data on other environmental attributes, on the grounds that the data is too heterogeneous and ephemeral to be useful.

Recommendations

Supporting interinstitutional and interdisciplinary research
Institutional reorganization can be disruptive, and it often encounters resistance, which takes time to overcome. The workshop therefore recommends that while such change is taking place, NARS be encouraged and assisted by IARCs and donors to introduce funding mechanisms that would encourage more institutional collaboration and interdisciplinary research. By allocating funds to programs and projects, rather than to organizations or departments, organizations can be persuaded to focus more on joint efforts to solve complex resource management problems.

Strengthening linkages between research and policy
The workshop recommends that more attention be given by NARS and IARCs to making policymakers aware of the opportunities for improving the overall productivity of land, water, and biological resources that can result from applying systems methods and NRM approaches. NARS and IARCs have an obligation to participate more directly in the policy dialogue. High priority should be given by NARS to strengthening linkages between natural resource users, scientists, planners, and policymakers.

Supporting long-term goals

The workshop recognizes that the vulnerability of NARS to externally funded priorities makes it imperative that they develop a capability to determine their own priorities to negotiate the allocation of investments with donors.

Natural resource and socioeconomic databases

The workshop endorses the continuing efforts by the CGIAR and UNEP to harmonize the environmental-information-management activities throughout the CGIAR, and to integrate natural resource and socioeconomic information required for sustainable agricultural development at the international, regional, and national level. As noted above, little progress was made on the last item at two previous workshops where it was discussed. But the issue arises with increasing frequency, as it did at this workshop, as attention to NRM and work at a systems level expands and the technology for information management improves. For this reason, it should be high on the list of current priorities in the UNEP/GRID and CGIAR project.

It urges that in the course of these efforts, early attention is given to the role and the needs of the national services that are the traditional sources of much of this information.

SECTION D

Experience at a national level of use of systems analysis methods

Multiple-goal planning as a tool to analyze sustainable agricultural production options in Mali

E.J. BAKKER

DLO Research Institute for Agrobiology and Soil Fertility, P O Box 14, 6700 AA Wageningen, The Netherlands

Key words: linear programming, regional planning, Sahel, sustainable agriculture

Abstract

In Mali, as in other countries in the Sahel, degradation of natural resources threatens future production possibilities in animal and crop husbandry There is a need for research that focusses on how sustainable agricultural production could be brought about This paper describes a multiple-goal linear programming model for analyzing options for sustainable land use The model is tailored to the possibilities for fertilizer use, as this seems to be a necessary ingredient of agricultural production that is both sustainable and able to satisfy the increasing demand for food The paper also discusses how the approach can be used to explore feasible options for sustainable production, for agricultural research planning, and to indicate the most appropriate price-policy instruments The discussion further focusses on the trade off between the various goals as well as the influence of economic parameters on the possibilities for ecologically and economically sustainable agricultural development Finally, the paper discusses the merits and limitations of the approach

Introduction

Mali is a vast west-African country in which agriculture (both crop and animal husbandry) is the most important sector in the economy. Traditionally, arable farming is practiced mainly in the south. Because of climatic conditions, the most productive animal husbandry systems are found in the north. Low soil fertility and highly variable climatic factors are the main causes of low agricultural productivity in Mali. The resulting high level of food insecurity is well known (Sijms 1992). High population growth aggravates this problem and has introduced another one: a large part of the nation's land resources run the risk of becoming unproductive soon due to nutrient depletion of the soil. In crop production, fallow periods have been shortened, while insufficient manure or chemical fertilizers are being used to restore soil fertility. Natural rangelands suffer from overgrazing, causing increased degradation through erosion, loss of perennial grass species, and a decreasing tree population.

An important element of the degradation problem (Breman and Traoré 1987) is the conflict between pastoralists and arable farmers. Arable farmers have continuously expanded the cultivated area, thus reducing the space available for rangelands. At the same time, the number of ruminants has increased, in part because farmers have purchased significant numbers of animals. Chaos in land tenure and a lack of responsibility for maintaining the land resources resulted. The lack of clear and enforceable regulations concerning land property and land use, together with scarcity and poverty, have led people to use the land beyond its long-term production capacity.

189

P Goldsworthy and F W T Penning de Vries (eds), Opportunities, use, and transfer of systems research methods in agriculture to developing countries, 189 - 198
© 1994 *Kluwer Academic Publishers*

Thus, even if one can speak of growth in agricultural output in Mali, it is not sustainable growth. A crucial question is, therefore, how sufficient food can be produced without (further) compromising tomorrow's agricultural production possibilities.

As far as the technical side of the question is concerned, answers could be found. Van Keulen and Breman (1992) show that higher agricultural productivity could be achieved by increasing the use of fertilizers, notably nitrogen and phosphorus. The use of these inputs would also increase the sustainability of the production systems, provided that other equilibria in the soil (e.g., organic matter) are not disturbed. The limited availability of manure and the high prices of chemical fertilizers, however, have so far made this solution impractible. In the Production Soudano-Sahélienne project, a Dutch-Malian research team, of which the author is a member, is investigating the technical options for sustainable agricultural growth and their economic feasibility for the Malian Sudano-Sahelian zone. This zone has an average annual rainfall of between 300 and 900 mm. In our research, we follow an approach that enables us to analyze technical options and economic conditions in an integrated manner. A central part of the approach is use of an analytical tool known as Interactive Multiple-Goal Programming (IMGP). The following section explains what IMGP is and describes the way in which we plan to implement it. The third section contains a discussion on the possibilities, the limitations, and the possible extensions to the approach.

Interactive multiple-goal (linear) programming

IMGP is a method to analyze a linear optimization problem with not just one but a number of objectives. The method has been described, used, or illustrated by Nijkamp and Spronk (1980), Spronk and Veeneklaas (1983), Van Keulen (1990), Veeneklaas et al. (1991), Van Keulen and Veeneklaas (1993), De Wit et al. (1988), Spharim, Spharim and De Wit (1992), and Romero and Rehman (1989). They all use IMGP for multiple-criteria decision making in a linear programming context, most of them in regional agricultural planning.

Our first version of the model is in the final stages of development. Each of the principal components of this model will be discussed briefly.

Resources
The region under study (the Sudano-Sahelian zone of Mali) is divided into four climatic zones. Using administrative border lines as a second criterion, 15 subregions are distinguished. Within each subregion, the land resources are subdivided according to soil type. Land availability is given per subregion per soil type. Availability of human resources (labor) is also calculated per subregion.

Activities

An activity is used here as a technical term denoting a quantitatively defined combination of inputs and outputs, such that the proper use of the inputs would lead to the production of the outputs. Five types of activities are distinguished: cropping activities (food crops, cash crops, and fodder crops), rangeland activities, tree plantation activities, fish production activities, and other animal production activities (cattle, sheep, and goats).

All activities are defined in such a way that they can be supposed to be sustainable, which means that they could be undertaken year after year without deteriorating the physical or chemical soil properties. In the model, sustainability is translated for cropping and tree activities in such a manner that the amounts of nitrogen, phosphorus, potassium, and organic matter in the soil are in equilibrium, i.e., the input in terms of fertilizers or in number of years of fallow is such that the amount of nutrient that is harvested or otherwise lost is returned to the soil. Moreover, it is assumed that rainfall runoff is controlled over time by sufficient investment in anti-erosion measures, if required. For animal-production activities, sustainability is enforced by limiting the rangeland production available to the animals in such a way that the rangeland is not degraded by the grazing of the animals. In practice, this means that the number of grazing animals has to be restricted.

For each production goal (e.g., beef, millet, cotton), a number of activities are defined. These activities should reflect the different levels of intensity at which the product can be obtained. For crops this means different levels of crop yield per unit area, with matching levels of input (especially fertilizers). For animal, levels of production vary with the quality and quantity of the feed intake. For a given product, the model does not include the entire production function but should include a sufficiently large number of points to be representative of the function.

For the quantitative description of the activities, we use models that reflect the available knowledge in different disciplines. For the crops, the average potential, water-limited yields are calculated for each crop-climate combination using crop-growth simulation models and actual rainfall data for 30 years. Next, some yield levels are chosen, e.g., 90 percent, 60 percent, and 30 percent of the potential yield. Finally, the necessary input levels of nitrogen, phosphorus, potassium, and organic matter needed to guarantee sustainability are estimated using nutrient-balance models. For animal production, models link feed quality (expressed as protein content and digestibility) with quantity of feed intake. These two are then linked with meat and milk production. For other types of activities, the definition of the inputs and outputs is based on data from literature and from local research institutions.

Final and intermediary products

Final products, such as cereal grain, meat, and milk, are the main results of the activities undertaken. They will generally be used for specifying the goals. Some of the inputs are the outputs of other activities. Such inputs are termed intermediary products (figure 1). Their use and production are calculated endogenously in the model. The most important intermediary product is animal feed, which is an input to animal production activities and an output from range-land and fodder-production activities and from byproducts of cropping activities. The most important resources and products are listed in table 1, which also indicates the types of activity in which each resource and each product (final or intermediary) is used as an input, as well as the types of activity from which each product is derived as an output.

Restrictions

Three types of restrictions can be distinguished in the model. They apply to the balance of resources, the inputs, and the outputs. Table 2 shows the principal restrictions used in the model. The first type makes the model internally consistent. The second type of restriction reflects constraints to be applied to factors that are not incorporated in the model, such as migration. The input-output matrix does not contain costs of migration, and therefore, according to the model, migration is free. For example, if one wants to allow a migration of up to 5 percent of the labor force between subregions, and if one wants to exclude the possibility of migration from or into the region, a restriction can be applied that states that total labor use in each subregion should not surpass available labor in the region by more than 5 percent. However, an extra restriction should be added to ensure that total labor use in the region cannot be more than the total available labor in the region. The third type of restriction, which is discussed in the following paragraphs, reflects the aspiration level of a goal variable.

Goals

As indicated above, several goals can be incorporated in the model. They all need to be defined as linear functions of the decision variables. Goals that are to be included in our model are increases in the amount or efficiency of food production, of milk and meat production, and increases in regional income, and employment. They can be included at the regional level, but also for a few or all of the subregions. The latter has the drawback that the number of goals becomes very large. Goals can be included either as aspiration levels (e.g., self-sufficiency in cereals) or as objective functions to be maximized (e.g., regional income) or minimized (cereal shortage). A combination is also possible. One may, for example, prefer a situation of full employment (maximization), while accepting unemployment only up to a certain level (aspiration level).

Prices

An important component of some of the goal variables are the prices of the inputs and the final products. These also have to be fed into the model. They may have a substantial impact on the benefits of the various activities and hence on the economic attractiveness of intensified fertilizer use.

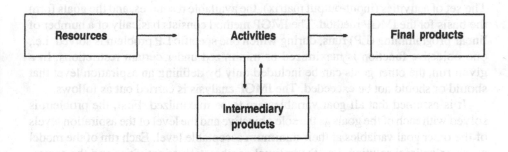

Figure 1 The production process in the LP model

Table 1 Resources, final, and intermediary products produced by five types of activity (I=input, O=output)

Resource/ product	Subdivided by	Types of activity				
		Crop production	Animal production	Fishing	Tree production	Rangeland use
Land	Soil type	I			I	I
Water surface				I		
Labor	Period	I	I	I	I	
Animal feed	Quality	O	I		O	O
Draft animals		I	O			
Manure		O	O		I	
Fertilizer		I			I	
Milk/meat/fish			O	O		
Food		O				
Money		I/O	I/O	I/O	I/O	

Table 2 General formulation of the restrictions on the use of resources and of the balance restrictions for intermediary products

Land, water surface	I available
Labor	I available + Immigration - Emigration
Feed	I O + Import - Export
Manure	I O
Draft animals	I O

The IMGP procedure

The set of activities (input-output matrix), the available resources, and the goals form the basis for the IMGP method. The IMGP method consists basically of a number of linear programming (LP) runs, during which one specific LP problem is solved, i.e., one objective function is maximized or minimized under certain restrictions. In a given run, the other goals can be included only by defining an aspiration level that should or should not be exceeded. The IMGP analysis is carried out as follows.

It is assumed that all goal variables are to be maximized. First, the problem is solved with each of the goals as the sole objective and the level of the aspiration levels of the other goal variables at their minimum acceptable level. Each run of the model gives an optimal solution, i.e., the optimal combination of activities and the corresponding values of the goal variables. The results define the maximum level of each of the regional and subregional goals, given the technical constraints and the minimum goal levels. The results also quantify the windows of opportunities in that region.

Next, in a step-like manner, the aspiration levels of the various goals are increased or decreased. Each time an aspiration level is increased, the corresponding goal restriction becomes more inhibiting and leaves less room for the fulfillment of the others. In the process, one obtains insight into the costs of one goal relative to the others. If the number of goal variables is high, it may be impractical to calculate the solutions for the whole range of possible combinations of aspiration levels. In that case, the interactive approach is preferred: each time a run has been carried out, one of the goals is given a higher aspiration level. If the resulting solution is preferred over the previous solution, i.e., if the gain in one goal was not too expensive in terms of the others, the next step is taken. If the solution is not preferred, one returns to the previous set of aspiration levels and tries to increase the aspiration level of another goal. This interactive phase should be carried out preferably with one or more of the parties interested in the results. Instead of using the full interactive approach, one may use a number of policy views that dictate the relative importance of the goals. An exploration of the solution space for each of the views can be made and presented before the actual interactive phase starts.

Figure 2 illustrates the trade-off principle between goals in the IMGP method. Figure 2a depicts a situation in which a regional income of at least I_1 is required. The maximum attainable level of cereal production under this restriction is calculated in one LP run and is denoted here by C_1^*. In this way, point $I_1 C_1^*$ is obtained. Similarly, point $C_1 I_1^*$ can be obtained, with C_1 as the minimum acceptable level of cereal production and I_1^* the corresponding maximum attainable regional income. If in a following run the minimum acceptable cereal production is increased to a level C_2, the maximum attainable regional income decreases to, say, I_2^*, as less land will be available for cash-crop production. The resulting situation is shown in figure 2b. Figure 2c, finally, shows how the solution space is further reduced if the regional income that is minimally required increases to I_2; the maximum attainable cereal production decreases to C_2^*. It can thus be seen that the optimal solution depends on

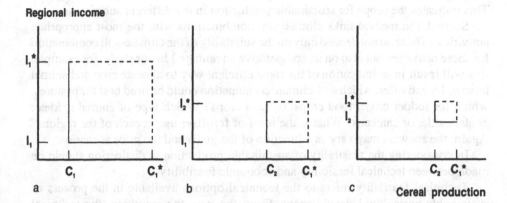

Figure 2. Illustration of two IMGP steps. For i=1,2 C_1, I_1 represent aspiration levels of cereal consumption and regional income and C_1^*, I_1^* the maximum attainable levels given the aspiration level of I_1, C_1 respectively.

the restrictions as much as, or even more than, on the specification of the objective function.

Possibilities and limitations of the method

Before embarking on a discussion of the use of the model and the analyses described above, it is worthwhile to consider the character of the model, as this will help explain the possibilities of and the limitations to the use of the model.

The model is a mathematical description of an allocation problem. This allocation problem, however, does not exist in reality. Although it is tempting to call the model a regional land-use planning model, in fact, it is not. The model is meant to describe the potential of sustainable agriculture for the region under study, as a function of a regional goal and a set of prices. If one wants to call it an allocation problem, it is that of a super farm, where a decision maker has the freedom or the power to allocate the existing regional labor force, the land, and other resources to the sustainable land use and animal-husbandry activities defined in the model. In reality, decisions are not made centrally but at a much lower level, in a situation where part of the land is already degraded and where sustainable activities are the exception, not the rule.

What then can be expected of the model and the IMGP method? The general answer is that it helps the user *explore the feasibility* of sustainable agricultural production.

First, it answers the question of whether sustainable production is economically feasible for the various price scenarios and the aspiration levels of the goal variables.

The economic balance is made for the region as a whole as well as for the subregions. This indicates the scope for sustainable production in the different subregions.

Second, the method links climate-soil combinations with the most appropriate activities. (These do not depend only on the suitability of the climate-soil combination for these activities, but also on its comparative advantage.) In our study, for example, this will result in an indication of the most efficient way to allocate crop and animal husbandry activities: which soil-climate combination could be used best for pastures, which for fodder crops, food crops, or cash crops? Which type of animal is ideal: goats, cattle, or chickens? What is the level of fertilizer use in each of the regions? Again, the answers may vary as a function of the goals and the price scenarios.

In considering the feasibility of sustainable production, a distinction should be made between technical feasibility and economic feasibility.

Technical feasibility refers to the technical options available in the process of sustainable agricultural development. From the way the activities (the technical options) have been defined, it is clear that hardly any of them have been tested for most of the climate-soil combinations in the region. In fact, testing is difficult and expensive, the more so because many of the processes that determine the level of sustainability of an activity are long-term processes. In the input-output model, all available knowledge is put together in a systematic way. The IMGP approach may now serve technical agroecological and zootechnical research. It is highly probable that during the IMGP analyses, certain combinations of activities are consistently chosen by the model where others are not. The chosen activities, then, would be preferred candidates for further testing on station and, eventually, on farm. This could considerably reduce the amount of money and effort spent for agricultural research.

Economic feasibility deals with the economic attractiveness of the combinations of technical options. The economic feasibility of a sustainable activity depends on price developments and can be regarded as an indicator of the potential for adoption of the activity by farmers. Running the model for various input and output prices makes it possible to analyze the effects on the goals and optimum set of activities. This could indicate the effectiveness of various price policy options. Policymakers can thus be helped to identify instruments aimed at stimulating sustainable production. One should be careful, however, not to attach too much value to predictions of the effects of price policy based on a regional model on the choices of farmers. Farm-level analysis is needed as a means of verification.

The existing knowledge of crop cultivation, processes in the soil, animal production systems, etc. is used to define the technical production possibilities. The model is thus a good basis for multi- and interdisciplinary research, in which specialists of various "technical" disciplines fill in their part of the input-output matrix. One drawback is that the user of the model may not be aware of the uncertainties involved in some of the technical coefficients. This can and should be avoided by involving the technical specialists in discussions about the results. The user of the model (the economist) should also be involved in the work of the specialists, if only to prevent them from going into too much detail and hence spending too much time on producing the input-output coefficients. In short, the process has to be interdisciplinary, not just multidisciplinary.

One limitation of the approach is that the analysis is static. It is assumed that the combination of activities that form the optimum solution can be carried out year after year. It thus describes a future sustainable situation, but it does not indicate the road that leads to that situation, eventhough the analysis provides technical and economic policy suggestions. It would be interesting to extend the approach to an analysis over longer periods of time. This would necessitate the inclusion of non- or less sustainable activities as well, with their influence on land resources and their consequences for future production possibilities. Such a framework could be used to analyze the trade off between actual and future production.

Another limitation is the absence of any risk assessment. The variation in input-output relations as a consequence of climatic variation can be quite large, but the activities in the model are designed to describe the averages. The risk of insufficient food production in a very dry year, for example, is not considered. We plan to include this aspect in the way Veeneklaas et al. (1991) have done. The input-output relations are defined for a typical "disaster year," for example one with only 60% of normal rainfall. It is then possible to set limits to certain goals for a disaster year, but at the cost of a large number of extra variables and coefficients in the model. In an exploratory analysis it may therefore not be worthwhile to include the risk aspect. The more the model is to be used for actual planning, the more it is necessary to include risk.

Acronyms

IMGP   interactive multiple-goal programming
LP   linear programming

References

Breman H, Traoré N (Eds.) (1987) Analyze des conditions de l'élevage et propositions de politiques et de programmes. Mali. Sahel d(86)302. Club du Sahel/CILSS/OECD, Paris, France. 243 pp.

Van Keulen H (1990) A multiple goal programming base for analysis of agricultural research and development. Pages 265-276 in Rabbinge R, Goudriaan J, Van Keulen H, Van Laar H H, Penning de Vries F W T (Eds.) Theoretical production ecology: Reflections and prospects. Simulation Monographs, PUDOC, Wageningen, The Netherlands.

Van Keulen H, Breman H (1990) Agricultural development in the West-African Sahelian Region: A cure against land hunger? Agriculture, ecosystems and environment 32:177-197.

Van Keulen H, Veeneklaas F R (1993) Options for agricultural development: A case study for Mali's fifth region. Pages 367-380 in Penning de Vries F W T, Teng P, Metselaar K (Eds.) Systems approaches for agricultural development. Kluwer Academic Publishers, Dordrecht, The Netherlands.

Nijkamp P, Spronk J (1980) Interactive multiple goal programming: An evaluation and some results. Pages 278-293 in Fandel G, Gal T (Eds.) Multiple criteria decision-making theory and application. Springer Verlag, Berlin, Germany.

Romero C, Rehman T (1989) Multiple criteria analysis for agricultural decision-making. Elsevier, Amsterdam, The Netherlands.

198

Sijms J (1992) Food security and policy interventions in Mali. Tinbergen Institute, Erasmus University of Rotterdam, The Netherlands.

Spharim I, Spharim R, De Wit C T (1992) Modelling agricultural development strategy. Pages 159-192 in Alberda Th et al. (Eds) Food from dry lands. Kluwer Academic Publishers, Dordrecht, The Netherlands.

Spronk J, Veeneklaas F R (1983) A feasibility study of economic and environmental scenarios by means of interactive multiple goal programming. Regional Science and Urban Economics 13:141-160.

Veeneklaas F R, Cissé S, Gosseye P A, Van Duivenbooden N, Van Keulen H (1991) Competing for limited resources: The case of the fifth region of Mali. Report no. 4. Development scenarios. CABO/ESPR, CABO, Wageningen, The Netherlands.

De Wit C T, Van Keulen H, Seligman N G, Spharim I (1988) Application of interactive multiple goal programming techniques for analysis and planning of regional agricultural development. Agricultural Systems 26:211-230.

Needs and priorities for the management of natural resources: a large country's perspective

CHENG XU

Beijing Agricultural University, Beijing 100094, People's Republic of China

Key words crop simulation models, decision support systems, environment, GIS, natural resource management, planning, sustainable agriculture, systems methods

Abstract
Modern quantitative tools have been used in China since they were introduced into the country in the late 1970s They have served mainly as a means to change the decision-making process, which until then depended on the views and experience of a few individuals or on unrepresentative information from an experimental site A natural resource inventory and regional planning at every level, based on survey data and systems science with its companion methodology-system engineering, has been widely used in formulating the general programming of social, economic, and environmental development With the focus of this work at the county level, more than 500 counties have set up their programming system, using various mathematical methods such as LP, SD, and dynamic simulation, as well as GIS technology At the same time, a decision support system that includes macroeconomic policy analysis as well as microlevel applications such as a "crop cultivation model" and growth simulation models, is either being developed or is already in use From the viewpoint of resource management, all this work has played an important role in reallocating resources to gain the optimal economic, social, and environmental benefits

Introduction

Integrated resource management (IRM) has become a vital tool in facilitating sustainable agriculture In China, since the formulation of the "Four Fields' Modernization" strategic goal, including agricultural modernization, and the adoption of the "Open and Reform" general guiding principle, two basic tasks related to IRM have been completed. The first is an agricultural resource survey (inventory) and the preparation of regional plans, and the second is a procedure for regional development programming at the level of province, prefecture, or county.

These measures are intended to serve leaders at every level, so that they can make informed decisions about the use and protection of resources. An outstanding feature of the implementation of these two huge projects has been the widespread application of modern science theory and methodology, including methods that used to be regarded as serving "capitalism," such as econometrics and development economics. Among the many examples of science theory and methods, the application of quantitative analysis and systematic analysis may be among the most notable. The prime example is the insight provided by the application of agricultural system engineering.

When commenting on the differences between the ways of thinking in the Orient and the Occident, the founder of synergetics theory, Dr. Haken, greatly appreciated the indigenous system thinking contained in ancient Chinese agriculture. It is a pity

P Goldsworthy and F W T Penning de Vries (eds), Opportunities, use, and transfer of systems research methods in agriculture to developing countries, 199 - 211
© 1994 *Kluwer Academic Publishers*

that such traditions have been seriously neglected. For example, in agricultural circles, almost everyone's attention has been focused on grain production. To increase grain production, people made every effort to reclaim various kinds of marginal land, such as hilly and mountainous land, grassland, forestland, and even the bottoms of drained lakes. People spared neither labor nor money to adopt an exhaustive system of cropping, mainly by hand work. In addition, they have usually unquestioningly overused chemical fertilizers, pesticides, and irrigation. The adverse effects on other industries, the stress upon resources and the environment, as well as the stagnation of farmers' incomes at a low level were all ignored. Another example of the developments that took place was the campaign to follow the example of the village of Dazhai, in the 1970s. Although there are major differences in natural conditions across the country and although the model of Dazhai was designed to tackle an unfavorable drought climate, the single agricultural model of Dazhai had to be adopted everywhere. The model included the building of terraced fields, adding organic fertilizer to enrich the soil, and deep ploughing, all with the aim of growing higher-yielding crops such as corn and sorghum. So it is no surprise that sustainable development of agriculture was seriously obstructed.

In 1979, a large-scale national project was started to survey natural and agricultural resources and to provide a basis for regional plans. At the peak of the project, the number of people involved had increased to 400,000. Following the initial success of the project, a new initiative was started to find out how the results could be used to benefit the formulation of agricultural development programs at every level. At the same time, top Chinese leaders began to realize that most economic failures and setbacks during the past 40 years could be attributed to faulty decisions, as for example, the "Great Leap Forward" at the end of the 1950s and "Centering on the Key Link of Grain" in agriculture. The leaders realized that the decisions behind these movements were made without any quantitative analysis or consultation with specialists. They were merely the result of an official's political will or a leader's experience, or based on unrepresentative information from an experiment site. The leaders introduced a guideline called democratic and scientific decision making. Under these circumstances, system science and system engineering have been introduced and applied on a large scale.

The following paragraphs will describe how in the past decade, Chinese agricultural scientists and technicians have introduced modern quantitative tools, including system engineering and geographic information systems (GIS), and how they applied these tools in research and analysis from the macrolevel down to the microlevel, and from policy research to regional programming to crop production simulation. This paper will analyze development trends and the perspective of popularizing systems methods, and finally offer some suggestions.

Challenges

The first challenge comes from the acute problems that exist in the current agricultural system. Agriculture is the foundation of China's national economy, with the agricultural system as a subsystem of the national economic system. It undertakes the tasks of providing food for the nation and raw materials for light industry. China's agriculture has achieved a great deal: it has fed a population that accounts for 22 percent of the world from an area of arable land that is only seven percent of the world's total. However, it should be pointed out that further development of agriculture has been restrained by the long-term and fundamental conflict caused by a serious shortage of land and water and a rapidly increasing population with a low level of education. Thus the first and foremost issue is how to make rational use of the limited resources. This means that scientists must actively take part in the search for optimal solutions for agriculture. The major components of their responsibility will be:

■ to analyze the potential for increasing yields and the opportunities to realize this potential in the near, medium, and long run;

■ to provide the scientific information required to guide the state and local government in their formulation of long- and medium-term plans as well as annual plans;

■ to conduct research on approaches to the rational utilization of every key production element and to find the optimal combination of key elements;

■ to analyze systematically each unit of the agrotechnological system as well as its general structure, so that the most reasonable and complete set of agrotechnologies can be built up.

The second challenge is related to concerns about the environmental consequences of agricultural activities. Agricultural development must not only meet the demands of the present generation, but it must also guarantee the necessary conditions for the welfare of future generations. Thus the relationship between sustainable development of agriculture, population growth, and the conservation of resources and the environment should be considered seriously.

China's population has increased rapidly from 540 million in 1949 to 1,150 million today, while the area of arable land has dropped from 110 million ha to 96 million ha. The erosion-stricken areas have increased from 1.16 million km^2 to 1.63 million km^2. The area ruined annually by disaster has increased from 9.26 million ha to 19.43 million ha. Forest cover has dropped from 13 percent to 11.5 percent. All these figures show that if effective and timely measures are not taken, by the turn of century, China will be faced with very urgent environmental problems. This would threaten the sustainable development of agriculture, as well as the foundation of China's existence and development. It is therefore urgent to apply the newly adopted theory and methods of system engineering to do the following:

■ address the issues of population growth, environmental damage, and energy supply with the aim of conserving the environment and ecological system, so that China can solve its own problems and contribute to the global environmental issue;

■ provide ecosystem analysis and ecosystem engineering designs for state and ecoregions;

■ design projects for the optimum management and use of resources in mountainous areas;

■ systematically examine China's rivers and lakes and propose projects for the
optimum use of these resources;
■ conduct a systematic analysis of natural catastrophes and countermeasures.

Efforts and achievements

An inventory of natural resources and a national cross-sectoral plan was prepared
during the 1980s. In 1979, the first session of the Chinese National Commission of
Agricultural Regional Planning was established, headed by Vice Premier Wan Li.
The commission consisted of 26 members, who were high-ranking officials of
ministries, state commissions, and state academies of sciences. A consultant group
of 30 senior scientists also served on the commission. The mandate of the commission
was:
■ to promote and lead natural resource surveys and regional planning;
■ to organize and coordinate related research;
■ to provide scientifically sound suggestions for the rational use and protection of
management of natural resources as well as give scientific guidance to relevant
ministries or provincial governments on the programming and organization of
agricultural production.

Until the end of 1980, regional planning commissions had been established all over
the country at every level, from province to prefecture to county. The total number
of professionals involved amounted to 12,000 people. According to the statistics in
1984, there were 2,108 counties that had conducted natural resource surveys and
regional planning. After 1985, a lot of the results of this work were published,
including "Chinese comprehensive agroregional planning," "China's land bearing
capacity research", "China's soil erosion area calculation and regional planning,"
"Investigation of natural predators of insect pests of crops", "Primary evaluation of
China's water resources," and a series of regional plans for crops, cropping systems,
animal husbandry, feed, forestry, fisheries, agromechanization, township-run enterprises, rural construction, and the use of chemical fertilizer and green manure.

In the past decade, an outstanding achievement in the application and extension
of system science has been the formulation of long-, medium-, and short-run general
programs mainly at the county level. The main quantitative tools have been linear
programming (LP) and systems diagnosis (SD). The principal outputs have been the
readjustment and optimization of various structures. So far, more than 500 counties,
or about one-sixth of the total counties in China, have implemented agricultural
programs, and the majority of them have been successful in getting approval from
the county congress, thus effectively avoiding the problem of the past when new
administrative leaders would often casually abandon or shift existing programs. A
few counties, such as Taoyuan, Jinyu, and Changqing, have reported the benefits they
have derived from a program that is respected by everyone in the county. At higher
levels of decision making, system engineering has played a decisive role in the
drafting of general programs at the prefecture, province, and even national level. At
present, Shanxi, Shandong, Jiling, and Xinjiang Provinces (or Autonomous Regions)

have completed their agricultural programming. At lower, local levels of decision making, the experiences gained during the process of county agricultural programming have served to guide program planning at the township and village level. Some of the publications on agricultural and natural resource management (NRM) planning are listed at the end of this paper.

The typical procedure for planning the county agricultural program includes five steps: environment identification, system diagnosis, design of general goals, determination of development strategy, and general design and optimization of resource allocation. Figure 1 illustrates the method of system diagnosis. The diagnostic procedure was derived from a series of activities, including brainstorming by a group of specialists, charting the relationships between main determinants in the agricultural system, preparing and entering data into computers, analysis, and development of a diagram of hierarchical constraints. From such a diagram the constraints to system function, as well as their direct and indirect causes, were clearly shown. This model did not aim at providing solutions, but at pointing out the way to achieving a solution.

Systems methods can play an important role in resolving conflicts between multiple goals that are part of the overall development goal. For this reason, almost all the general programs, especially those for the county level, take into account the social and ecological factors that affect the agricultural system. Figure 2 shows the procedure for paying attention simultaneously to social and environmental benefits, aside from the principal economic benefit.

Apart from their use in program planning, modern quantitative tools are also being applied to assist in the analysis of national agricultural problems. The following are examples of some of these applications in the 1980s:

- the application of multivariate linear regression, developed by the Research Institute of Systems Science of the Chinese Academy of Science, to predict national annual grain production;
- the application of LP methods, developed by the Chinese Academy of Agricultural Science, to optimize the national structure of grain production;
- a method of assessing the overall potential production capacity of China's agricultural sector, developed by the former State Council's Research Centre for Rural Development;
- the application of GIS in national and regional planning.

In the early 1980s, agricultural scientists in Loudi Prefecture Research Institute, Hunan Province, introduced systems methods in their agricultural research. They successfully designed a technique for optimizing factors in hybrid rice cultivation, using a specialized form of multivariate analysis and other statistical techniques. Subsequently, this method has been extended to many other crops and has proved successful for wheat, soybean, millet, watermelon, and sugar beet. The method has had a significant impact on the northern-China wheat region and the southern-China rice-producing region, raising yields and reducing costs.

Research has also been conducted on the development of crop simulation models and expert systems for crop production. Following the introduction of well-known simulation models such as CERES, COMMAX, and others, Chinese agricultural

204

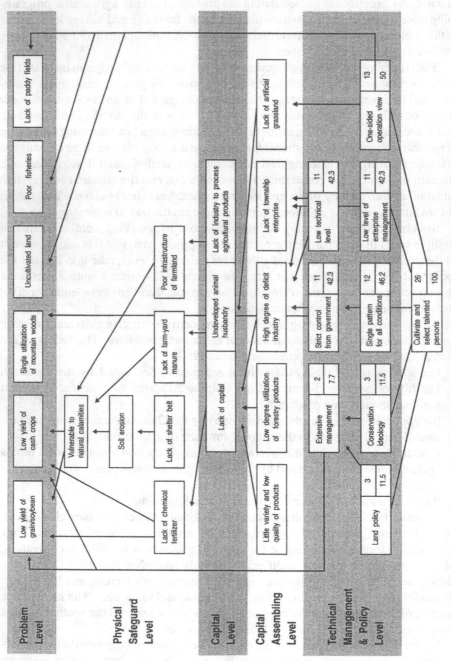

Figure 1 Diagram for structure diagnosis of territorial development and utilization of Hailun Country, Helongjiang Province
Numbers in boxes upper influential region, lower influential weight

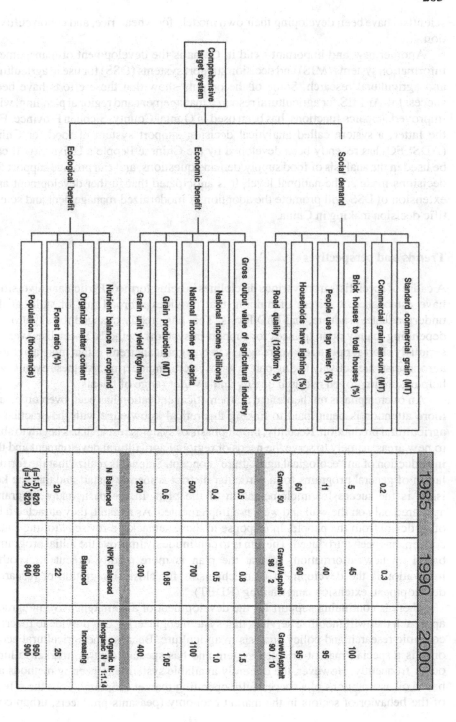

Figure 2 Target tree of Hailun Country general goal *) Numbers of pregnancy

	1985	1990	2000
Social demand			
Standard commercial grain (MT)	0.2	0.3	
Commercial grain amount (MT)	5	45	100
Brick houses to total houses (%)			
People use tap water (%)	11	40	95
Households have lighting (%)	60	80	95
Road quality (1200km %)	Gravel 50	Gravel/Asphalt 98 / 2	Gravel/Asphalt 90 / 10
Economic benefit			
Gross output value of agricultural industry	0.5	0.8	1.5
National income (billions)	0.4	0.5	1.0
National income per capita			
Grain production (MT)	500	700	1100
Grain unit yield (kg/mu)	0.6	0.85	1.05
Ecological benefit			
Nutrient balance in cropland	200	300	400
Organize matter content	N Balanced	NPK Balanced	Organic N: Inorganic N = 1:1.14
Forest ratio (%)	20	22	25
Population (thousands)	(l=1.5) 820 / (l=1.2) 820	860 / 840	950 / 900

scientists have been developing their own models for wheat, rice, and cotton cultivation.

Another new and important trend in China is the development of management information systems (MIS) and decision support systems (DSS) for use in agriculture and agricultural research. Some of the outputs show that these efforts have been successful. An MIS for agricultural resource management and regional planning, with improved graphics functions, has been used in Qunlai County, Sichuan Province. For the latter, a system called analytical decision support system of food for China (ADSSFC) has recently been developed by the Chinese People's University. It can be used in the analysis of food supply/demand questions, and can provide support for decisions made at the national level. It is anticipated that further development and extension of DSS will promote the adoption of modernized management and scientific decision making in China.

Trends and perspectives

A cadre of specialists from various disciplines is being formed. Particular universities have established a system of training that provides a broad mix of skills at the undergraduate, graduate, and PhD levels. This has greatly changed the situation of depending mainly upon trainees for support in systems methods. Specialists were in seriously short supply a few years ago. A very significant trend is that more and more agronomists are actively taking part in agrosystem engineering. All these events will help ensure improved research quality and a wider range of talents.

An overemphasis on theoretical mathematical education has been overcome, and more attention is being paid to linking theoretical knowledge with the practice of agricultural production. Recently, the emphasis of systems research has begun to shift to new areas, namely to serve the needs of regional agricultural development and the introduction of an "ecological agriculture" concept. Scientists realize that the formulation of general programs for a particular district is merely a start and that the key issue is the successful implementation of the plan. In the past, many programs appeared only on the wall and were not implemented. As a result, they attracted a lot of criticism from the people. In response to these setbacks, a revised "rolling-plan" concept has been introduced. The aim is to continuously improve the initial programs, based on new information, so that the plan is more readily executed. Another innovation is the development of mechanisms for closer integration of research, development, extension, and training (RDET).

There is continuing support for the development of a comprehensive integrated approach that will mobilize services that system engineering can provide to macroeconomic research and policy analysis in agriculture. Because the agricultural economy is a special, market-oriented system and has special features, market failures occur frequently. However, the currently available system-engineering methods are based on assumptions of a successfully operating market economy. Thus the effects of the behavior of sectors in the market economy (peasants-producers, urban consumers, grass-roots cadres, and scientists) should be taken fully into account to modify the policy proposals that are derived from system-engineering methods.

The application of system engineering to NRM will continue. At present the concept and the terminology of resource economics and environmental economics is still very strange to most Chinese, including a lot of agronomists. Because most people do not understand concepts such as "internalization of costs," the "tragedy of communal wealth," and "market failure", they think only of a versatile market and do not consider the negative effects caused by overuse of inputs into agriculture. For example, in conducting the study of land's bearing capacity, opinions largely representing those of some senior agronomists who advocated more and more inputs were particularly influential. They used a series of quantitative analysis methods, including a fuzzy system prediction, and a logistics growth model to extrapolate the yield and production of grain in the year 2000. Their conclusions about the land's bearing capacity were optimistic; they estimated that by the turn of this century, China's arable land would be able to support a population of 1.28 billion. However, their estimate of the corresponding increase in the demand for energy was 40.4 percent, based on 1986. This is not realistic because their target for the annual growth of energy inputs for the following 10 years was about 3.5 percent, whereas the actual overall growth rate of energy production was below 2 percent. Thus the extrapolation method they used appears to be inadequate.

Furthermore, they did not take into account concerns about the environment. For example, according to their estimate, the total consumption of chemical fertilizer in 2000 will reach 32.53 million ton, which suggests that the annual application of NPK per hectare will rise to 325 kg! If we consider the very uneven use of chemical fertilizer within a given area, the consumption level in some relatively developed areas may be more than 500 kg per hectare, with the obvious associated risk to the environment. Although such a predictive analysis clearly has some shortcomings, those responsible nevertheless did a lot of excellent systems research by collecting and analyzing the real numbers. Using a modified agroecological zoning method, they got much more valuable indications of land productivity from the measurement of key variables than had been available previously. Figure 3 illustrates the constraints and the resources required for the food-production system. People are gradually beginning to understand that the ultimate objective of reform is the rational reallocation of resources, including natural resources. Against this background, it will be necessary to develop the knowledge and the logic for a system of managing natural resources, as well as the methodology for implementing it.

The movement to introduce a concept of a Chinese ecological agriculture that is believed to be a unique form of sustainable agriculture in China has progressed rapidly. More than 500 counties are recognized as experimental "ecological agriculture counties" by the Ministry of Agriculture. Since the introduction of criteria for evaluating the success with which development programs are implemented and operated, some systems analysis methods, such as multiple-objective programming, have been found to play an influential role in formulating the general program. For multiple-objective programming, the methods permit the user to consider three

208

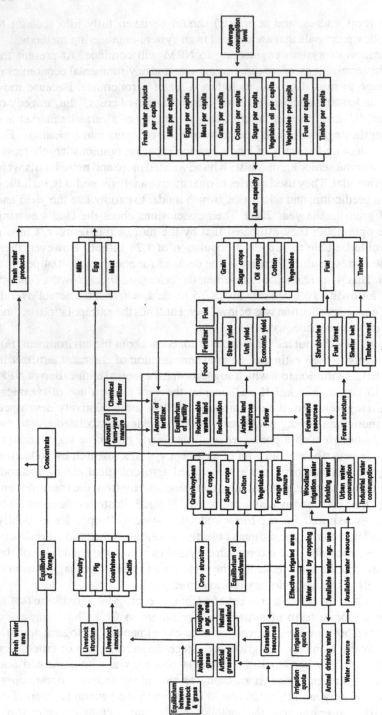

Figure 3. Research framework for land resource-bearing capacity and productivity in China

groups of objectives: economic, social, and ecological (more usually expressed as efficiency, equity, and security objectives). Today, system engineering has been more closely linked with the concept of Chinese ecological agriculture and its practice. An additional and outstanding feature of this movement is that it has involved more social scientists. This is important because agriculture is merely one part in the country's overall development. In solving problems of population, social equity (in China usually linked to areas of poverty), and social welfare, sociologists are playing an increasingly important role.

Problems and suggestions

In order to sharpen the existing tools of system research as an aid to decision making at different levels as well as to build institutional and human resource capacity in this field, it will be necessary to review and find solutions to the following problems:

- Data collection and survey: These are prerequisites for any successful quantitative analysis, especially for crop growth simulation and the use of expert systems. However, until now, the inventory of natural resources that is available to Chinese agricultural scientists has been much more limited than those available in many other parts of the world.
- Human resource development: To make the benefits of system research more accessible for agricultural development, more specialists have to be trained, particularly those who have an agronomy background and understand system science and technology.
- Transfer of technology: Considering the fact that China has experience in using resource inventories, in applying system engineering, expert systems, and GIS, as well as having had a good start toward the establishment of sustainable agriculture and rural development and ample talent in developing computer software, it is appropriate to suggest that if given the chance, Chinese scientists could play a special role in the transfer of modern quantitative analysis technologies and resource management experiences to other developing countries.

Conclusion

China has done a lot of the groundwork in applying systems research methods to resource management and agricultural/rural development. The development and management of a large natural resource inventory, as well as regional planning for agriculture and industry at all levels, have made progress over the past decade. Their effects can readily be seen in the manner in which decisions are made, a process that was earlier dominated by empiricism and estimation.

However, it should be remembered that most of the methods and approaches referred to above are still at an early stage of development and capability. It should also be noted that in China, there is a group of agronomists who are taking a "mainstream" position in influencing policymakers. Though they have recognized the need for and have even actively taken part in activities directed toward developing an inventory of resources for regional planning, they still have only a limited awareness of sustainable development. They still advocate increasing amounts of physical inputs, in particular fertilizer, to agriculture. To overcome these difficulties, it is suggested there be an international effort to combine economic principles and quantitative analysis methods with sustainable development principles into procedures for system research as early as possible. After that, the international agricultural organizations and institutes would have a lot of work to do to extend knowledge about the application of these procedures and to train developing-country scientists in using them.

Acronyms

AEZ	agroecological zoning
DSS	decision support system
GIS	geographic information system
IRM	integrated resource management
LP	linear programming
MIS	management information systems
NPK	nitrogen-phosphorus-potassium
SD	systems diagnosis

Publications on agricultural and NRM planning

Cheng Baiming et al (1991) Researches on China's land resource productivity and its bearing capacity for population Chinese People's Univ

Chen Li (1981) Agricultural system engineering China Agricultural Publishing House

Cheng Xu (1992) Sustainable agricultural development in China World Development 20(8)

He Kang (1992) To create a new situation for developing agricultural resource surveys and regional planning Proceedings of the first symposium on China's agricultural resources survey and regional planning

Li Jianbai (1990) Pilot experiment for agro-modernization by means of system engineering System sciences and comprehensive studies in agriculture no 1

Pei Xingde (1990) Linear programming, goal programming and their application Science and Technology Literature Publishing House

Qian Xueshen et al (1990) A new sphere of science Open and complicated giant system and its methodology Scientific Decision and System Engineering In Chinese Science and Technology Publisher

Qunlai County Regional Planning Office Research report on county-level agroresources economy information and decision consultant system (unpublished)

S&T Committee, Ministry of Agriculture (1989) Forty years review of China's agroscience and technology China Science and Technology Publishing House

Tao Mingjiang et al System engineering programming for demonstrated countryside/township of Zhelimu League, Inner Mongolia (1986-2000) (unpublished)

Xinjiang S&T Commission (1989) Research and application of rural development programming model in Xinjiang autonomous region (unpublished)

Yang Tinxiu (1987) General designing of agro-system engineering Shandong S&T Publishing House

Zhang Renwu et al (1990) Techniques and design for ecological agriculture, Hebei Science and Technology Publishing House

Zhang Xiangshu (1991) Systematic analysis on China's food demand and supply, and the research of its decision support system. Adopted from Medium-long term research on China's food development strategy. Agriculture Publishing House.

Zhang Xiangshu (1993) Market-oriented economy and system engineering for natural resources management. Ecological agriculture research 1(3).

Note: All but Cheng Xu (1992) are in Chinese.

The use of systems analysis methods—the experience at a national level (India)

S. SANKARAN
Tamil Nadu Agricultural University, Coimbatore 641 003, India

Key words: agroecological zones, climate change, geographic information system, India, rice, simulation models, systems analysis, systems research

Abstract
Although systems approaches in agricultural research have been introduced in India only recently, the Simulation and Systems Analysis for Rice Production (SARP) project has helped expand the country's activities in this area Systems thinking, analysis, and simulation have been used in agroecological zoning, estimating the potential production of crops, crop improvement, assessing the effects of climate change, nutrient and water management, and crop protection To get quantitative measures on the behavior of rice in different environments, a substantial amount of work has been done The data and human skills required for systems research appear to be major constraints Continued international cooperation is needed to strengthen work in this field and to extend systems research to other areas of agricultural research, especially rainfed agriculture

Introduction

Out of the 329 million hectares that covers the geographical area of India, 136 million hectares are cultivated, 33 percent (43 million ha) of which is irrigated and 67 percent (93 million ha) rainfed. Agricultural research in India has helped farmers increase production to a level that is sufficient for the current population of 870 million. The food production growth rate (2.4 percent) is slightly higher than the population growth rate (2.1 percent). However, with an expected population of 997 million by the year 2000, and with limited opportunity to expand the cultivated area or extend access to irrigation, researchers and farmers face substantial challenges. To manage the country's natural resources in a productive and sustainable manner has become the most important aspect of agricultural development in India. It will not be enough to just satisfy the demand for food; the task is also to protect the natural assets of the soil and water for future generations.

Rainfed agriculture supports 40 percent of the human population and about 60 percent of the cattle and accounts for 44 percent of the total food production. During 1988-89, 91 percent of the coarse cereals, 90 percent of the pulses, and 81 percent of oilseeds production came from rainfed agriculture (Katyal 1992). The dryland farmer depends on erratic and uncertain rainfall, soils of poor nutrient status and low productivity, and very limited economic resources. A large part of the cultivated land suffers from salinity, erosion, or other constraints to production. There were about 50 million farms in the 1940s, but the fragmentation of land holdings has doubled that number to 100 million now. One of the results of this fragmentation is that the

P Goldsworthy and F W T Penning de Vries (eds), Opportunities, use, and transfer of systems research methods in agriculture to developing countries, 213 - 225
© 1994 *Kluwer Academic Publishers*

spread of new technology is very slow. There is an acute shortage of green fodder, and the country's 448 million cattle has to live on only about half of the dry fodder they require.

India has built up a sound infrastructure for agricultural research. 27 Universities (out of a total of 250 in the country) are devoted exclusively to agriculture. Agricultural universities have between 20 and 70 research stations under their supervision each. The Indian Council of Agricultural Research (ICAR), the principal body for agricultural research in the country, has more than 60 research institutions throughout the country. ICAR also coordinates more than 50 network projects with agricultural universities. The agricultural scientists of India are confident they can meet the challenge of ensuring the agricultural prosperity of the country.

Intensifying production, achieving sustainability, and optimizing the use of scarce resources are the key issues for the future of agricultural research in India, especially for rainfed agriculture. Systems thinking and systems simulation have become important tools in dealing with the complex agricultural systems.

Simulation and systems research

Systems analysis and crop growth simulation are relatively recent techniques that provide a means to quantify the effects of climate and soils on crop growth and on the productivity and stability of agricultural production from different classes of land. These techniques can reduce the need for costly and time-consuming field experimentation. Since national agricultural research systems are responsible for developing new production technology, simulation and systems analysis have become important tools to gain insight into the complex crop-production environment. Simulation and system analysis can be used to extend the results of research conducted in one environment to other larger areas with a similar environment. The inductive logic employed in the systems approach helps overcome the failures of the conventional deductive approach.

The central idea of systems research is to identify, define, and understand a system sufficiently to be able to influence it in a predictable manner (Spedding 1990). The systems researcher has to define the system boundaries and try to understand the system with all its variables and the processes related to the system's behavior. The modelling that is involved serves to highlight gaps in knowledge and to point the direction for further research. The systems analysis, crop modelling and simulation begun by De Wit in 1958 has now become a worldwide endeavor.

Conventional research has depended on statistical regression models to analyze environment-yield relationships. In systems research, however, the models describe the processes that govern the behavior of the system in a quantitative manner. These descriptions are explicit statements based on scientific theory and hypotheses. For a crop model, the components of the system are first analyzed, and the growth processes are quantified on the basis of knowledge of how they are influenced by the environment. The model is then built by integrating the knowledge about the individual components of the system to produce a description for the entire system. In these models, simultaneous differential equations are used to simulate dynamic processes of crop growth and development (Penning de Vries et al. 1989).

Implementation of systems approach in agricultural research in India

Systems analysis methods were first used in Indian agricultural research only recently, thanks to the knowledge acquired from participation in the project Simulation and Systems Analysis for Rice Production (SARP). Jointly carried out by the Centre for Agrobiological Research (CABO-DLO) and the Wageningen Agricultural University (WAU) and by the International Rice Research Institute (IRRI), the SARP project trained four teams of scientists in India in using systems analysis methods. The teams came from two national institutes (the Indian Agricultural Research Institute, New Delhi, and the Central Rice Research Institute, Cuttack) and two agricultural universities (Tamil Nadu Agricultural University and G.B. Pant Agricultural University). In collaboration with the SARP project, ICAR used the SARP-trained manpower to extend the knowledge to another six institutes in the country in 1991. The continuing contacts with the SARP project personnel and an informal network of collaborative research programs have further improved the scientists' capacity to use the tools and methods of systems research.

In India, the systems approaches support research on agroecological zonation, the potential production of crops, climate change and rice production, the evaluation of rice genotypes, plant breeding, plant nutrient management, water management, and crop protection. The tools and methods employed are simulation languages (CSMP, FORTRAN), crop growth models (MACROS, SOYCROS, ORYZA), databases (from national agricultural research centers, IRRI, and CABO), and a geographic information system (GIS). Our experience with GIS is very limited and there is scope for widening its use. Our experiences with systems research are briefly discussed in the following sections.

Agroecological zonation
Crop and environment data collected at specific sites, together with systems analysis and crop growth simulation, can be used to characterize the variation within a region, and thus to identify different agroecological regions. Agroecological zoning provides a basis to determine research priorities and the regional allocation of resources, to plan technology transfer and regional development, and to study the impact of climatic variability on agricultural production. Crop simulation can examine the practicability of a wider range of crop production/environment options than conventional methods can. The combination of systems analysis and simulation with a GIS database that contains environmental and socioeconomic information offers even greater opportunities to identify homologous agroecological environments that satisfy criteria selected by the investigator. These criteria may include any number of factors, such as climate, soil, land use, population, and incomes.

216

An example of the application of these approaches is the work done by Aggarwal (1993), who demonstrated that simulation models can be used to determine the productivity of wheat at different locations in India. These locations were selected by climate and the availability of water (figure 1). Similarly, using a simple water-balance model and the rainfall data from 115 locations, Ramaswami and Selvaraju (1993) identified the areas suitable for rainfed rice cultivation during the Northeast monsoon (November-December) in Tamil Nadu. In another example of this kind of application, Jeyaraman et al. (1993) analyzed the influence of agrometeorological parameters on the productivity of rice in the *kuruvai* season (June-September) in Aduthurai, India, during 1991 and 1992. They found that grain yields were positively associated with maximum and minimum temperatures and solar radiation, and that rainfall during flowering negatively affected the grain yield.

Systems research for crop improvement
Although maintaining stability of yield and quality of economic produce are also important, improving the yield potential of crops is one of the major responsibilities of agricultural research in India, and it continues to be the biggest challenge. Crop simulation provides a means to assess the probable performances of new genotypes

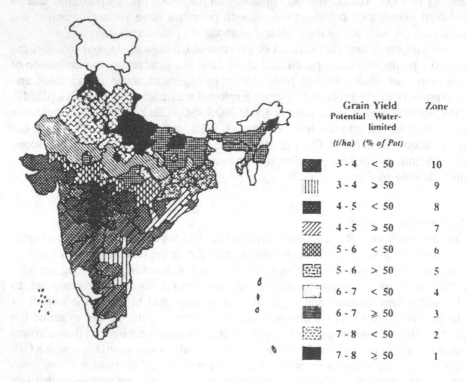

	Grain Yield		Zone
	Potential	Water-limited	
	(t/ha)	(% of Pot)	
■	3 - 4	< 50	10
‖‖‖	3 - 4	≥ 50	9
■	4 - 5	< 50	8
▨	4 - 5	≥ 50	7
▨	5 - 6	< 50	6
▨	5 - 6	> 50	5
▢	6 - 7	< 50	4
■	6 - 7	≥ 50	3
▨	7 - 8	< 50	2
■	7 - 8	≥ 50	1

Figure 1. Iso-yield wheat zones of India based on potential and water-limited productivity (*Source:* Aggarwal 1993)

in different environments. The parameters and response functions in the model that reflect the variation between genotypes can also serve to show what modifications would lead to better crops (Penning de Vries 1991).

In Tamil Nadu, high-yielding genotypes are identified in a series of time-consuming yield tests. Palanisamy et al. (1993) used the crop-growth simulation model MACROS L1D to reduce the time and cost of variety improvements in rice. The model was able to predict the performance of rice genotypes at Coimbatore and Madurai, and for selection purposes it gave results comparable to those obtained from field trials (figure 2). Simulation models can be used to explore options (e.g., a hypothetical new genotype) before it is possible to do so in practice. This is a promising aspect for future system research (Mohandass et al. 1991, 1993).

Potential production of crops
In environments with irrigation and an optimum supply of nutrients and without significant pest or disease damage, the growth rate of the crop depends only on the current state of the growth of the crop and on the weather, in particular the radiation receipt and the temperature (Penning de Vries et al. 1989). Simulation models and

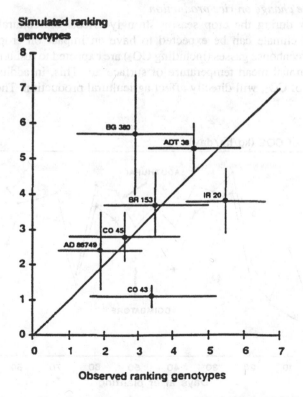

Figure 2. The average ranking of seven genotypes in the simulation and in the Multi Location Trials, with the standard deviations (*Source:* Palanisamy et al. 1993)

218

systems analysis can estimate the potential production, the optimum sowing date and the inputs required to achieve the potential yields in different environments.

Thus, Mohandass et al. (1991) found the potential production of IR 50 rice was 9.7 t ha^{-1} at Coimbatore and 9.5 t ha^{-1} at Aduthurai. Using simulation methods, they were able to attribute the difference between these yields and observed yields to the benefit derived from soil CO_2 produced by the decomposition of added green manure. It was assumed that above-normal CO_2 concentration contributed to the photosynthesis of the rice crop. They also simulated the production of CO_2 at Aduthurai and Coimbatore. The rate of production was generally highest from 15 days before flowering to 20 days after (figure 3). In the field, the soil also supplies CO_2; upland soils reportedly supply from 1.5 to 52.0 g m^{-2} day^{-1} of CO_2 for photosynthesis (Yoshida 1976).

In another example of the application of simulation methods to estimate potential yield of a crop, Thiyagarajan et al. (1993) used the MACROS model to show that rice yields in six locations of Tamil Nadu could vary by 16 up to 31 percent, depending on the planting dates.

Effect of climate change on rice production

As the weather during the crop season strongly affects agricultural production, changes in the climate can be expected to have an impact on crop production. Emissions of greenhouse gasses (including CO_2) are expected to result in an increase in the global annual mean temperature of surface air. This, in addition to higher concentrations of CO_2, will directly affect agricultural production. The higher CO_2

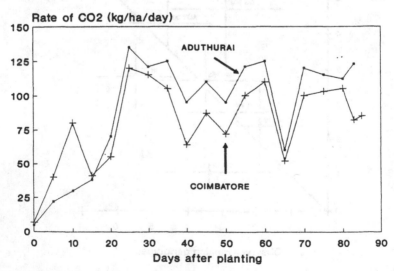

Figure 3. Simulated rate of soil CO_2 production due to green manure application to IR 50 rice at Aduthurai, DS, 1990 (*Source:* Mohandass et al. 1991)

concentrations will lead to higher assimilation rates, an increase in stomatal resistance, a decline in transpiration, and an improve in water-use efficiency. At the Tamil Nadu Rice Research Institute, Aduthurai, attempts have been made, under the auspices of the IRRI/SARP project, to estimate changes of these climate changes on the potential yield in the major rice growing areas in India. When temperatures rise, yields decline. The study showed that the decline reached a maximum of 12 percent for a 4^0C rise in temperature. Even so, for a rise in temperature of less than 2^0C, the effect on yield was small (Mohandass et al. 1993). Increased concentrations of CO_2 in the atmosphere generally led to an increase in simulated estimates of the potential rice yield. The average yield increase for the entire country was 11, 20, 28, 40, and 48 percent for 390, 440, 490, 590, and 680 ppm atmospheric CO_2 concentrations respectively. Within these general trends there were differences between locations.

Plant nutrition

With the rising cost of fertilizer, more research is needed to improve the efficiency of fertilizer inputs. Nitrogen plays a key role in crop nutrition, and its role in determining crop yields has been studied extensively (Sinclair 1990). The response of a crop to fertilizer can vary greatly, depending on crop management and on the weather during the growing season. Since crop simulation models make it easier to assess this variability, they are important tools in nutrient management (Dahnke and Olson 1990).

Field experiments conducted at Thanjavur, India, with rice variety CR 1009 during the wet season (September-January) of 1991-92 and 1992-93 (Sivasamy et al. 1993), provide an example of the application of simulation methods to a crop nutrition problem. The experiments showed that for the same nitrogen fertilizer strategy (150 kg N ha^{-1} split into six equal applications), the grain yields in 1992-93 were 18 percent lower than in 1991-92. A simulation analysis of the behavior of the two crops provided the explanation; nitrogen uptake ceased during the period between maximum growth of tillers and panicle initiation of the crop in 1992-93, and the crop was unable to recover from the setback later (figure 4). Further analysis showed that a lack of rain during the critical period between the tillering and panicle initiation interrupted the nitrogen uptake in 1992-93. Another simulation study (Thiyagarajan et al. 1993) to examine the response of rice to 100 kg N ha^{-1} (four times more than normal) showed that the nitrogen-use efficiency was higher for the nitrogen applied at the P1 growth stage when no nitrogen had been applied to the crop before that stage (table 1). The study also indicated that in fertile soils that can support a good crop growth in the early stages, the best strategy for increasing nitrogen-use efficiency is to apply more nitrogen during the later part of the vegetative growth of the crop.

A systems approach enabled Thiyagarjan et al. (1993) to propose a new concept concerning the changes over time in the concentration of leaf nitrogen required to attain a given yield (in this example 8 ton ha^{-1}). The proposal, based on the analysis of results from a number of experiments, indicates that if the concentration of leaf nitrogen falls below a specified level at any stage of the crop, the yield will fall to below 8 ton ha^{-1}. They also found that for a yield of 10 ton ha^{-1}, the required

concentrations at any stage of growth are higher and must be maintained longer—
from tillering to flowering. The experiment also indicated that the levels of leaf
nitrogen required before flowering to attain these yields could only be achieved
through green manuring.

Water management

In the semi-arid tropical regions of India, the specific adaptation of crops to the local
environment is important. Dynamic crop-growth models can be used effectively to
compare different strategies for cropping on land use and water management (Huda
et al. 1978). Since the attainable yield is determined mainly by the amount and
distribution of rainfall, an efficient management of water-use is vital for crop
production. A simulation approach can be used to extend research findings to
different environments, provided that the quantitative relationships between the crop,

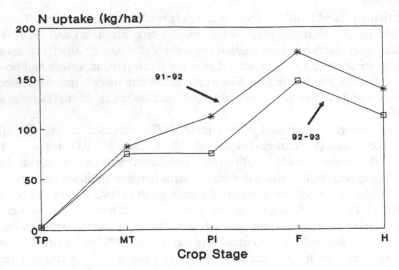

Figure 4. Nitrogen uptake by rice (CR 1009) in the wet season of 1991-92 and 1992-93, Thanjavur, India.
N uptake stalled during maximum tillering (MT) and panicle initiation (PI) stage resulted in 18 percent
yield loss in 1992-93 (*Source:* Sivasamy et al. 1993).

Table 1. Nitrogen application strategy and nitrogen use efficiency in rice (CR 1009), WS, Thanjavur, India

N applied kg ha^{-1}			kg grain kg^{-1} N		
At 10 DAT	At PI	Total	1991-92	1992-93	Mean
0	100	100	22.0	8.0	15.0
50	100	150	18.7	6.0	12.4
150 (6 splits)		150	23.3	10.0	16.7

(*Source:* Thiyagarajan et al. 1993)

soil, and weather are understood, and that changes in any component of the system can be evaluated adequately.

The work of Palanisami (1993) is an example of the application of a simulation model for water-use management. The model was developed to help determine optimum cropping systems for tank irrigation in Tamil Nadu, where water is scarce. The model simulated the effects of water stress on yield and net incomes for different years in which the filling of the tank varied. The results indicated that the optimum cropping system consists of a 25:75 percent division between rice and non-rice crops. Kalra et al. (1993) also used a simulation model to evaluate the effects of different levels of water and nitrogen on the growth and yield of wheat in India. In a different application, Ramasamy (1993) used a simulation approach to show that an effective drainage in wetland rice soils increases the activity of crop roots and improves the nitrogen uptake and grain yield.

Research on rainfed agriculture needs greater emphasis, however. In addition, a lot of work is needed to improve our understanding and control of the relations between crop water use and nutrient uptake. Systems analysis and simulation can provide valuable help in this domain of research.

Ecological farming

Since the plant nutrient status of soils in India is often extremely poor, soils require plant nutrient inputs to achieve acceptable levels of production. At the same time, however, the dependence on purchased fertilizer inputs needs to be reduced (Swaminathan 1991). Although the argument for a greater emphasis on ecological concerns is widely accepted, there are not yet proven technologies to replace the current practices. Systems of integrated nutrient management (with crop rotations, green manures, and biofertilizers) that farmers can readily adopt are yet to be developed. Studies conducted at Thanjavur (Sivasamy et al. 1993) and Aduthurai (Budhar et al. 1993) illustrate how systems research methods could improve the understanding of how organic sources of nitrogen can be used more effectively in crop production. The studies show that only part of the nitrogen required to achieve high crop yields can be obtained from organic sources (green manure, azolla, farm yard manure). The rest is obtained from inorganic sources. The pattern of nitrogen uptake by the crop depends on the nitrogen source. The entire crop requirement may be met by organic sources, provided that the native fertility of the soils is first improved and then maintained by using organic manures regularly. Mohandass et al. (1991) also showed that green manure application delayed the leaf senescence in rice IR 50—an effect that could not be accounted for by improved nutrition alone.

Pest management

Pest management in crops is complex and involves many components. It is made even more complex by man's production-oriented interventions, such as plowing and the use of pesticides, and by weather variables. In the early years of pest management, researchers made attempts at mathematical modelling and computer simulations, but without explicitly recognizing the processes involved. Pest models commonly simu-

222

late the dynamics of a single disease or insect and how the disease is affected by its host and the physical environment. Pest zoning is useful in particular for pest-management problems that cover a large area in which a common strategy and tactics can be applied. Thus, predictive models can be used to define zones with similar pest risks, so that pest management technology from key sites can be extended to a broader area. It can also be used to identify areas of adaptation to deploy resistant varieties.

In an example of a crop protection study of this kind, Narasimhan and Abdul Kareem (1993) studied the effects of bacterial leaf blight (BLB) on leaf function and the leaf-area index of the canopy of a rice crop. They introduced these effects as forcing functions in the MACROS L1D model (figure 5). The studies indicated that it was possible to use this modified version of MACROS L1D model for assessing the crop growth and yield loss due to BLB infection. To validate its use in this manner, the model requires further testing with more cultivars of different degrees of susceptibility in multi-location experiments over a number of seasons.

Weed management
Weed management is a key element in crop protection management in most agricultural systems. Integrated Weed Management (IWM) is based on how different methods of control complement each other, while reducing the dependence on chemical control methods that involve environmental risks. Developing such weed management systems requires quantitative information about how weeds behave in agrosystems and how they affect crops. Systems methods can then be used to help management find solutions for given situations. One example is a new and simple descriptive regression model for the early prediction of crop losses arising from weed competition. The model, introduced by Kropff and Spitters (1991), was derived from

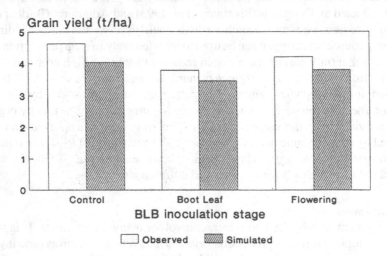

Figure 5. Simulation of effect of bacterial leaf blight infection on IR 50 rice grain yield, Aduthurai, India (*Source:* Narasimhan et al. 1991)

the well-tested hyperbolic yield-density model. This model relates yield loss to relative weed leaf area soon after crop emergence.

Bottlenecks in the application of systems research

The pattern of agriculture is determined largely by climate, while it is the weather that strongly influences the variations in crop production from year to year and from place to place. Solar radiation on the surface of the earth, influenced by both climate and weather, is a major factor that determines the productivity of crops. The reliable measurement of solar radiation is therefore an important input into crop simulation models. Until recently, problems in measuring this radation was a serious bottleneck to systems analysis. Thanks to the SARP project, however, we now have radiation integrators in six of our research stations in Tamil Nadu. Other stations will receive similar equipment.

As expected, simulation has increased the awareness of research scientists that they often lack reliable data, and that their understanding of crop behavior is incomplete. In this sense, systems methods can help point to where further research is needed. One example is the uptake of nitrogen by rice after flowering. It appears that there can be at least four possible situations: (a) the crop does not absorb nitrogen from the soil, (b) it continues to take up nitrogen from the soil, (c) it loses nitrogen to the soil, or (d) absorption and loss of nitrogen take place simultaneously. At present, our knowledge on this is very limited, but simulation methods have helped show where research is required.

Another difficulty is collecting the data required for a reliable estimate of the soil-water balance. Though it is appreciated that acquiring realistic data on soil moisture characteristics is difficult, easier and more universally acceptable ways are needed to measure it if we are to make faster progress in improving technology for dryland farmers.

A final, well-recognized, and important bottleneck in systems research is the present shortage of people with the expertise and skills required. The SARP project has done a tremendous service to agricultural research by imparting this expertise to scientists in countries such as India. More efforts are required to introduce other researchers to systems approaches and thinking.

Conclusions

Systems research, conducted by inter-disciplinary teams, can overcome some of the limitations of conventional research and technology transfer. More basic research is required to fill gaps in the understanding of quantitative relationships in a crop production environment. In India, the environmental database should be strengthened and developed; so also should linkages for collaborative research. Measures are needed to improve the levels of expertise in systems methods in the education system and in research institutions.

224

Acronyms

BLB	bacterial leaf blight
CABO-DLO	Research Institute for Agrobiology and Soil Fertility
DAT	days after transplanting
GIS	geographic information system
ICAR	Indian Council of Agricultural Research
IWM	integrated weed management
SARP	simulation and systems analysis for rice production
WAU	Wageningen Agricultural University

References

Aggarwal P K (1993) Agro-ecological zoning using crop growth simulation models; characterization of wheat environments of India. Pages 111-125 in Penning de Vries F W T, Teng P S, Metselaar K (Eds.) Systems Approaches for Agricultural Development, Kluwer Academic Publishers, Dordrecht, The Netherlands

Budhar M N, Palaniappan S P, Thiyagarajan T M, Ten Berge H F M (1993) Influence of organic and inorganic sources of N on the growth and yield of lowland rice — a simulation analysis. SARP Workshop on Nitrogen Management and Modelling in Irrigated Rice, 1-10 November 1993, Suweon, South Korea.

Dahnke W C, Olson R A (1990) Soil test correlation calibration and recommendation. Pages 45-71 in Westerman R L (Ed.) Soil testing and plant analysis. Soil Science Society of America, Madison, USA.

Huda A K S, Sivakumar M V K, Virmani S M (1980) Interdisciplinary research needs of agroclimatological studies: Modelling approaches and minimum data set. Pages 197-202 in Proceedings of the International Research Needs of the Semi-arid Tropics, 22-24 November 1978, ICRISAT, Hyderabad, India.

Jeyaraman S, Thiyagarajan T M, Mohandass S, Abdul Kareem A (1993) Influence of agrometeorological parameters on productivity of rice. Paper presented at the International workshop on simulation of potential production of rice, 25-28 January 1993, TNRRI, Aduthurai.

Kalra N, Aggarwal P K, Sinha S K (1993) Production functions evaluating the responses of water and nitrogen on growth and yield of wheat. SARP Workshop on Nitrogen Management and Modelling in Irrigated Rice, 1-10 November 1993, Suweon, South Korea.

Katyal J C (1992) Dry land farming; Corporatisation indispensable. Pages 17-22 in The Hindu — Survey of Indian Agriculture 1992. The Hindu, Madras, India.

Kropff M J, Spitters C J T (1992) A simple model for crop loss by weed competition on basis of early observation on relative leaf area of the weeds. Weed Research 31:97-105.

Mohandass S, Thiyagarajan T M, Palanisamy S, Abdul Kareem A (1991) Influence of leaf N status on grain characteristics and yield in IR 50 rice — A simulation analysis. Paper presented at the International Workshop on Water, Nutrients and Roots, 6-11 May 1991, the Universiti Pertanian Malaysia, Serdang, Malaysia.

Mohandass S, Thiyagarajan T M, Palanisamy S, Abdul Kareem A (1993) Impact of increased temperature and CO_2 on rice productivity in Cauvery Delta Zone, India—A simulation analysis. J. Agr. Met. (Jpn)48:791-793.

Mohandass S, Abdul Kareem A, Jeyaraman S, Thiyagarajan T M (1993) Rice production in India at current and future climate. Page 31 in Project Country Report of Simulation of Impact of Climate Change on Rice. (unpublished).

Narasimhan V, Abdul Kareem A (1993) Simulation of effect of bacterial leaf blight diseases on yield reduction in rice. Paper presented at the SARP Workshop on Mechanism of BLB disease and their effect on yield, 3-5 March 1993, Central Rice Research Institute, Cuttack, India.

Palanisami K (1993) Optimization of cropping patterns in tank irrigation systems in Tamil Nadu, India. Pages 413-425 in Penning de Vries F W T, Teng P S, Metselaar K (Eds.) Systems Approaches for Agricultural Development, Kluwer Academic Publishers, Dordrecht, The Netherlands.

Palanisamy S, Penning de Vries F W T, Mohandass S, Thiyagarajan T M, Abdul Kareem A (1993) Simulation in pre-testing of rice genotypes in Tamil Nadu Pages 63-75 in Penning de Vries F W T, Teng P S, Metselaar K (Eds) Systems Approaches for Agricultural Development, Kluwer Academic Publishers, Dordrecht, The Netherlands

Penning de Vries F W T, Jansen D M, Ten Berge H F M, Bakema A (1989) Simulation of ecophysiological processes of growth in several annual crops Simulation Monographs, PUDOC, Wageningen, The Netherlands and International Rice Research Institute, Los Baños, Philippines

Penning de Vries F W T (1991) Improving yields, designing and testing VHYs Pages 13-19 in Penning de Vries F W T, Kropff M J, Teng P S, Kirk G J D (Eds) Systems and Simulation at IRRI IRRI Res Paper Series No 151 International Rice Research Institute, Los Baños, Philippines

Raju N, Abdul Kareem A, Palanisamy S (1992) A simulation model for rice yellow stemborer in relation to growth and yield of ADT 39 Paper presented in the SARP workshop on Mechanism of stemborer damage, 3-5 August 1992, Khonkaen, Thailand

Ramaswamy S (1993) Integration of drainage, plant population and nitrogen levels and simulation modelling in lowland rice PhD Thesis Tamil Nadu Agricultural University, Coimbatore, India

Ramaswami C, Selvaraju R (1993) Agroclimatic zoning, a water balance approach for rainfed rice in Tamil Nadu State of India Proceedings of the Workshop on Agroecosystems, April 1993, Zhejiang Agricultural University, China

Sinclair T R (1990) Nitrogen influence on the physiology of crop yield Pages 41-56 in Rabbinge R, Goudriaan J, Van Keulen H, Penning de Vries F W T, Van Laar H H (Eds) Theoretical production ecology Reflections and prospects Simulation Monographs PUDOC, Wageningen, The Netherlands

Sivasamy R, Thiyagarajan T M, Ten Berge H F M (1993) Nitrogen and rice Uptake and recovery of applied Nitrogen Paper presented at the SARP Workshop on Nitrogen Management and Modelling in Irrigated Rice, held at Suweon, South Korea, 1-10 November 1993

Spedding C R W (1990) Agricultural production systems Pages 239-248 in Rabbinge R, Goudnaan J, Van Keulen H, Penning de Vries F W T, Van Laar H H (Eds) Theoretical production ecology Simulation Monographs 34, PUDOC, Wageningen, The Netherlands

Swaminathan M S (1991) Sustainability beyond the economic factor Pages 10-15 in Survey of Indian Agriculture, 1991 The Hindu, Madras, India

Thiyagarajan T M, Jeyaraman S, Budhar M N (1993) Rice production situation in different agroclimatic zones of Tamil Nadu Paper sent for the International Workshop on Agroecosystems, held at Zhejiang Agricultural University, Hangzhou, P R of China (in press)

Thiyagarajan T M, Sivasamy R, Ten Berge H F M (1993) Time course nitrogen concentration in rice leaves Paper presented at the SARP Workshop on Nitrogen Management and Modelling in Irrigated Rice, 1-10 November 1993, Suweon, South Korea

Thiyagarajan T M, Sivasamy R, Ten Berge H F M (1993) Nitrogen and rice Influence of N application levels and strategy on growth, leaf nitrogen and N use efficiency Paper presented at the SARP Workshop on Nitrogen Management and Modelling in Irrigated Rice, held at Suweon, South Korea, 1-10 November 1993

Yoshida S (1976) Carbon dioxide and yield of rice Pages 211-222 in Climate and Rice, International Rice Research Institute, Philippines

Opportunities for use of systems approaches in agricultural research in developing countries—the Senegal experience

D.Y. SARR
Institut Sénégalais de Recherches Agricoles, Boîte Postale 3120, Dakar, Senegal

Keywords: multidisciplinarity, reorganization, systems research, training

Abstract
In the last decade, Senegal created systems research projects in its agricultural research, where sociology and economics were added to the biophysical sciences and agronomy. Its success stimulated the foundation of a Systems Department with structured field and office activities. This approach is now being implemented in all research departments.

Introduction

In most developing countries production takes place in sectors as diverse as agriculture, livestock and forestry. In this context, a systems approach has been shown to be an effective way to clarify the complexity of production systems in different agroecological zones. It helps to highlight the importance of interaction between the sectors mentioned and, at a lower level, between the elements within each production system. As a result of many years of research and a determination to focus its research on the concerns of the producers, ISRA has given greater importance to a systems approach in its research plan of action by establishing the Department for Research on Production Systems and Technology Transfers, which was later renamed Department for Research on Agrarian and Agricultural Economics (DRSAEA).

Without elaborating on the Senegal experience in this field, we present here the issues that we feel play a major role in the design of an efficient research tool that can promote economically strong production systems that will have the support of producers and that are ecologically feasible.

Systems approaches in agricultural research organizations in Senegal

The experience of the Experimental Units (Unités Expérimentales) Project laid the foundation for a systems research program in Senegal. One of the main features of the Experimental Units has been that they involved sociology and economics, disciplines that until then were considered to be of secondary importance to agricultural research. These disciplines were included with the aim of:

227

P. Goldsworthy and F.W.T. Penning de Vries (eds.), Opportunities, use, and transfer of systems research methods in agriculture to developing countries, 227 - 232.
© 1994 *Kluwer Academic Publishers.*

- a better understanding of the problems and needs of the farmers;
- identifying the constraints that obstruct or limit the adoption of technologies proposed.

The experience of the Experimental Units helped to identify clearly the position of farmers and the major role they play, not only in the implementation of the technologies proposed to them, but also in the identification of constraints and in the search for possible technical and economical solutions. In other words, the aim was actually to let them participate in the search for solutions to the extent that they were willing and able to do so.

As a consequence, the Experimental Units were seen as the forerunners of systems approaches in Senegal. On this basis a Systems Department was set up, with a mission not only to identify the socioeconomical constraints associated with a proposed technology, but also to put agricultural activities in their proper context, and to highlight what is required to:

- better understand the functioning of production systems used by farmers and farmer organizations;
- better understand the problems and needs of the producers;
- improve the adaptation of technologies to match farmers' requirements and, thereby, also ensure that more efficient agricultural research is done to resolve the constraints that have been identified and to serve the basic needs of the farming population;
- better understand the influence of agricultural policies on the performance of the farming units in order to help define development strategies;
- work on effective communication between the research department and its various partners (extension, non-governmental organizations (NGOs), individual farmers and/or via their organizations) within the scope of the mission, which aims to improve the well-being of the populations through increased and sustainable production.

To meet these objectives, the Systems Department has focussed on an overall research systems approach. I shall not discuss the methodological side of such an approach, because a considerable amount of information is already available on this subject.

Furthermore, the Department has implemented an organization, which is particularly effective in strengthening structures that are needed within the various programs of the institute. For this purpose, the Department included:

- interdisciplinary regional teams, to deal with regional problems;
- the Office for Macroeconomic Analysis (BAME), which handles at a macrolevel the information and problems collected by the field teams at a microlevel;
- the assignment of thematic researchers, working under the authority of the research/support division of each center that has appointed a team for research on technical solutions to the problems identified;
- finally, and above all, the Central Analysis Group (GCAS), charged with administrative coordination and scientific support to the field teams. This group consisted of experienced researchers, able to organize, conduct, and maintain, within an agroecological context, a consistent approach as well as proper coordination.

Frequent site visits by the teams and management coordination meetings of the department have contributed in a very effective way to the smooth running of both the administrative and scientific services within the department. The visits and meetings made it possible to study and comment on all aspects of each others' programs in a most efficient way. Such interaction was all the more necessary because of the difficulties encountered, particularly at the level of the field team.

Constraints in the execution of the department's programs for research on agricultural systems

Especially during the first few years after the programme started, it appeared that there were a number of constraints:

- Some of the researchers were unable to devote their time entirely to the functioning of the programs because colleague researchers were on training. As a consequence, it was up to researchers involved in other programmes to organize the work of the absent researchers until they themselves could concentrate fully on a new program.
- The team researchers lacked the experience and skills required for the systems approach. The researchers had not received proper training in the subject beforehand, and it became clear that they needed to be taught the basic elements of concepts and methodology.
- The team encountered problems in creating a team spirit and accomplishing full integration, because the field team members were more prone to the thematic approach they had always followed.
- Contacts and communication between the teams outside the research management coordination meetings were limited.

Although these problems affected the program during the first few years after its implementation, it should be noted that in due course significant improvements were made. This resulted in commendations from external observers (mainly in mission reports of donors) and from the Committee for Science and Techniques of ISRA, which was appointed to assess the relevance of the research programmes of the institute, and the level and quality of the execution of these programmes.

Contribution of the systems research department

Without elaborating on the various research themes nor on the approach, one could say that the Systems Department ushered in a new era for both the way research priorities are set and the way programmes are managed. The Systems Department has:

- made a major contribution to establishing a strong awareness among the researchers, even those outside the regional teams, of the need, when determining the research thrusts, to adopt an approach which encompasses the social, economic, ecological, and political aspects of the constraints which affect production;

- focused on alternative technologies that are more in harmony with the environment and the circumstances of the farmers, thus contributing to the reorientation of the research programs;
- promoted the establishment of new relations between researchers and their partners (extension agents and farmers).

Thus, by the time of its dissolution in 1990, the Systems Department had left its mark on the global approach of the institute. The same year, this induced the Scientific and Technical Council to recommend generalization of the systems approach to all research divisions within the institute.

The Systems Department within the ISRA organization

The Systems Department was set up in April 1982 as the Research Department on Production Systems and Technology Transfer, in the same way the research departments on crop production, forestry production, etc. were set up. At a later stage, in order to give a clearer definition of its activities, the Department was renamed the Research Department on Agrarian Systems and Applied Economics (DRASAE). Its mission was to act as a kind of training-school because a new methodological approach to research, e.g., the research systems approach, was implemented in Senegal. A main feature of this new approach was that it involved the different production sectors, taking into account not only the specific issues of each sector, but also the interactions between different sectors and between the sectors and their environment.

The Scientific and Technical Council's recommendation to all research departments to generalize the implementation of such an approach has in fact a threefold significance. First, it confirms that the Systems Department successfully accomplished its training-school mission. This could also explain its dissolution as a specialized unit. Second, it expresses the clear need for an overall research approach that is more in harmony with its environment. Finally, the recommendation reflects recognition of the feasibility of systems approaches, despite the problems that may sometimes occur.

Until 1983, most efforts were put into the restructuring and organization of the Department, in particular the installation of research teams in three regional ISRA centers: the Djibelor (Ziguinchor region), Saint-Louis, and Kaolack centers. As a next step, scientific activities were identified and defined by the regional teams, with the help of researchers from the Research/Support Division.

A major step forward has been made with the event of an internal workshop specifically devoted to the conception and implementation of the programs on production systems. This workshop, which was attended by all the teams, made it possible to:

- define the guidelines of the systems approach;
- present the major stages in the scientific approach of the department;
- make suggestions with regard to the links that should be installed between systems research and thematic research, as well as between research and development.

Finally, the responsibilities of the Central Group for Analysis were defined during this workshop.

Although during the implementation phase of the programs, the Department may have faced difficulties, this did not prevent the regional teams from gradually refining their approach and obtaining the results that brought the Department recognition from external evaluators and encouragement from the Scientific and Technical Council of ISRA.

Possibilities for the application of the systems approach in developing countries

As explained earlier, in most developing countries, the agricultural production systems are multisectoral. This makes the systems approach the most appropriate for funding ways to make better use of production potentialities and their complementarities.

The balance that could result from such an approach should not be limited to production only. It should also take into account the ever growing problem of damage caused to the environment, and try to find the balance between the demand for increasing productivity and the need to preserve the environment. In other words, for developing countries it is not possible to put into practice the slogan "nourished in harmony with the environment", without a broader view of the problem. The systems research approach gives them this opportunity.

There is absolutely no doubt that the implementation of research systems in agriculture has its advantages. It allows the followers of agricultural research to:
■ better understand the production constraints in relation to the environment;
■ better understand the research objectives in the solving of problems that were encountered and identified with the communities concerned;
■ put in place new technologies or alternatives to existing technologies that are in better harmony with agricultural production potentialities and the particular environment.

In view of these advantages, introduction of systems approaches to research in developing countries seems desirable. However, it has to be remembered that most of these countries do not have sufficient funds, material, nor manpower to conduct an approach of this kind effectively. This is in addition to the institutional problems that research organizations face in some of the countries.

These constraints mean that, in order to establish an effective systems approach to research, these countries depend very heavily on donor funding and support. As far as the developing countries are concerned, their lack of means would be an obstructing factor in the implementation of the systems approach.

232

Conclusion

Given the results, one could say that the implementation of a systems approach in Senegal has been successful. However, to maintain the standard of work that has been accomplished so far, there are problems that still remain to be solved. The need to solve them has become even more crucial since the dissolution of DRSAEA, the special unit designed for this purpose.

Acronyms

DRSAEA   Department for Research on Agrarian and Agricultural Economics
ISRA   Institut Sénégalais de Recherches Agricoles

The SANREM CRSP: a framework for integration of systems analysis methods in a sustainable agriculture and natural resource management research and development agenda

W.L. HARGROVE[1], J.W. BONNER[2], E.T. KANEMASU[3], C L. NEELY[1] and W. DAR[4]

[1] SANREM CRSP, University of Georgia, College of Agricultural & Environmental Sciences, Georgia Station, Griffin, Georgia 30223-1797, USA
[2] USAID Washington DC, USA
[3] University of Georgia
[4] Bureau of Agricultural Research of the Department of Agriculture, Diliman, Quezon City, Philippines

Key words GIS, interdisciplinary research, natural resource management, participatory research, Philippines, simulation methods, sustainable agriculture

Abstract

The mission of the USAID-funded Sustainable Agriculture and Natural Resource Management Collaborative Research Support Program (SANREM CRSP) is to implement a comprehensive, farmer-participatory, interdisciplinary research, training, and information exchange program that will elucidate and establish the principles of sustainable agriculture and NRM on a landscape scale

Systems analysis and modelling can be useful tools in analyzing and integrating research results from many different disciplines, identifying research gaps, and recommending further experimentation Two other features of systems analysis are repeatability across space and repeatability over time The judicious use of simulation models will permit predictions that we can use to evaluate practices that are most likely to be sustainable GIS will be used to store data from various sources with one common descriptor, that of geographic location

The SANREM activities in the Philippines are described as an example of how systems analysis is integrated with farmer participation and action and experimental work The goal of simulation modelling is to determine how human activities affect soil and water quality and quantity and to provide guidance to field-oriented programs in the development of effective and sustainable land-management practices

Through the participatory identification of constraints and research questions, the research agenda is "demand driven", resulting in appropriate and more readily adoptable solutions than traditional "top-down" approaches The results of systems analysis can be used to aid communities and national programs to identify and test alternative strategies for sustainable agriculture and NRM

The framework of the SANREM CRSP

Background and general approach

Our concept of the landscape is illustrated in figure 1. The approach being used is to describe, as part of the ecology of a landscape, the complex processes within and between the individual ecosystems of a toposequence which transect two or more agroecological zones. This includes social, physical, and biological dimensions of ecosystems. Improved management techniques are being designed with participation by farmers and their appropriateness is evaluated by farmers and other users of natural

233

P Goldsworthy and F W T Penning de Vries (eds), Opportunities, use, and transfer of systems research methods in agriculture to developing countries, 233 - 246
© 1994 Kluwer Academic Publishers

234

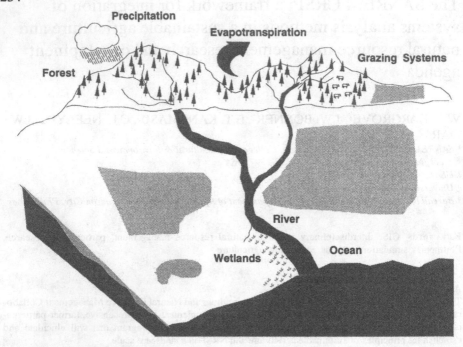

Figure 1. Conceptual model of a landscape

resources in terms of agricultural, environmental, economic, and social sustainability. This farmer-participatory approach has been described by Rhoades (figure 2). Thus, landscape/lifescape analysis and participation of all those concerned form the cornerstones of our approach.

The approach also includes a mix of experimental and systems analysis techniques focussed on a "demand-driven" research agenda, aimed at developing, testing, and implementing improved strategies for sustainable agriculture and natural resource management (NRM). The wide applicability of these principles and methodologies in fragile environments will be demonstrated. Through training, institutional strengthening, and networking, local and regional contributions to agricultural sustainability and improved NRM will be enhanced.

We are focussing on two broad climatic zones: the humid tropics and the seasonally-arid tropics. Both of these climatic zones have characteristics that make them of interest in a study of sustainable agriculture and NRM. In mountainous regions of the humid tropics, these characteristics include:

- reserves of plant and animal genetic diversity;
- centers of both cultural and biological diversity;
- high rates of soil loss and sedimentation;

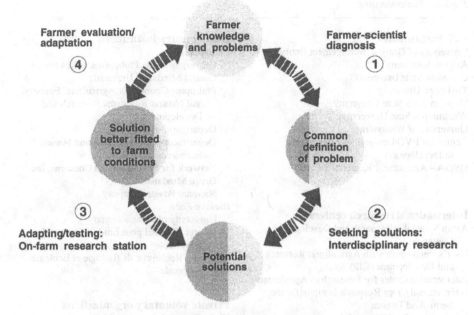

Figure 2. The farmer-back-to-farmer model assumes that the final decision on the appropriateness of a technology rests with the people who will ultimately use the technology, not with the scientists. It involves four major activities, each with a goal. The shaded areas in the circles indicate an increasing understanding of the technological problem as research progress. Note that the process can recycle.

- significant downstream impacts (pesticide contamination, salinization, siltation, destruction of coastal resources and living aquatic resources);
- zones of human migration; and
- significant forest resources.

In the seasonally arid tropics, the features of special interest include:
- desertification;
- high spatial variability;
- human and livestock populations that generally exceed the carrying capacity of the land;
- food supplies that are highly dependent on the vagaries of rainfall and tropical storms; and
- critical vulnerability to famine.

Currently, we are implementing our program at two sites: one in the Philippines and one in Burkina Faso. For each of these experimental sites, transects or watersheds have been identified in which a sequence of these agroecological zones are being studied. We are planning additional activities in Honduras, Ecuador, and Morocco.

We have built a consortium (table 1) that has the breadth of disciplines and expertise and the participation of a broad array of agencies, universities, and service groups with diverse interests but one common goal: the development of a new

Table 1 The consortium

U.S. institutions	Host-country institutions
University of Georgia, Management Entity	*Philippines*
Auburn University	University of the Philippines at Los Baños
Colorado State University	Central Mindanao University
Tuskegee University	Philippine Council for Agriculture, Forestry,
Virginia Tech State University	and Natural Resources Research and
Washington State University	Development
University of Wisconsin	Department of Agriculture
Center for PVO/University Collaboration	Department of Environment and Natural
in Development	Resources
USDA - Agricultural Research Service	Network for Environmental Concerns, Inc
	Green Mindanao
	National Power Company
	Burkina Faso
International research centers	University of Ouagadougou
Asian Vegetable Research and Development	Institut National pour Etude et Recherche
Center	d'Agricole
User's Perspective with Agricultural Research	Institut Recherche de Biologie et Ecologie
and Development (CIP)	Tropicale
International Center for Research in Agroforestry	
International Crops Research Institute for the	
Semi-Arid Tropics	**Private voluntary organizations**
International Rice Research Institute	Heifer Project International (all sites)
	Christian Children's Fund (Philippines)
	Plan International (Burkina Faso)
	CARE (Honduras and Ecuador)

paradigm for agriculture and NRM that includes an integrative, interdisciplinary approach to research, training, and farmer adoption activities, centered around the end-users and an improved understanding of the landscape. It is anticipated that this program will serve as a new and lasting model for rural development activities.

Research goals and methodologies
The following is a broad list of goals and a brief description of the methodological approach, particularly the role of systems analysis techniques, to reaching these goals. The goals are broadly applicable to all sites.

1. Identify and describe the problems relating to sustainability using farmers' and other end users' goals and perspectives.
To utilize the vast storehouse of end user knowledge and skills in problem-solving, we have a diverse toolkit of innovative, participatory methods for working with resource-poor farmers and other end-users. By using community meetings, end-user group interviews, key informants, group treks, village preference rankings and other methods, the end users themselves have the opportunity to help design the research agenda from the earliest stages. Scientists (expatriate and local), public officials, community leaders, PVO/NGO representatives, extension agents, and end-users

regularly and systematically come together in the watershed context to identify goals, problems, and solutions to problems. Instead of focusing on individual farmers in isolation, diverse actors in the food and land-use systems (e.g., farmers, fishermen, processors, traders, consumers, and resource managers) are involved in this strategy. Precisely because a landscape is a mosaic of interacting ecosystems, achieving sustainability on this scale requires that all end-user constituencies represented within the landscape collectively define broad goals and seek solutions to the problems that are common to all; otherwise, the divergent and conflicting interests of separate user groups are likely to undermine the sustainability of the landscape as a unit.

2. Identify and collate existing biophysical and socioeconomic baseline data, including relevant indigenous knowledge and determine the need for additional baseline data collection.
Our approach to this process involves conventional systems analysis, simulation modelling and geographic information systems (GIS) (figure 3) and incorporates indigenous knowledge gathered in formal and informal meetings with end-users. The collective understanding of the landscape units under study will be refined continually as SANREM CRSP develops and matures. For the purpose of information exchange, meetings and interaction of this sort are scheduled on a regular and systematic basis, adding new information and fresh perspective to the collective knowledge base, weaving the threads of indigenous knowledge throughout.

3. Collect and integrate additional physical, biological, and socioeconomic baseline data.
Baseline information gaps, identified by integrating existing knowledge bases, could limit our understanding of the ecosystem and their component processes and interactions. These gaps are being filled with monitoring and/or survey data. For example, biotic surveys (e.g., natural vegetation, aquatic biota) are being used to establish or supplement measurements of biodiversity, and serve as an index of environmental conditions prior to system manipulation, and social surveys are being conducted to assess the role of women in agricultural production and NRM.

4. Recognize and understand the cultural, socioeconomic, political, and institutional framework.
All NRM and agricultural-production decisions are made within an existing social, political, economical, and institutional framework. It is imperative to recognize, evaluate, and understand this framework as an early step toward identifying and overcoming the constraints to sustainability. An important tool in the analysis of the human structural framework is modelling. Through simulation modelling, incorporating both qualitative and quantitative information, multi-level constraints can be accounted for and possible alternative management practices can be evaluated in a comprehensive manner.

238

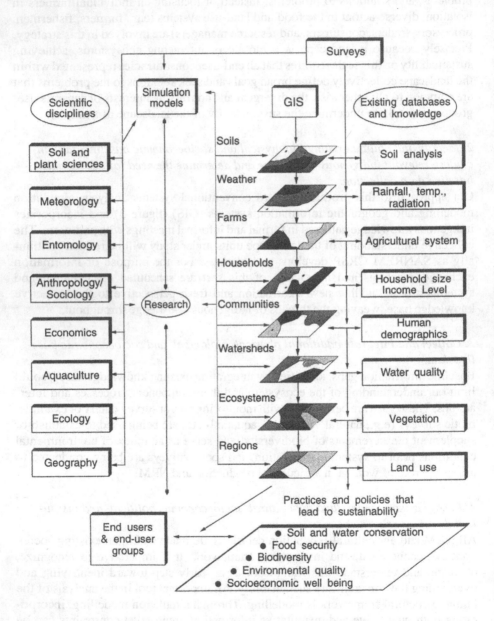

Figure 3. Conceptual model of the roles of interdisciplinary research, simulation in the development o
practices and policies that lead to sustainability

5. Improve the understanding of important ecosystem processes and critical ecosystem linkages in a landscape setting.

Critical physical, chemical, biological, socioeconomic, and cultural landscape linkages are being identified employing aspects of agroecosystem analysis. Prime goals of this project are improved understanding of: 1) the hydrologic cycle at each site, 2) how ecosystems are linked through the hydrologic cycle, and 3) how agricultural practices impact other ecosystems in the landscape. In particular, it is necessary to quantify rainfall, irrigation, runoff, run on, evapotranspiration, infiltration, deep percolation, soil water storage, and ponding on sites of contrasting land use. A thorough knowledge and understanding of nutrient cycles and their component processes are essential also, as nutrients are often the source of negative off-site impacts as well as being key to long-term soil fertility and productivity. The tools for accomplishing this include not only experimental field studies of varying scale (plots, fields, watersheds), but also simulation modelling and GIS. Simulation models provide a framework for understanding complex processes; GIS organizes existing information and facilitates its integration with newly acquired data.

6. Identify quantifiable "indicators of sustainability", measurable parameters that will indicate improvements in sustainability.

Indicators of sustainability can be grouped into two broad categories: biophysical indicators and socioeconomic indicators. The primary on-farm biophysical indicators of sustainability derive from measurements of edaphic and biotic factors that influence, or serve as indices of, the productive capacity of the farm. With respect to social indicators of sustainability, key demographic variables are being measured, and changes in demographic trends that result from the adoption of key interventions will be analyzed.

With respect to economic indicators, there are two principal types of economic indicators of sustainability. First, economics can provide an integrated indicator that combines diverse changes in components related to sustainability. One version of this descriptive economic indicator would measure from year to year, or project to future years, the combined annual value of resource-based goods and services. Second, economics provides methods for comparing the long-run benefits of increased sustainability to the more immediate costs of implementing the alternative measures and strategies required for their improvement. Models will be key to the testing and evaluation of such indicators.

7. Develop and evaluate viable management strategies for achieving sustainability in agricultural and natural ecosystems.

To address this goal, we are employing the farmer-first methodology combined with a cadre of expertise in innovative technologies such as integrated resource management, integrated pest management, and agroforestry and integrated social forestry systems. The goals of integrated resource management are the restoration of productivity of degraded lands, maintenance of long-term productivity, and conservation and preservation of natural biodiversity. At the core of integrated resource management is improved soil, water, and nutrient management, which can lead to an alleviation of the pressure to expand onto marginal lands. Because pesticides can be a source of human health risk and loss of beneficial biological resources, management strategies that reduce dependence on pesticides are needed also. These strategies include the coordinated use of multiple tactics, including biological control, to assure stable crop and animal production and maintain, pest damage below the economic injury level. Agroforestry strategies will also be evaluated. Some of the benefits of agroforestry systems include: soil stabilization; fodder production; supply of mulch for improving soil; improving microclimate; recycling of nutrients; and production of wood for firewood, pulpwood, charcoal, and boards for local construction.

8. Promote education, training, and information exchange in sustainability issues.
The emphasis in training is on community-based education, farmer-back-to-farmer information exchange, and formal training. A major emphasis in community-based education is the "training of trainers". Formal training is at the undergraduate, graduate, and visiting scientist levels.

Integration of systems analysis

Systems analysis and simulation modelling
Achievement of the research goals will require an interdisciplinary approach, utilizing a range of research tools. Systems analysis and modelling can be useful tools in analyzing and integrating research results for many different disciplines, identifying research gaps, and recommending further experimentation (figure 4). Two other features of systems analysis and models also stand out: repeatability across space and repeatability over time.

The concept of sustainable agricultural production incorporates problems with long timeframes, uncertain and changing environments, and variable social and economic settings. The judicious use of simulation models will permit predictions which we can use to evaluate practices which are most likely to be sustainable.

Once models have been validated and tested they will be applied at different levels to study issues related to sustainability in agriculture and the interaction with both the biophysical and social environment (figure 3). Models will be used:
1. at the field level to study the interaction between the crop and its climatic and edaphic environment and to determine optimum crop management practices;
2. at the farm level to study the interaction with the various farm enterprises;
3. at the household level to study issues related to the economics and net-income of the households;
4. at the community level to study the traditional farming practices, indigenous knowledge, and the effect of modified farming practices on the economics and social well-being of the entire community;
5. at the watershed level to study the effect of current and modified farming practices on soil erosion and water pollution;

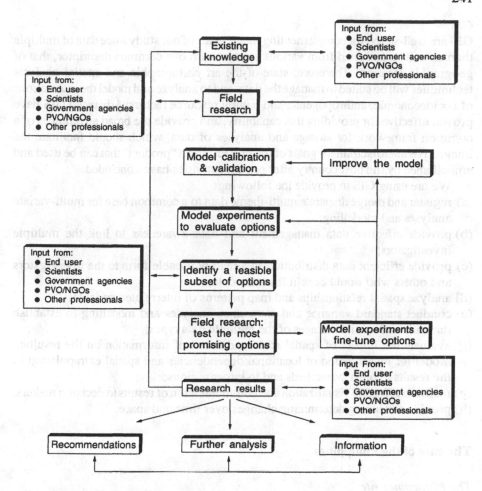

Figure 4. Modelling and the research process

6. at the ecosystem level to study the interaction of farming practices with the environment;
7. at the regional level to integrate the individual farm responses and to study regional production patterns; and
8. at the national level to provide information on long-term policies and practices for policy- and decision makers.

At each level, these models will be used by interdisciplinary research teams as a tool to link with current data bases, to analyze existing knowledge and information, to determine knowledge gaps, to screen research options, and to further test the most promising results from experiments, trials, and surveys.

GIS

GIS are well-suited for complementing major goals of our study since data of multiple themes will be collected from various sources with one common descriptor, that of geographic location. Therefore, state-of-the-art cartographic and spatial analysis techniques will be suited to manage the data, and to analyze and model the interactions of socioeconomic, anthropogenic, and natural resource factors. Advanced GIS have proven effective for providing this capability, and provide the broad advantages of a common framework for storage and analyses of data, which should facilitate the integrative/interdisciplinary goals of the project, and a "product" that can be used and embellished by the host country after project activities have concluded.

We are using GIS to provide the following:

(a) register and merge disparate multi-theme data to a common base for multi-variate analyses and modelling;
(b) provide effective data management and data awareness to link the multiple investigators;
(c) provide efficient data distribution in a readily useable form to the investigators and others who would benefit from the data;
(d) analyze spatial relationships and map patterns of interrelated variables;
(e) conduct standard variance and covariance analyses and modelling to establish thematic and spatial linkages of the multiple data types;
(f) evaluate the impact of spatial scale and detail of information on the resultant model for investigation of locational dependencies and spatial extrapolation of the results to other watersheds and to larger regions;
(g) facilitate effective visualization and communication of results to decision makers;
(h) provide a meaning determining changes over time and space.

The case of the Philippines

The Philippines site

The watershed selected for research and development in the Philippines is the Manupali River which drains approximately 80,000 hectares and is a tributary of the Pulangi River Basin. The Manupali watershed is located on north-central Mindanao (longitude 125"W, latitude 8"N), 20 kilometers south of Malaybalay and 85 kilometers south of Cagayan De Oro City (see figure 5).

The Manupali watershed is characterized by the ecosequence outlined in figure 6. The watershed has moderate to steep slopes, with greater than half of the land area having a slope of 15 percent, fragile soils and severe erosion. Soils in the area are well-drained, highly erodible clays of low pH. A rapid increase in small-scale farms with insecure land tenure and a necessity to provide food for the family, has resulted in unsustainable agricultural practices on the uplands. Farm sizes are typically 0.25 to 19.5 hectares with an average size of 3.0 hectares. Farm sizes less than 3.2 hectares are typically tenant or leasehold farms. The Manupali watershed has a diversity of agricultural crops. There are large tracts of irrigated wetland rice, sugarcane, and

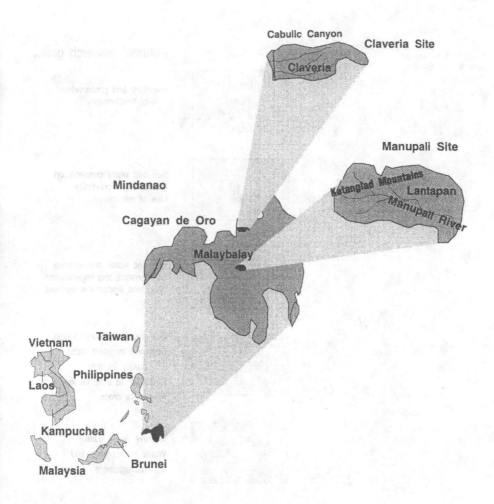

Figure 5. Location of research sites in the Philippines

pineapple which utilize high external inputs. In addition, production of lettuce, potato, cassava, and tomato, and some small vegetable gardens are present in the uplands. The upper reaches of the Manupali watershed is occupied by the Katanglad Philippine and National Park, a primary forest bio-preserve.

The Philippines is fertile ground for encouraging an integrative approach to the research of SANREM. Currently there are large numbers of non-governmental organizations (NGOs) and national programs working to improve the ecological and agricultural sustainability in the Philippines. Although the necessary expertise and willingness to address these problems are present in the Philippines, current efforts are fragmented. Coordination of various groups is poor and typically the farmer or

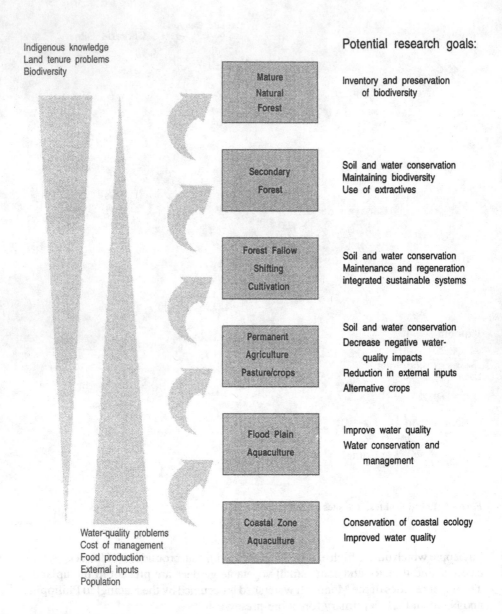

Indigenous knowledge
Land tenure problems
Biodiversity

Potential research goals:

Mature Natural Forest

Inventory and preservation
of biodiversity

Secondary Forest

Soil and water conservation
Maintaining biodiversity
Use of extractives

Forest Fallow Shifting Cultivation

Soil and water conservation
Maintenance and regeneration
integrated sustainable systems

Permanent Agriculture Pasture/crops

Soil and water conservation
Decrease negative water-
quality impacts
Reduction in external inputs
Alternative crops

Flood Plain Aquaculture

Improve water quality
Water conservation and
management

Coastal Zone Aquaculture

Conservation of coastal ecology
Improved water quality

Water-quality problems
Cost of management
Food production
External inputs
Population

Figure 6. Characteristics of and potential research goals for the typical ecosystems sequence in the humid tropics

local community voice is not heard. The landscape approach to sustainability of the SANREM CRSP seeks to bring the efforts of many groups involved in sustainable agriculture and NRM together in a coordinated manner. Within the Philippines, there exists enormous potential for researchers and development workers to break the old paradigm of reductionist thinking, and begin to take part in the mosaic of an integrative, people-based system.

The Philippine Council for Agriculture and Resources Research and Development (PCARRD) coordinates agricultural and forestry research at the national level. In addition to government agencies such as the Department of Agriculture, Department of Environment and Natural Resources, University of the Philippines at Los Baños, and Central Mindanao University, several NGOs, both local and international, are active in the watershed. The SANREM CRSP has developed strong linkages between these sectors through an explicitly integrative team research approach. By targeting the research to specific watersheds and institutionalizing a coordinated research design and implementation, the project will evolve and test a unique model suited to the efficient conduct of systems research at the landscape level.

Integration of systems analysis into the Philippines agenda
The SANREM site and activities in the Philippines will serve as a good example of how systems analysis is integrated into a demand-driven research agenda. One of the constraints to sustainability identified through the participatory landscape/lifescape analysis is that streams are becoming more "polluted" and carrying higher sediment loads, while stream levels are decreasing. Soil erosion, water quality, and water quantity are principal concerns of the community.

During our workshop held near the site, the following research question was identified relevant to this constraint: How do outputs of sediment, water, and chemicals (nutrients and pesticides) relate to land-use systems (natural, production, fallow) by agroecozone, topography, and soil type? Our approach to this question is one in which farmer participation and action, experimental work, and simulation modelling will be integrated. The goal with respect to simulation modelling is to use simulation models of soil and water processes coupled with GIS to determine how human activities impact soil and water quality and quantity across the landscape and to provide guidance to field-oriented programs in the development of effective and sustainable land-management practices.

The planned activities include building a database of thematic layers (soils, land use, hydrology, digital terrain map) which will be the inputs to the HUMUS model (Hydrologic Unit Model for the US). This model is under development and testing at the Texas A&M University Research Center at Temple, Texas, under the leadership of Drs. Alan Jones and Paul Dyke. A field survey of the land-use practices will also be conducted as input to the HUMUS model. The HUMUS model will be tested and adapted to local conditions of the Manupali watershed, and then used to evaluate the effects of present land management on water quality and quantity. These modelling activities will be coupled with examining the communities' perceptions of practices which critically affect soil and water quality and using these perceptions and the results of the modelling activities to aid in the communities' development of alternative land-management strategies.

This work will be done collaboratively with the local community, the University of the Philippines at Los Baños, the Network for Environmental Concerns, Inc. (a local NGO), IRRI, and Texas A&M University and the University of Georgia. It should provide a useful model for integrating systems analysis, experimental research, and local people action to solve NRM problems.

Summary and conclusion

The SANREM CRSP provides a framework for integrating systems analysis in a participatory research and development project to solve problems in sustainable agriculture and NRM. Through the participatory identification of constraints and research questions, the research agenda is "demand driven", resulting in appropriate and more readily adoptable solutions than traditional "top-down" approaches. Simulation modelling and GIS are appropriate tools to address the identified constraints to sustainability in a cost-effective manner. The results of systems analysis can be used to aid communities and national programs to identify and test alternative strategies for sustainable agriculture and NRM and thus can provide a platform for initiating needed policy changes.

Acronyms

GIS   geographic information systems
HUMUS   Hydrologic Unit Model for the US
NGO   non-governmental organization
NRM   natural resource management
PCARRD   Philippine Council for Agriculture and Resources Research and Development
SANREM CRSP Sustainable Agriculture and Natural Resource Management Collaborative Research Support Program
USAID   United States Agency for International Development

Discussion on Section D: Experience at a national level of use of systems analysis methods

From the experience of some of the NARS represented at this meeting it is evident that resistance to institutional change can be a formidable obstacle to the introduction and adoption of systems approaches in research. Where such approaches have been introduced it is often as a result of donor influence and support, which makes their continued use uncertain when this support comes to an end. As a result, systems approaches remain marginalized in many national systems. The discussion on Section C has already referred to some of the policy, institutional, and funding constraints that contribute to this situation. The papers on the use of systems methods in China and India gave rise to additional comment on these issues, as reported here.

Institutional change

In the discussion on Section C, it was noted that the short-term perspective of uncertain funding is inconsistent with the requirements of a systems approach to research, in which agricultural productivity has to be linked to the overall productivity of natural resources. It is one of the main reasons why insufficient attention is generally given to the institutional changes that the broader research agenda and the establishment of systems approaches in NARS imply.

The ability to forge interinstitutional linkages and joint programs of research will depend to some extent on identifying common problems that require a systems approach, and on clearly defining the scope of the systems approaches proposed. Such programs are likely to consist of problem-solving interdisciplinary teams.

In general, NARS are not organized to accommodate effective interdisciplinary research. Commodity and disciplinary research is still the dominant mode of organization. This is not the best formula for tackling problems concerning NRM, and some reorganization is usually required. Like successful farming systems research, NRM has to be fully integrated with other research activities in the NARS, or it will not become a part of the institution. Measures have to be taken to guard against NRM becoming a separate activity, organizationally apart within the NARS, focussing solely on systems methods. Care is also needed to avoid giving any impression of models being developed for their own sake.

Too often in the past, failure to make the institutional changes that are needed resulted in scientists having become proficient in systems methods within the life of a project, but finding themselves without funding or support when the project ceases. There are very few examples of NARS that have made the institutional changes or that have committed financial resources needed to develop and maintain systems research programs unaided.

Apart from the institutional changes required, the wider research agenda calls for more investment in problem diagnosis and research planning. The analysis conducted by SARP in India is a good example of what is required (Sankaran); it enabled research needs to be clearly identified, and it highlighted the need for training and data exchange. Thorough planning is particularly important if the research involves different institutes, to identify common problems and objectives that require systems approaches. The level and scale of the system that is the object of a proposed study has to be clearly defined (see discussion on Section B). This may mean that systems approaches cost more time at the research planning stage than traditional commodity or disciplinary approaches, but the additional investment is often crucial to ensuring the successful implementation and eventual outcome of the research.

Standard terminology
Interdisciplinary discussion on systems methods would be helped if a less ambiguous vocabulary was developed for the subject of systems research. Interdisciplinary teams in IARCs and NARS need a common terminology.

One of the recommendations of the Rome workshop was that a glossary of common technical terms to describe weather and climate, soils, landforms, and vegetation should be prepared to facilitate the exchange of data about agricultural environments among centers and between centers and cooperators outside the CGIAR system. The task was undertaken by a working group of centers' staff together with other specialists.

This took care of some of the needs for standard terminology at the level of database management. The discussion at this workshop indicated that the work now needs to be extended to a more conceptual level. Concepts rooted in one discipline may translate poorly into others. Ways are needed to define more clearly how the terms system, level, input, and output are being used, and how system responses are measured.

In response to this need a proposal was made by one of the working groups for a classification of systems methods and approaches, based on the following characteristics:

- *The hierarchical level* at which the method operates. This could be global, macro (national), (eco)regional or provincial, farming system, or a farming system component (plant, animal, soil).
- *The objective* of the system method. The following categories of objectives were proposed: improved understanding, improved communication, development and testing of new concepts, decision support, and forecasting. Obviously, there is some overlap between the objectives, and at this point they are intended for illustration, not as a definitive list.
- *The client groups* of the method/model. Principal clients would be: agricultural producers, researchers, and policymakers.

It was observed that some of these categories are linked, and that therefore some combinations of the proposed categories would be more likely to occur than others (e.g., at a national level, models are likely to serve a communication function and be directed at policymakers).

Some additional categories were suggested: process versus integrative models, expert systems, deterministic versus stochastic models, and optimization (normative) versus empirical (positive) approaches. The discussion brought out the enormous variability in systems methods and the impracticality of discussing them all.

Strengthening research-policy linkages

The weak linkage between research and policy levels of decision making (see also earlier discussion on Section C) is a major factor contributing to the marginalization of systems approaches, and to the difficulty of institutionalizing them. However, the experience of China (Cheng Xu) shows that systems models can be employed successfully for land use and research planning at a regional level. One result of the success is that there is now much stronger support from China's policymakers for systems-based research, and it has stimulated a demand for more training in the use of systems approaches.

This experience reinforces the views reported already that policymakers should be made aware that systems methods can improve the efficiency of research by making it possible to explore a wider range of scenarios that will satisfy environmental and sustainability objectives while also meeting agricultural production and development goals; options that are often missed in purely commodity- and production-type research.

By providing quantitative information on the technical and perhaps economic impacts of the different courses of action available, systems methods also provide valuable support to the policy decision-making process.

It was suggested that one way to gain greater policy support might be to demonstrate the use of systems research on well-defined benchmark sites, representative of key ecological zones, and use the results to illustrate some of the options for the zone as a whole. Clearly, by whatever means they choose, NARS leaders and researchers must give special attention to making policymakers more aware of the role of systems approaches in the broader agricultural research agenda.

Adapting models to NARS needs

At present, NARS often lack the information technology to use the available models effectively, though as the papers in this and previous sections of this volume have shown, their capacity to do so is developing rapidly.

More attention and support needs to be given to adapting systems models and methods to specific users' needs. NARS representatives reported that some of the biophysical models available are unsuitable for their needs, for two major reasons. First, the data requirements of the models limit their utility in many NARS where the datasets available are sparse. However, this probably reflects the need for establishing adequate natural resource databases rather than inherent inadequacies of the systems models. Second, most of the models were designed originally for applications in a developed country, whereas the objectives of users in developed and developing environments are often different. Differences in biophysical and socioeconomic conditions may lead to irrelevant data-collection efforts on the one hand and unrealistic behavioral and biological assumptions on the other. Consequently, the models available will not always provide answers to questions that are being asked by the

NARS. Obviously, such discrepancies reduce the utility of the models, but perhaps even more important, they damage the credibility of the systems scientists in the countries where the models are being used.

There was a plea from NARS representatives that those who develop models should validate them under a wider range of conditions; it was felt that too much of the responsibility for doing this is at present left to users. It should be a matter for concern that this echoes a view expressed eight years ago at an earlier workshop in Rome, that validation must be taken more seriously than in the past.

NARS representatives also requested better documentation of the models, and that it be made easier for users to reset the model parameters to correspond with the situations in which they are being applied. However, as Harris et al. emphasize in their paper in the next section that there are real dangers unless those applying a model have an adequate knowledge of the principles on which it is structured.

These user demands might appear exacting to the developers of models, but among the users, expectations are generally realistic. There was a realization at this meeting that there are no shortcuts to the introduction of more quantitative systems methods by NARS, and that the data requirements of models will always be demanding.

The workshop was told that models were sometimes unnecessarily complex and that this caused problems. The comment was directed at the largely mechanical assembly of models from lower levels in a system hierarchy into more comprehensive, higher-order models. It was felt that apart from increasing the complexity, such combinations were also likely to add to imprecision. A danger then is that the focus of attention on modelled components may well be at the cost of non-modelled components that may be equally relevant.

Recommendations

Institutional change
The workshop recommended that research managers in IARCs and NARS exercise caution to ensure that the work of groups of systems scientists is integrated into the programs of the remainder of the institute in which they are based. The need to guard against any tendency to become isolated is particularly important in institutional environments without an established culture of systems research.

The workshop recognizes that apart from the institutional changes required, the increased complexity of NRM research calls for more investment in diagnosis and research planning.

Strengthen research-policy links
The workshop recommends that IARCs and NARS consider a proposal to establish well-defined benchmark sites and data sets, which represent key agroecological zones, as a way to show practical applications of systems research, and so gain greater policy support. The sites would not be "dry-labs", but key locations within a broader

agroecological context, which would allow individual NARS to contribute to a wider regional objective while focussing on agroecological systems and issues of direct relevance to national interests. Further reference to this idea is made in Section F, in relation to the ecoregional work of IARCs.

Adapting models to NARS needs
The workshop wishes to draw attention to the plea from NARS that those who develop models should validate them under a wider range of conditions. NARS have also requested better documentation of the models.

SECTION E

Training in the use of systems approaches

Requirements for systems research in agricultural and environmental sciences

GAJENDRA SINGH[1], B.K. PATHAK[1] and F.W.T. PENNING DE VRIES[2]
[1] Asian Institute of Technology (AIT), G P O Box 2754, Bangkok 10501, Thailand
[2] DLO Research Institute for Agrobiology and Soil Fertility, P O Box 14, 6700 AA Wageningen, The Netherlands

Key words agriculture, decision support, developing countries, multidisciplinarity, rural development, simulation, systems research, training, university

Abstract

With the increasing complexities of agricultural and environmental issues, there is a growing consensus that systems approaches are essential to understand and handle these issues properly Although systems research has attracted attention in developed countries, developing countries have shown little interest in it so far This paper identifies five key requirements for carrying out systems research skilled users and developers of system research tools, adequate research organization, basic data about agroecosystems, access to computer technology, and opportunities for multidisciplinary cooperation Researchers as well as users need to be trained in the contemporary technologies and methods Asia in particular lacks properly trained scientists Emerging worldwide electronic networking provides additional stimuli for the systems-research community, as data and programs can be shared much more easily

With a primary task in regional training in technology for agriculture, environment and rural development, the Asian Institute of Technology can help answer the demand for systems-research training at pre- and post-doctoral levels by adapting and implementing training programs

Developments in the use of systems approaches

In an effort to feed their growing populations in an equitable and environmentally safe way, developing nations are changing their agriculture dramatically. However, changes in one part of a system can dramatically affect other parts—human, economic and environmental. If the system fails, the results in human terms may be catastrophic. The central idea behind systems research in agriculture, therefore, should be to foster an awareness of such interactions and to develop the skills required to quantify these reactions in a clear manner.

Systems theory is used in almost every field of science and technology, and its use is growing with the increasing demand for a more exact understanding of the real world. The application of the systems approach in agriculture has grown in parallel with its application in other areas. The different approaches to systems research that have been discussed over the years may be categorized as follows:

- surveys and analyses that are performed on a logical and systematic basis, leading to a descriptive outline of the system;
- modelling, where essential processes are represented by mathematical equations and where the model can "behave" as the real system.

255

The first approach is characterized by key words such as "farmer oriented" and "bottom up", and is commonly referred to as farming systems research and development (FSR&D). Judging from the impressive amount of literature that has been published on the subject since the mid-1970s, few research approaches in the agricultural sector have been so popular among scientists, in particular in the developing world. It is becoming increasingly clear, however, that FSR&D has not always been able to live up to the high expectations (Stroosnijder and Van Rheenen 1993).

The second approach is "science oriented", and has produced many mathematical models of agricultural production systems. Models can help understand the structure and functioning of a system and can integrate different experiences with the system. Mathematical models can also help extrapolate our knowledge to locations and conditions where experiments were either difficult or impossible to carry out. Using this approach is most compelling when experimentation is expensive and time consuming and where it provides only local answers (Dent 1993). Dynamic simulation modelling and linear programming have been important tools for many years. Both still suffer from serious limitations, however. Simulation models cannot include all the factors that play a role in reality, and they often comprise empirical data that need to be calibrated for site-specificity (Jones et al. 1987). Mathematical programming, in particular linear programming with a single objective, has also been often used for optimizing. But in reality, decision makers must handle several, often conflicting objectives. To enhance the applicability of the models, these shortcomings should be rectified in future research on agricultural systems. In addition, an unequal distribution of research funds may have slowed the development of understanding of agricultural systems (Dent 1993). In spite of that, systems research has already acquired an important role, which can be expected to grow in future.

A system cannot be properly understood through ad hoc studies of its separate elements. The interrelationships among the components and the instability of the total environment in which a system operates create a whole that is more complex than the sum of its individual parts (Dent and Blackie 1979; Doyle 1990). To understand the complexities of a system, therefore, systems researchers in agriculture, agroecology and land-use planning need to have some knowledge of systems theory, mathematics, statistics, economics, and social science, besides a basic knowledge of the agricultural sciences. This multidisciplinary approach is also essential among the various branches of agricultural science, i.e. agronomy, soil science, agrometeorology, crop protection, and agricultural engineering. Multidisciplinarity is essential, because all critical environmental problems caused by agricultural practices and policies are systems problems—not disciplinary ones.

Models have become very common tools in science. Although collectively, they take many forms and serve different purposes, individually, they are designed for specific purposes, each with its own requirements (Penning de Vries 1983). Some models are designed for training and education, others are primarily research tools, while yet others can be applied to predict and extrapolate. Here, we will discuss the last category in particular.

Proper systems research requires the following (Penning de Vries et al. 1991; Kropff et al. 1994):

- models and methods (the technical tools) that represent the system under study reasonably well and that support the analyses effectively;
- basic data that characterizes the (bio)physical and socioeconomic environment of the locations where the model is to be applied;
- computers to carry out simulations, optimizations and analyses;
- trained staff to specify research objectives, adapt the models and methods, select and inspect the basic data, and interpret the results of the analyses critically;
- an organizational structure that facilitates the functioning of multidisciplinary teams.

Achieving these requirements is difficult. Our theoretical knowledge is far from complete and tools need to be improved. In addition, basic data is generally scarce, while computers are not yet as commonly used in developing countries as in industrialized countries. We regard the shortage of trained staff as the major bottle-neck. Figure 1 shows how these five requirements have to be met to overcome the bottleneck in systems research into agricultural and environmental issues in developing countries. Most scientists in developing countries have been trained in a disciplinary fashion, instructed to extract and solve mono-disciplinary issues from complex problems. Since they are generally not used to working in multidisciplinary

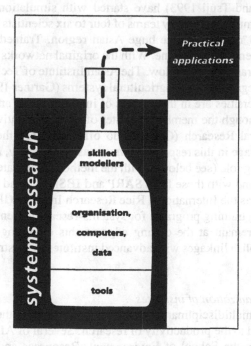

Figure 1. Shortage of scientists skilled in systems research in agricultural and environmental sciences is the bottleneck for the application of systems research to practical problems.

teams, they find it difficult to apply their agricultural knowledge in such teams. In addition, many research institutions in developing countries are organized to operate in a monodisciplinary way, which adds to the problem of multidisciplinary cooperation.

The role of AIT in meeting requirements for systems research

Training

Users need to be fully aware of the major features and pitfalls of planning and analysis methods, including systems modelling and simulation, and of their advantages and their disadvantages, if they are to use such methods effectively. Users also have to interpret the results of the models in the context of the problem, and need to be aware of the underlying assumptions. Without such an awareness, there is a risk that users may either be misled by the apparent precision of the results, or become unduly biased against mathematical methods.

Since there are very few scientists in developing countries with modelling skills, training programs in systems research and crop simulation are badly needed (Penning de Vries et al. 1991). However, few training programs are being provided; we do not know of any Asian university that formally teaches systems research. Two international research networks that have recently been created (SARP, Ten Berge 1993; IBSNAT, Uehara and Tsuji 1993) have started with simulation training courses. Although about 25 multidisciplinary teams of four to six scientists have been formed, this is a very small number for the huge Asian region. Trained teams now teach courses, but much remains to be done. With the original networks having made their mark, national programs should follow. The Asian Institute of Technology (AIT) has started a master's degree course in agricultural systems (Gartner 1993), while several SARP-related universities are in the process of including long and short courses in their curricula. Although the member institutes of the Consultative Group on International Agricultural Research (CGIAR) do offer training, they do not have a comparative advantage in this respect. However, as a university, AIT is well placed to fulfill this training role (see below). With the methods and materials developed in its own programs, and with those from SARP and IBSNAT, and with courses from organizations such as the International Rice Research Institute (IRRI), AIT plans to offer short and long training programs for systems-research scientists and end users of the results. To remain at the cutting edge of this emerging technology, such programs will establish linkages with advanced institutes in Australia, the USA, and Europe.

Changes in the organization of institutes

Problem-oriented, multidisciplinary collaboration can add to the efficiency of research planning and to the productivity of research. Several of AIT's divisions have been integrated into the School of Environment, Resources and Development, in order to deal with agricultural and rural development issues in an interdisciplinary manner. Although AIT has no mandate to restructure organizations, it contributes to change by setting examples for its students. The way in which IRRI was restructured recently to make full use of systems research can also serve as an example (Kropff et al. 1994).

Access to systems-research technology

Users of systems technology include government policy makers, researchers, extension workers, and farmers. All these user groups will need access to computer technology, although users of systems methods usually require less software and hardware than do researchers and those who develop the methods. Although access to computers in developing countries is improving, computers continue to be expensive items in many countries. On the positive side, developing countries that make a late start in introducing computer technology may benefit from the rapid advances that are being made in more developed countries.

Electronic networks could become very valuable for systems research in all countries. Internet, the global network of networks, is gaining popularity in the developing countries and may soon modify the exchange of information in a revolutionary way. Electronic networking has many ramifications for the worldwide academic and research community and hence for agricultural systems researchers. Through an electronic network, it is possible to access databases irrespective of their location, and to transfer almost any type of binary data or text, whether databases, documents or executable programs, from one part of the world to another within minutes. The systems-research community in developing countries should be encouraged to adopt this new way of information exchange. Electronic mailing is another advantage of networking: it is rapid, inexpensive, and reduces the use of paper. Electronic mail could facilitate the setting up of common-interest discussion forums for the systems research and the users community. AIT is willing to initiate a forum for systems research in Asian agriculture and rural development.

Data

AIT has no immediate role in collecting and storing data. Its systems-research training program, however, underlines the importance of gathering minimum sets of weather, soil, crop, pest, sociological, and economic data, thereby stimulating the development of national databases. Using this data in various applications will also encourage the development of relevant databases. Through its Regional Computer Center (RCC) and UNEP's Global Resource Information Database (GRID), AIT can facilitate the linking of national databases to international ones.

Technical tools

In its research programs and as part of its students' programs, AIT develops, modifies, and adapts simulation models, decision support systems, expert systems, and geographic information systems (GIS). These modules are usually targeted to specific problems in the students' home country. AIT's capacity to use models and methods developed in CGIAR-institutes and other advanced institutes is still underdeveloped

(Gartner 1993). Since the range of technical tools for systems research is broad, we will discuss only some of them, and indicate the involvement of AIT:

GIS. Considered a technological breakthrough when it was introduced, a GIS can store large volumes of geographically referenced data, manipulate and analyze the data, and present the data in the form of maps. The attractive display features allow patterns to be recognized and grasped easily, and messages to be conveyed readily. GIS provides researchers with a powerful modelling tool, and it is becoming popular as an exploratory tool for resource planners and managers (FAO 1988). At AIT, GIS has been shown to deal effectively with the large arrays of data that are required for agroecological modelling. Attempts have been made to develop procedures to combine simulation modelling and GIS. Sabbagh et al. (1993) discussed two examples. In the first, GIS was linked to a crop simulation model, a linear programming model, and a ground-water flow model. The objective was to study the long-term impact of agricultural management practices on the ground-water levels. In the second application, GIS was linked to three simulation models for studies of a hydrological basin. Combinations such as these can be used to determine the agricultural potential of a region and to identify major constraints on productivity.

The Sioux Falls Center in USA has carried out a series of studies, using models and GIS, of the agricultural potential of Senegal, by crop and region, in order to estimate the number of people that could be supported by rain-fed agriculture (GRID 1993). Studies such as these provide a valuable perspective on options for future land use. We want to emphasize, however, that individuals with a knowledge of the systems methods used and of the environment in which they are being applied need to examine the results thoroughly, for there are often pitfalls in interpreting them.

Expert Systems. Expert systems are computer programs that embody the expertise (facts, definitions, theories, as well as rules of thumb) of one or more experts in some specialized problem. These systems can be used to retrieve expert knowledge for making useful inferences at a level comparable to a human expert (Barr and Fiegenbaum 1981). The rapidly emerging technology of artificial intelligence for expert systems holds great promise for applications in agricultural systems. Simulation models can be incorporated in expert systems.

Decision Support Systems (DSS). AIT has carried out a number of research studies on DSS. Three of these were aimed at agricultural-production and -processing systems in Asia:

■ *Agricultural machinery and power selection for farms in western India*. Farmers face difficulties in chosing farm equipment needed to produce agricultural commodities both effectively and economically. A decision involves factors such as the weather and its uncertainties, the timeliness of operations, the soil types and their conditions, the types of crops and crop rotations, management practices, labour availability, and fuel supply. It is critical to take the right decision because machinery costs are high. Purchases of machinery are made only very infrequently and are often irrevocable. Based on earlier work by Singh and Gupta

(1981), Butani and Singh (1993) developed a DSS designed to optimize the choice of farm machinery in western India. The DSS has the flexibility needed to accommodate regional variations in crops and cropping practices, farm characteristics, sizes of equipment, and in costs and expected outputs (figure 2). Two sources of mobile power—animals or tractors—and two sources of stationary

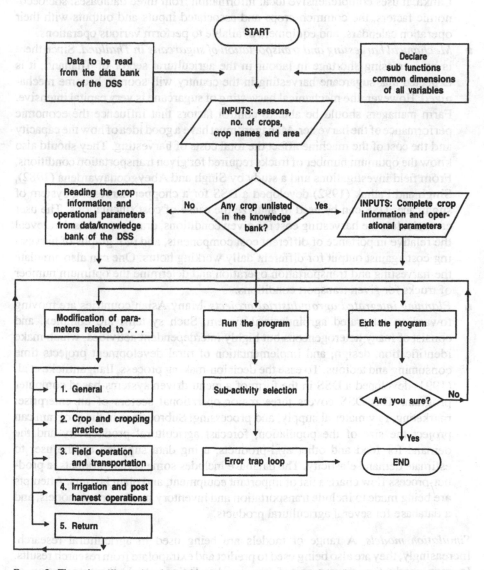

Figure 2. Flow chart illustrating the working principle of the main module of the decision support system

power—electric motors and diesel engines—were considered. With the DSS, one can determine the optimum power source and the matching equipment from locally available sources and equipment. The DSS takes into account the number of working hours required, energy used by the equipment, fixed and operating costs for each of the selected farm operations, and economic returns on area, crop, seasonal, and annual bases. Chandraratne (1993) developed an elaborated DSS for determining crop plans and selecting the ideal machinery for farming in Sri Lanka. It uses comprehensive local information from three databases: socioeconomic factors, the common crops and associated inputs and outputs with their operation calendars, and equipment available to perform various operations.

■ *Mechanical harvesting and transportation of sugarcane in Thailand.* Since there is an increasing shortage in labour in the agricultural sector of Thailand, it is expected that sugarcane harvesting in the country will soon to become mechanized. However, the mechanical harvesting of sugarcane is very capital intensive. Farm managers should be able to control factors that influence the economic performance of the harvester. Managers must have a good idea of how the capacity and the cost of the machine affect the total costs of harvesting. They should also know the optimum number of trucks required for given transportation conditions. From field investigations and a study by Singh and Abeygoonawardena (1982), Singh and Pathak (1992) developed a DSS for a chopper harvesting system of sugarcane. DSS can be used to analyze the cause of cost fluctuations. The user can calculate the harvesting cost for given conditions, draw pie charts that reveal the relative importance of different cost components, and plot graphs of harvesting cost against output for different daily working hours. One can also simulate the harvesting and transportation operation and determine the optimum number of trucks for given transport conditions.

■ *Planning integrated agroindustrial projects.* Many Asian countries are moving toward an integrated agroindustrial system. Such systems are complex, and consist of many heterogeneous but highly interdependent activities, which make identification, design, and implementation of rural development projects time consuming and tedious. To ease the decision-making process, Ilangantileke et al. (1991) developed a DSS in the form of a menu-driven systems-based computer program. This DSS covers three major operational sectors of an enterprise: marketing, raw material supply, and processing. Subroutines in the program can project the size of the population, forecast agricultural productivity and the demand for food and other agri-products, using data supplied by the user to estimate demand elasticity. The database includes some cereal products, a product-process flow chart, a list of important equipment, and plant layouts. Attempts are being made to include transportation and inventory management models, and a database for several agricultural products.

Simulation models. A range of models are being used in agricultural research. Increasingly, they are also being used to predict and extrapolate from research results. In crop production, four areas of use can be distinguished: characterization of agroecological zones, matching crops with suitable environments, crop and soil management studies, and crop protection. Some models, in particular those from IBSNAT (Uehara and Tsuji 1993), have focused at crop and soil management, and aimed at identifying optimum strategies for farming in risky environments. Models have also become instruments in designing new plant types (Dingkuhn et al. 1993) and in agroecological zoning studies (Aggarwal 1993). They are also being used in animal-production, ecology, genetics, hydrology, climate studies, economics, demography, and many other purposes. A broad selection was presented at a symposium at AIT on Systems Approaches for Agricultural Development (December 1991).

Multicriteria optimization. Farmers and governments may differ in their agricultural objectives, in particular with respect to economic profitability and environmental sustainability. Dealing with such multiple and sometimes conflicting objectives in decision-making in agriculture can be facilitated by using multiple-goal optimization techniques and crop-simulation models (Alocilja and Ritchie 1993). Van Keulen (1993) described a technique of interactive multiple-goal linear programming, which can be employed for evaluating the various development options in a region with a wide range of technical and socioeconomic conditions.

The Asian Institute of Technology (AIT)

The Asian Institute of Technology (AIT) is an autonomous, international postgraduate technological institute near Bangkok, Thailand. It has an international community of teaching faculty and research staff from 30 nations, and it fosters the exchange and dissemination of advanced technological knowledge and expertise. Founded in 1959, it is funded by a range of donors and national programs. AIT provides advanced education in engineering, science, and allied fields through academic programs. Agricultural and Food Engineering (AFE) is one of the five academic programs of the School of Environment, Resources and Development. Agricultural systems is one of the four fields of research and training in AFE. AIT offers master- and doctoral-level education and provides special short-term programs as mid-career training for professionals from the government and the private sector. Excellent facilities permit AIT to enroll more than 1000 students in its long-term training programs, and to train about 3000 professionals from developing countries annually in short courses. These training programs are funded by international donor agencies and the national governments of the participants.

With the basic facilities and infrastructure, AFE can conduct training and research in agricultural and environmental systems for Asian students. Multidisciplinary team work is encouraged. There are also good opportunities to interact with the other academic programs, such as computer science, environmental engineering, and natural resources management. AIT has full Internet connectivity, and all computers on its campus are linked in a network. One of the major centers of GRID is located at its campus, and serves as one of the outreach centers of the Institute. GRID archives, collates, and disseminates information in digital format, and encourages collaborative

264

projects with national and international organizations, especially in the Asian and the Pacific region.

Acronyms

AFE	agricultural and food engineering
AIT	Asian Institute of Technology
CGIAR	Consultative Group on International Agricultural Research
DSS	decision support systems
FSR&D	farming systems research and development
GIS	geographic information system
GRID	Global Resource Information Database
IBSNAT	International Benchmark Sites Network for Agrotechnology Transfer
IRRI	International Rice Research Institute
RCC	Regional Computer Center
SARP	Simulation and Systems Analysis for Rice Production
UNEP	United Nations Environment Programme

References

Aggarwal P K (1993) Agroecological zoning using crop growth simulation models Characterization of wheat environments of India Pages 97-109 in Penning de Vries F W T, Teng P S, Metselaar K (Eds) Systems approaches for agricultural development Kluwer Academic Publishers, Dordrecht, The Netherlands

Alocilja E C, Ritchie J T (1993) Multicriteria optimization for sustainable agriculture Pages 381-396 in Penning de Vries F W T, Teng P S, Metselaar K (Eds) Systems approaches for agricultural development Kluwer Academic Publishers, Dordrecht, The Netherlands

Barr A, Fiegenbaum E A (1981) The handbook of artificial intelligence, Volume 1 Kaufmann, Los Altos, California, USA

Ten Berge H F M (1993) Building capacity for systems research at national agricultural research centers SARP's experience Pages 515-538 in Penning de Vries F W T, Teng P S, Metselaar K (Eds) Systems approaches for agricultural development Kluwer Academic Publishers, Dordrecht, The Netherlands

Butani K M, Singh G (1993) A decision support system for the selection of agricultural machinery with a case study in India Computer and electronics in agriculture (in press)

Chandraratne I W D T (1993) A decision support system for crop planning and equipment selection MEng thesis, AIT, Bangkok, Thailand (unpublished)

Dent J B (1993) Potential for systems simulation in farming systems research Pages 325-340 in Penning de Vries F W T, Teng P S, Metselaar K (Eds) Systems approaches for agricultural development Kluwer Academic Publishers, Dordrecht, The Netherlands

Dent J B, Blackie M J (1979) Systems and simulation in agriculture Applied Science Publishers, London, UK

FAO (1988) Geographic information systems in the Food and Agriculture Organization (FAO) of the UN, Rome, Italy

Gartner J A (1993) Post-graduate training in agricultural systems The AIT experience Pages 485-503 in Penning de Vries F W T, Teng P S, Metselaar K (Eds) Systems approaches for agricultural development Kluwer Academic Publishers, Dordrecht, The Netherlands

GRID (1993) On bridging the gap Information for decision making GRID, United Information Environment Program, Nairobi, Kenya

Ilangantileke S et al (1991) A decision model for planning integrated agroindustrial projects Paper presented at the workshop on Grain Post Harvest System Analysis using Microcomputers ASEAN grain post harvest program 25-27 June 1991, Singapore

Jones J W et al (1987) Combining expert systems and agricultural models Transactions of the ASAE 30(5) 1308-1314

Van Keulen H (1993) Options for agricultural development A new quantitative approach Pages 355-365 in Penning de Vries et al (Eds) Systems approaches for agricultural development Kluwer Academic Publishers, Dordrecht, The Netherlands

Kropff M J, Penning de Vries F W T, Teng P S (1994) Capacity building and human resource development for applying systems analysis in rice research (pages 323 - 339 of this volume)

Penning de Vries et al (1991) Systems simulation at IRRI Page 5 in Penning de Vries et al (Eds) Systems simulation at IRRI Research paper series no 151 IRRI, Philippines

Penning de Vries F W T (1983) Modeling of growth and production Pages 118-150 (Chapter 4) in Encyclopedia of Plant Physiology, New Series, Vol 12 D, Springer Verlag, Berlin, Germany, Germany

Sabbagh G J et al (1993) Two applications of GIS in agricultural modeling Pages 250-259 in Heatwole C D (Ed) Application of advanced information technologies Effective management of natural resources Proceedings of the ASAE Conference, 18-19 June 1993, Spokane, Washington, USA

Singh G, Pathak B K (1992) Final report on evaluation of austoft sugarcane harvester in Thailand AIT Research Report No 259 Division of Agricultural and Food Engineering, AIT, Bangkok, Thailand

Singh G, Abeygoonawardena K A R (1982) Computer simulation of mechanical harvesting and transporting of sugarcane in Thailand Agricultural Systems 8(1982) 105-114

Singh G, Gupta M L (1981) Machinery selection method for farms in north India Agricultural Systems 6(1980-81) 93-120

Stroosnijder L, van Rheenen T (1993) Making farming systems a more objective and quantitative research tool Pages 341-354 in Penning de Vries F W T, Teng P S, Metselaar K (Eds) Systems approaches for agricultural development Kluwer Academic Publishers, Dordrecht, The Netherlands

Uehara G, Tsuji G Y (1993) The IBSNAT project Pages 505-514 in Penning de Vries F W T, Teng P S, Metselaar K (Eds) Systems approaches for agricultural development Kluwer Academic Publishers, Dordrecht, The Netherlands

Discussion on section E: Training in the use of systems approaches

Training in the use of systems methods in agroecological and resource management research was a subject that came up more frequently than any other in discussion throughout the three-day course of the workshop. The importance that participants attached to this topic reflects the comparative scarcity of people with a systems perception and systems way of thinking in the NARS (McCown et al., Section B) and to some extent also in the IARCs (Harris et al., Section F).

Centers and nations have added to understanding agroecology through farming-systems research (FSR), but few specialists are competent to handle different components of agroecological characterization. Individuals who can integrate the concepts, methods, and techniques of the different fields are even more rare. To remedy the situation, an understanding of the principles involved has to be introduced into the educational background. Starting in the universities and at the post-graduate level, there is a need to train more scientists to study and evaluate the environment and the effects of environmental stresses on agriculture and the use of natural resources, particularly in marginal environments (Harris et al., Section F). Building capacity at graduate and post-graduate levels takes time, and therefore one of the implications is that the widespread adoption and effective use of systems approaches will be a comparatively slow process.

Although there are some training opportunities in subjects related to agroecology at institutions in the developed nations (e.g., at WAU and the University of Reading), they will also be needed at institutions in developing countries.

IRRI with CABO at Wageningen has organized training in crop modelling and systems analysis for teams from Southeast Asian countries, under the auspices of the Systems Analysis Simulation in Rice Production (SARP) project (Kropff et al., Section F). The aim has been to build the capacity of the national programs by training groups of their scientists, who, on return, continue to operate as a team. There are short courses available for managers and project leaders. An additional aim in recent courses has been to train trainers, who on their return can train other groups.

IRRI and IITA are also working to establish multidisciplinary teams to work on agroecological characterization. The International Benchmark Sites Network for Agrotechnology Transfer (IBSNAT) has provided training in the use of crop simulation models (CERES and SORGF).

One of the purposes of the intercenter workshop in Rome was to determine how to develop more rapidly the capability of the centers to use systems approaches, and how to make this technology more readily available to the research systems of developing countries (Bunting 1987). The SARP project and some of the other training initiatives have contributed substantially to this end, as indicated here in the papers by Singh et al. and Kropff et al. (Section F).

In the last four years the IARCs have reduced their commitment to some forms of training. Singh et al. observe that the centers do not have a comparative advantage in training in the application of systems methods to agroecological research. Organizations other than the CGIAR centers will have to provide most of the training that is required. However, the centers can contribute by offering an environment for training of this kind with a broad background of active interdisciplinary research in many related topics, and an emphasis on field studies, as well as facilities, equipment, and libraries. The centers should therefore continue to promote competence in this field by providing opportunities for graduate and other forms of training for national cooperators.

Training courses should reflect the interdisciplinary nature of resource management research. Some courses or parts of courses should be oriented towards the needs of managers, to increase their awareness of the potential contribution that systems methods can make to improving the efficiency of research. The courses should contain provision to consider what reorganization of an institute may be required on introducing resource management concerns into the research agenda. Issues to consider include, for example, the cooperation between institutions and disciplines that is required, and what special measures might be needed to keep systems groups intact (see Kropff et al. and Torres, Section F).

Recommendations

The workshop recognized that a shortage of scientists trained in systems approaches is currently one of the most acute constraints to the introduction and adoption of systems methods in developing countries. It recommends that high priority be given to training, and recommends to the CGIAR, NARS, and donor agencies the following:

- Understanding of the principles involved in systems approaches has to be introduced into the educational background, starting at the universities and at post graduate level. The CGIAR, NARS, and donors should seek to create more opportunities for graduate and post-graduate training in systems methods and in subjects related to agroecology in the developed nations and in developing country universities.
- Reinforce efforts to build the capacity of the national programs by providing additional short-term training courses in systems approaches, similar to those that have been provided for rice scientists by the SARP project, and by IBSNAT, but with particular emphasis on the needs of rainfed agricultural systems. Where appropriate, train groups of scientists who on return will continue to operate as a team. Include short courses that are designed for managers and project leaders.
- CGIAR centers should continue to promote competence in this field by providing, in cooperation with the universities, opportunities for graduate research and other forms of training for their national cooperators. These training sessions should be conducted in an interdisciplinary environment at the centers.

SECTION F

The use of systems methods in research at an international level

Systems research methods and approaches at CIAT—current and planned involvement

F. TORRES and G. GALLOPIN
Centro Internacional de Agricultura Tropical (CIAT), Apartado Aereo 6713, Cali, Colombia

Key words: aggregation level, decision making, expert system, GIS, natural resource management, research planning, simulation models, systems approach, training

Abstract
The paper describes the way CIAT is using systems approaches in its research activities. Systems research concentrates on different levels of the hierarchy of agroecosystems ranging from the soil subsystems to the regional level, through the cropping, farming and landscape levels.

GIS applications are used in CIAT to address the spatial dimension of research problems. Previously, those techniques were used to characterize research sites, and lately to delineate and select the three major Latin American agroecosystems upon which CIAT's strategic plan concentrates.

CIAT is applying simulation models to investigate soil nutrient and organic matter dynamics in crop pasture systems, to analyze cropping systems and to examine responses of legume-grass pastures to management alternatives. Simulation models for economic evaluation at the farm level have also been used, and their application to decision making is at the development stage.

CIAT is initiating a line of research aiming at the development of systemic frameworks and methodologies for the understanding and treatment of causal cross-scale linkages relevant to land management in tropical America.

Introduction

In partnership with other institutions, especially national agricultural research organizations, the mission of the Centro Internacional de Agricultura Tropical (CIAT) is:
To contribute to the alleviation of hunger and poverty in tropical developing countries by applying science to the generation of technology that will lead to lasting increases in agricultural output while preserving the natural resource base.

The emphasis of this mission is on growth, equity and enhancement of the resource base. As we move toward the 21st century, the world community has realized the urgency of finding lasting solutions to the widespread deterioration of the natural resource base for agricultural production, including the loss of genetic diversity, depletion of water resources, soil erosion, deforestation, and environmental pollution. These solutions must also preserve the opportunities for achieving economic growth and, in developing countries, the production of food to meet the rapidly expanding demand. Research provides the scientific basis for improved understanding of the causes of environmental degradation and options for policy and technology to contribute to sustainable resource management.

In this problem-solving environment, CIAT will pursue its mission through research on germplasm development and resource management.

271

Research at the frontiers

Research on the management of natural resources is a perilous undertaking. One risks falling into a bottomless pit of analysis that has no product other than new knowledge. To avoid this hazard requires not only a commitment to development, but careful planning and management. Among other things, it compels us to make hard choices at the outset about where to work. Almost two years ago, CIAT made a good start toward answering this question through a rigorous exercise in strategic planning.

As part of this process, we first divided all of Latin America and the Caribbean into broad environmental classes. Then, using a set of measurable socioeconomic criteria, we began a process of elimination that eventually left us with six agroecologies. Based on the potential for achieving growth in production, encouraging equity, and contributing to sound resource management in these environments, we selected three to serve as the focus for CIAT's resource management research: 1) the cleared margins of rain forests, 2) well-watered hillsides, and 3) tropical savannas.

An important feature of the three groupings is that all are frontiers of agricultural development. Working in such areas makes a lot of sense to us. In the first place, they contain the region's most valuable natural resources. Moreover, since the conditions that govern land use in these agroecosystems are still in flux, we have a better chance of shaping current patterns in the day-to-day practice of farm communities and of influencing policy decisions.

The forest margins in the Amazon and Central America are the line of battle in the struggle to preserve the region's remaining forest. An important challenge for research in this environment is to develop and promote practices that enable "slash-and-burn" farmers to continuously derive a decent living from the same land over the long term. We must also study policy options to discourage other activities, such as large-scale ranching, that lead to deforestation.

Many of the immigrants clustered at the forest margins are refugees from another disaster unfolding on the hillsides of Central America and the Andean zone.

Inappropriate farming methods have increased water runoff and soil erosion on sloping land, causing extensive damage in the watersheds of hilly areas, and affecting soil structure and fertility. As in the forest margins, a large part of our challenge is to design acceptable technologies that permit more efficient management of natural resources, while still giving farmers acceptable returns on their investments. Moreover, the technology initiative must be accompanied by research on policies that are conducive to more appropriate resource management and on instruments for implementing these policies.

Making hillside farming more viable should also relieve some of the population pressure on the forest margins. Another, less direct, way of accomplishing the same end is to devise a combination of technology and policies that will permit more effective use of the vast savannas of Brazil, Bolivia, Colombia, and Venezuela.

Best suited for mechanized production, these environments are unlikely to divert the flow of landless immigrants to the forest margins. But if properly developed for medium-scale production of livestock and crops, the savannas could at least lessen the incentive for commercial exploitation of tropical forests. More important, farmers in the savannas could steadily increase supplies of staple foods—especially milk, beef, maize, rice, and other crops—thus lowering their prices. That would benefit poor urban consumers in particular, another group that contributes heavily to the ranks of small-scale farmers at the forest margins. But the savannas ought not be sacrificed for the benefit of more valuable tropical forests. They, too, are highly susceptible to degradation, so ways must be found to farm them on a sustainable basis.

The systems approach
The approach to natural resource management includes such components as agroecological characterization, analysis of land-use patterns and options, understanding of soils, water and plant nutrition relations, crops and cropping systems alternatives, and agricultural and forest policy alternatives. These components aim at creating the basis for sustainable development.

CIAT is highly conscious of the fact that agricultural production problems and opportunities cannot be resolved through unidimensional approaches that concentrate on isolated elements only. At the same time, we are aware of the pitfalls of a systems orientation *per se*, that could easily result in endless systems analyses and location-specific research activities. Thus, we seek to identify practical and relevant entry points into agricultural production systems. Work on these entry points is carried out with a clear perspective concerning the physical, biological, socioeconomic, policy, and land-use dimensions that may impinge on technology options. That is, throughout the process, a macro-agroecosystems perspective is combined with a micro-production systems perspective.

The following pages briefly discuss the major ways in which the systems perspective is being applied or planned in CIAT. This is not in the sense of an exhaustive listing. Most of the past and present involvement of CIAT with the systems approach can be addressed under three basic headings:
(a) a hierarchical systems perspective to research;
(b) geographic information systems (GIS) applications (spatial emphasis); and
(c) mathematical models.
Planned developments addressing cross-scale interactions are then discussed, followed by training needs.

A hierarchical systems perspective
As we seek ways to reconcile the potentially conflicting goals of productivity growth and prudent resource management, we must go beyond the conventional focus on farmers as producers of commodities to examine their behavior as land users and the circumstances that shape this behavior. The farmers' world may be viewed as a hierarchy of interconnected systems, whose effects are expressed in particular patterns of resource utilization across the rural landscape.

To understand and improve these patterns of land use, we must, at one level, look at what is happening in individual plots—that is, elucidate the biological and physical processes underlying agricultural production and resource degradation. From these insights we can then derive measures of sustainability by which to judge the merit of alternative technologies.

At a second level, we must gain new insights into the principles that guide farmers' deployment of resources and incorporate this information into the design of appropriate technologies. From the start, our efforts to generate new practices must be thoroughly grounded in the socioeconomic realities of the rural community.

The technology initiative will not go far unless it is accompanied by major efforts at a third level—that of decision makers in government. More effective land management requires greater flexibility and a decentralized approach in decision making. This research can contribute to change by bringing to light new information about the circumstances that shape decisions and by examining the links between these, focusing specifically on the way in which land users' choice of technology is influenced by the policy and organizational settings, and in the empowerment of local organizations that can then influence policymaking.

A general hierarchical framework embodying the most relevant levels of agriculturally managed ecosystems is presented in figure 1. The major levels of interest for the present discussion are:

- *The cropping system*, which includes all cropping patterns grown on the farm and their interaction with farm resources, other household enterprises and the physical, biological, technological, and socioeconomic factors.
- *The farming system*, which is the unit of production, consisting of a human group (often a household) and the resources it manages, involving the direct production of plant and/or animal products. Factors such as climate and weather, land tenure, land quality, and socioeconomic variables are included. A farming system is always a part of a larger social, political, economic, cultural, and political environment that impacts on everything happening within the farming system. Thus, it can be said that the next level of analysis upward can be a rural village, a compound, or some physical unit of space that includes several farming systems.
- *The landscape system*, which is defined primarily according to the criterion of spatial contiguity. Its boundaries may be traced according to different criteria. Thus, a watershed can be viewed as a landscape system delimited by hydrological boundaries, and a rural community as a landscape system delimited according to social criteria.

GIS applications

Although for some, GIS are not part of systems analysis, they are included here because they represent powerful tools for dealing with the spatial dimensions of complex systems, and because CIAT is planning to develop dynamic simulation models linked with GIS applications.

Target agroecosystem identification. During the preparation of the strategic plan, a GIS approach was used to determine which ecosystems were most appropriate for CIAT's attention. In addition to the criteria of growth, equity and resource preserva-

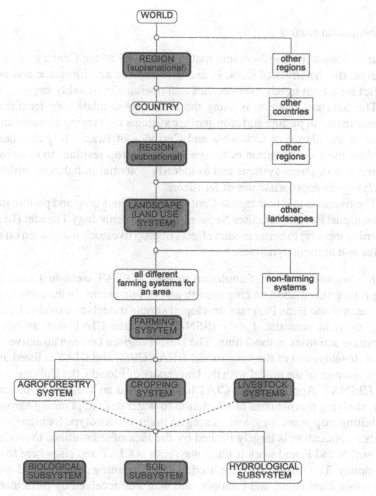

Figure 1. A hierarchy of agroecosystems. CIAT's systems activities concentrate on the shaded boxes.

tion outlined in the mission's statement above, those of feasibility and efficiency were added.

Last year saw the completion of an innovative classification of the Cerrados region of Brazil to determine appropriate study areas for joint research with Empresa Brasilera de Pesquisa Agropecuaria (EMBRAPA) and local agencies. Data from CIAT's climate database and the land-system study were used to provide images of climate, soils, and terrain for the region. These were complemented by data from the Brazilian agricultural censuses from 1970, 1975, and 1980. The study was used in Brasilia to select candidates for the final study area.

Mathematical models

Plant-soil models. A soil organic matter submodel of the Century model is used to simulate the dynamics of C, N, P, and S in organic and inorganic soil pools. It can predict long-term trends over decades and operates in monthly steps.

The Savanna Program is using the model to simulate long-term trends in soil organic matter in pasture and crop-pasture systems for varying climates and soil types found in the Llanos of Colombia and Cerrados of Brazil. It is also being used to simulate the decomposition of forage litter and crop residues to examine fluxes of nutrients in different systems and to identify potential imbalances which may need rectifying via appropriate use of fertilizers.

The overall aim is to integrate Century with various crop and pasture models (e.g., International Benchmark Sites Network for Agrotechnology Transfer (IBSNAT) and Thornley models) to better predict effects of crop livestock integration on soil organic matter and nutrient dynamics.

Crop simulation models. Simulation models at CIAT were first used as tools for integrating information on crop growth and development. In the early to mid-1980s, the Cassava and Bean Program developed simple models to examine factors contributing to yield potential. Lately IBSNAT provided CIAT with an opportunity to re-initiate activities in modelling. The Bean Program became an active collaborator in the development of the bean model BEANGRO, and CIAT is listed as one of the co-developers of the model with the University of Florida, the University of Georgia, and IBSNAT. Applications at CIAT have included an analysis of international trial data, studying mechanisms of adaptation to water deficit, planning agronomic trials, modelling crop phenology, and seeking an improved ideotype for high yield potential. Further application is largely limited by the lack of subroutines to model effects of low soil N and P, and work is underway both at CIAT and elsewhere to remedy this deficiency. The model has been used in CIAT training activities and workshops in Colombia, Zimbabwe, and Ethiopia, and was well received by participants.

BEANGRO has been distributed to researchers in over 40 countries, and the model has been applied for diverse problems including effects of global warming in Venezuela and zonification of potential bean production regions in Puerto Rico, which used a GIS interface to integrate modelling results with other data.

Pasture models. Pastures are much more complex systems than annual crops. No model exists that describes the relations between the components of a tropical pasture. However, the Hurley pasture model, a model of a temperate pasture written by Dr J.H.M. Thornley, (Institute of Terrestrial Ecology in Edinburgh), has been found to work well in southern England. Accordingly, the Savanna Program of CIAT has evaluated with Dr Thornley its suitability for application to grazed grass-legume pastures in the lowland tropics.

As the Hurley model is concerned with the dynamics of carbon and nitrogen pools and flows in a grazed ryegrass monoculture under temperate conditions and their relation to environmental and management variables, it has a number of limitations that prevent it from being used directly for mixed grass-legume pasture. To take account of these added complexities, considerable *de novo* program development is needed. In collaboration with the CIAT plant-soil group, Dr Thornley has written a soil submodel that takes account of phosphorus, nitrogen, and potassium, and has written a prototype legume submodel that includes meristematic tissue growth. Preliminary evaluations suggest that these submodels perform sensibly. Further development of the appropriate submodels is continuing.

In another research, a mathematical model of the nitrogen cycle was used to examine the response of legume-grass pastures to varying levels of utilization of the forage by grazing ruminants, and to determine the optimum legume composition of both temperate and tropical pastures. The model showed that the level of utilization of the pasture is important, in that at low levels of utilization, recycling occurs through plant tissue decay, with low losses, while at high levels of utilization, recycling is more through animal excrete, where losses are correspondingly higher.

Farm-level models. With this in mind economists at CIAT have either assembled or adjusted different simulation models to evaluate the costs and benefits of the proposed rice-pasture system. The models now in use are math-based models such as: Espadas, 1978, (The World Bank); Novillo, 1987 (CIAT); HATSIM, 1977 (CIAT); RUSH-MOD, 1987 (CIAT); BEEFMOD, 1985 (Inter-American Development Bank-IDB); and MULBUD, 1986 (International Centre for Research in Agroforestry-ICRAF).

All of these models are partial equilibrium models employing budgeting techniques (whole-farm and partial budgets) to estimate: a) the net income flow discounted over the production period being considered for integrated crops/trees/livestock production systems, and b) the internal rate of return (IRR) of investments in this land-use research. All models simulate the effect of variations in technical coefficients (input-output data), prices and costs over the IRR in the context of a sensitivity analysis. In that sense their predictive power heavily depends on quality of data (experimental data and farm survey data). The applications of the main model are: a) select the best alternative solution (land-use system/technology), and b) predict the economic consequences of several mutually exclusive solutions. The latter has been helpful in priority-setting exercises to focus experimental designs in rice-pasture systems.

In collaboration with the University of Florida, a farm-level economic decision model that incorporates daily time-step crop models is being developed in the Hillsides Program. The crop models selected are the suite of IBSNAT models which are fully operational. The econometric model explicitly links field-level processes, e.g., crop management, and higher level process, e.g., farmgate costs and prices. Biophysical inputs to drive the crop model will come from historical time series and will be supplemented and verified by intensive instrumentation.

Vertical interlinkages

Given the increasing global interdependence, the links between the micro and the macro, the local and the global need also to be examined to gain a multi-level perspective. Cross-scale interactions are likely to intensify in the future, requiring profound changes in strategies, policies, and institutions. A systems approach is essential in order to understand these linkages affecting agricultural sustainability.

There are at least three lines of argument for the need to consider the "vertical" causal linkages between agroecosystems at different levels of organization.

The first line of argument deals with understanding interactions across spatial and temporal scales. The use of land is determined by the interplay of a number of factors operating at different scales from plot practices to national policies, and even the international economy or the global ecology. The analysis of causal interactions between hierarchical levels in complex systems is a difficult methodological problem, but of great practical importance. The sustainability of land use may be influenced by events operating at different temporal and spatial scales, and far away from the fields.

The second one relates to institution-building, and the need for multi-level coordination. The permanent interplay between the micro and the macro requires institutional mechanisms that can operate simultaneously and in a coordinated way at different scales from the macro policies to the local community action. This has implications on governance and on levels at which decisions are made.

The third line of argument addresses the issue of complementarity between top-down and bottom-up approaches to rural development. To attain sustainable agriculture and eradicate rural poverty, both local strategies and macro policies have to play a role. Due to social and ecological specificities, strategies for sustainable development have to be context specific. The role of macro (national or international) policies cannot be neglected. They can either lead to the crashing of local efforts, or enable, amplify, and spread them. This means that local programs should be responsive to both the constraints and opportunities offered by national and international environments, but also that global programs must be assessed in terms of their effects on the current and future situation of the local populations and agroecosystems.

CIAT is currently defining its systems research strategy to address the issue of cross-scale interactions in land-use dynamics. A line of research will initially aim at developing a conceptual systemic framework for understanding "vertical" causal interlinkages between socio-ecological systems of land use at different levels of aggregation, and to identify the most critical cross-scale links affecting land-use patterns in tropical America. Expected outputs include: a methodology and framework for understanding causal cross-scale linkages; a set of cross-scale environmental, technological, and socio-economic factors and relationships critically affecting land-use patterns in selected agroecosystems; an analysis of the policy and institutional implications of cross-scale interactions for sustainable land use in the studied agroecosystems; and recommendations for policy and institutional changes to respond to cross-scale phenomena so as to foster sustainable land use.

CIAT will not attempt to build a single, detailed, multi-level, hierarchical simulation model, but it will adopt a flexible approach including a constellation of models at different scales and levels of resolutions. Conceptual models are a necessary step towards building mathematical models, but they are also powerful tools for the analysis of non-quantifiable variables and relations, which may critically affect the sustainability of agriculture and land use.

The outputs of some of the more detailed models may be used as inputs to the more aggregated ones. All models will have to operate, however, with only partial information on initial conditions and parameters, to accommodate to a typical condition in developing countries.

One invaluable quality of GIS and simulation modelling is that they have the potential for being much more interactive and "user friendly" for decision makers who are inexperienced and generally suspicious of classical statistical analysis. This interactive, "what-if" modelling ability is potentially an important tool in conflict resolution and consensus building for multi-objective, multi-stakeholder, resource-use decision. Another line of research will then focus around the development of decision support systems at different levels: the unit of production, the community or village, the local or national government. The decision support systems are expected to include GIS-linked databases, crop and farm simulation models, scenario analysis, userfriendly interfaces, and expert systems.

Training needs

Agricultural research in Latin America is usually conducted according to scientific disciplines or agricultural products. Scientific reductionism has been the approach to agricultural research. The majority of agricultural researchers belong to this school, characterized by isolating research topics from their context and from the interactions within the same systemic level (e.g., the production system) or with other hierarchical levels such as communities, micro-watersheds, agroecosystems.

The problem of sustainability, however, is usually complex and multi-dimensional. Frequently, there is no optimal solution within a systemic level. What is even worse, technological solutions may be developed without considering the farmer, his immediate context (social, cultural, institutional), or his more distant context (political, macro-economic). Ignoring these factors has led to two undesirable results: i) technologies that, when applied, have detrimental social and environmental effects, and ii) technologies rejected by potential users as inappropriate to their needs and circumstances.

To study problems of agricultural sustainability, the traditional approach is not good enough. It should be complemented with the holistic or systems approach, which incorporates systemic dimensions and levels relevant to solutions that are technically feasible, economically viable, ecologically sound, and socially acceptable.

There is a need, therefore, to introduce the systems approach into the training of scientists in a multidisciplinary framework. Towards this end CIAT is organizing

280

regional post-graduate courses aimed at training national scientists on research in sustainable agricultural land use.

Acronyms

CIAT	Centro Internacional de Agricultura Tropical
EMBRAPA	Empresa Brasilera de Pesquisa Agropecuaria
FAO	Food and Agriculture Organization
GIS	geographic information system
IBSNAT	International Benchmark Sites Network for Agrotechnology Transfer
IRR	internal rate of return

Systems approaches for crop improvement and natural resource management research in CIMMYT: past and future

L. HARRINGTON[1], J.CORBETT[1], S. CHAPMAN[1] and H. VAN KEULEN[2]

[1] Centro Internacional de Mejoramiento de Maíz y Trigo (CIMMYT), P O Box 6-641, Mexico 06600, D F Mexico

[2] DLO Research Institute for Agrobiology and Soil Fertility, P O Box 14, 6700 AA Wageningen, The Netherlands

Key words adoption, climate surfaces, crop improvement, crop simulation models, GIS, G x E interactions, maize, sustainable agriculture, wheat

Abstract

In the past, CIMMYT's contributions to a more productive and more sustainable agriculture were largely in crop improvement Some of its contributions were direct CIMMYT has developed maize and wheat varieties that resist insects and diseases with minimal use of pesticides, or that use water and nutrients more efficiently But it has contributed more in indirect or preventive ways new maize and wheat technology has averted or forestalled resource degradation by helping alleviate poverty, generate employment, and stimulate broad-based economic development, thus reducing the pressure on fragile agricultural lands

This paper describes CIMMYT's experiences in the use of systems methods to raise the efficiency of crop-improvement research CIMMYT has achieved higher efficiency by improving the identification and characterization of production environments (mega-environments), the identification and characterization of testing environments, and the characterization of genotype responses and interactions with environments The paper then describes CIMMYT's direct contributions to a sustainable agriculture through its endeavors in NRM research, and it suggests opportunities for the use of systems methods in improving the efficiency of these endeavors The paper closes with a word of caution about the relevance of the outcome, in terms of the link between systems methods (crop models, GIS) and a farming systems perspective (e g , factors affecting the adoption decisions by farmers)

Introduction

Established in 1966, CIMMYT has seen the world change dramatically. The population in developing countries has grown from 2.4 to 4.2 billion, maize production in these countries has risen from an annual 91 million tons to 217 million tons, and wheat production from 74 to 244 million tons. The fear of global famine has receded to a considerable degree. Widespread and unrelenting poverty continues to be a cause for concern, however, in particular in South Asia and sub-Saharan Africa.

In the next 20 years, maize and wheat farmers in developing countries will have to double current harvests to meet growing demand. It is important that they do so without degrading agricultural resources, because degraded agricultural resources threaten the well-being of future generations. New technologies are needed that both enhance the productivity and conserve resources.

P Goldsworthy and F W T Penning de Vries (eds), Opportunities, use, and transfer of systems research methods in agriculture to developing countries, 281 - 288
© 1994 Kluwer Academic Publishers

In the past, CIMMYT's contributions to a more productive and more sustainable agriculture were mainly through crop improvement. Some of these contributions were direct: CIMMYT has developed maize and wheat varieties that resist insects and diseases without requiring pesticides, or that use water and nutrients more efficiently. But it has contributed more in indirect or preventive ways.

New agricultural technology can avert or forestall resource degradation by helping ameliorate poverty, generate employment, and stimulate broad-based economic development, thereby reducing pressure on fragile agricultural lands (Harrington 1993). There is a tendency to dismiss these preventive contributions as esoteric or unimportant. Note, then, how Green Revolution technologies have contributed to preventing additional land from being cultivated:

"Without modern varieties, production of rice and wheat in the 90 countries with humid tropical lands may have been 20-30 million tons less than it is. To have made up this shortfall from non-irrigated lands with traditional varieties and management practices would have required an additional area under cultivation on the order of 20-40 million ha, probably an underestimate given the rapid degradation of newly cleared land." (CGIAR 1985, Chapter 14).

CIMMYT has set itself various goals for the future. It will continue to produce maize and wheat varieties that perform well with minimal use of pesticides; it will use biotechnology to accelerate breeding research; continue efforts to rescue, store, utilize, and share maize and wheat genetic resources; augment social-science research to ensure that environmentally friendly technologies remain profitable and in other ways "farmer friendly"; and add to research on natural resource management (NRM) and conservation, focusing on large ecosystems where maize or wheat is a major crop (CIMMYT 1992).

CIMMYT stresses that it is increasingly interested in expanding its use of systems research methods to improve the efficiency of conventional field research, and to conduct forms of experimentation—especially in NRM research—that are difficult or impossible to carry out in the field.

The remainder of this paper is divided into two sections. The first discusses CIMMYT's past and current investments in the use of systems research methods. These investments have been primarily in the use of geographic information systems (GIS) and crop models to target germplasm improvement to well-defined production zones or "mega-environments". The second part describes CIMMYT's current and future investment in NRM research and introduces possible roles for systems research.

Using systems research to increase the efficiency of crop improvement

Introduction

CIMMYT's global mandate for maize and wheat improvement compels it to deal with an extensive range of production environments. Farmers' germplasm requirements vary notably over these environments with regard to adaptation, maturity, and tolerance to biotic and abiotic stresses. Efficient management of research resources in global crop improvement requires an understanding of production and testing environments and of the types of germplasm being improved.

Identification and characterization of production environments
Within CIMMYT, the concept of the "mega-environment" (ME) has been developed as a means to organize and focus crop improvement programs. An ME is defined as a large geographic area (not necessarily adjoining and frequently transcontinental), characterized by similar biotic and abiotic stresses, cropping-system requirements, and consumer preferences. Each ME represents a minimum of one million ha, resulting in zones large enough to be of interest to an international center. MEs are most useful when they help target breeding efforts.

The use of GIS has enhanced CIMMYT's capacity to identify and characterize the diverse maize and wheat MEs throughout the world. Compared to earlier cartographically-based approaches to agroecological zoning (e.g., FAO 1981), the combination of GIS and simulation methods can characterize environments more accurately and assess changing production frontiers (e.g., genotypes) and constraints (ranging from diseases to resource access). So far, work at CIMMYT has included the development of robust, reproducible monthly climate surfaces based on a data set of climate normals (station or point data), a global digital elevation model, and robust algorithms for spatial interpolation of the climate data. These climate surfaces provide data for grid cells in a GIS and can then be used as input for grouping these cells on the basis of selected variables, such as temperatures and rainfall for the months of April through September.

A prototype study carried out a cluster analysis on seven months of monthly climate data (matching the maize growing season) in Mexico. The summary characteristics of the resulting clusters (environments) were then interpreted into maize adaptation zones. The criteria to form the zones from one or more clusters were derived from maize breeders' perceptions of the suitability of different maize germplasm to different environments. The results illustrate two advantages of this approach: (1) with the climate surface database in place, researchers can select only those elements of climate that best differentiate germplasm-specific zones, and (2) different environments are no longer combined on the basis of arbitrary, discrete boundaries.

These methods are being used in a collaborative project with the Kenya Agricultural Research Institute (KARI). Clustering techniques have been applied in combination with ground truth and production data to define germplasm-specific maize environments. Preliminary results indicate that the incidences and relative size of various maize production environments in Kenya differ substantially from what had been assumed (table 1). This information will be used to reassess priorities for maize improvement and agronomic research in Kenya.

Table 1. Estimates of the proportion of maize area found in Kenya in different maize environments: a comparison of two approaches

Maize environment	Estimate from 1988 maize megaenvironment database , based on subjective judgment[a]	Estimate from 1993 collaborative project with KARI featuring GIS and cluster analysis[b]
Lowland tropics	7%	3%
Mid-altitude tropics	56%	18%
Tropical transition zone	35%	43%
Highland tropics	1%	28%
Other	1%	8%
Total maize area ('000 ha)	1,425	1,075[c]

[a] CIMMYT Maize Program 1988.

[b] Preliminary results, KARI-CIMMYT collaborative maize GIS database project.

[c] National maize area under the GIS project was estimated from five years of data from aerial photography, hence is thought to be more reliable than the earlier estimate.

Identification and characterization of testing environments

Maize. The usefulness of the climate surface database increases as reliable mechanisms that relate monthly data to an array of agricultural issues are identified. Every year, CIMMYT's International Maize Testing Program distributes seed for evaluation trials all over the world. Climate surfaces can contribute to interpreting the results of international trials by characterizing the environments of test sites to identify those representative of Mes or for particular stresses.

The analysis of international trial results can identify the stations in "hot spots" for biotic stresses. CIMMYT pathologists can include susceptible genotypes in trials planted at various locations to test disease pressure empirically. With actual weather data from the sites, descriptive models can then be constructed. These models, combined with climate surfaces, can be used to extrapolate the relationships, and can estimate the potential frequency and intensity of disease pressure in different areas. The resulting expert diagnosis of a cluster analysis of climate variables specific to a disease can identify the spatial extent of areas where the disease is expected to be severe. This serves two purposes: (1) resistant maize varieties can be better targeted while priority can be given to improving the resistance based on potential demand, and (2) suitable test sites can be identified. Abiotic stresses are also considered a serious challenge to CIMMYT germplasm. Drought, for example, is thought to be responsible for about 15 percent of all maize yield losses in the lowland tropics. A proposed project aims to use climate surfaces to characterize the frequency and severity of drought in selected areas (e.g., southern Africa), so that we may better estimate the potential returns to continued investment in drought-tolerant germplasm.

Wheat. In an analysis of 26 years of multi-location field trials with 50 cultivars per year, CIMMYT's Wheat Program identified key characteristics of wheat production environments (DeLacy et al. 1993). The study differentiated wheat production environments on the basis of the performance of wheat germplasm itself. Future research can build on these results by assessing the extent to which environmental information can explain observed patterns in wheat germplasm performance. In addition, the location of trial sites can be rationalized in terms of identified wheat environments and their underlying characteristics. Finally, the spatial extent of climatically similar areas (with regard to wheat germplasm) can then be assessed, allowing more accurate targeting of wheat germplasm and its improvement. In general, analyzing environmental and trial data with a GIS can help identify two types of testing sites: those that best represent a particular ME and those that represent 'hot spots' for specific stresses. This information increases testing efficiency and improves the accuracy of the weight given to the data obtained.

Characterization of genotype responses and interactions with environments

Maize. An improved understanding of the characteristics of the range of CIMMYT germplasm can fine-tune environment definitions as well as efficiently interpret trial performance data where genotype-by-environment (GxE) interactions across locations are complex. Data from international testing trials and CIMMYT trials have been used to estimate parameters for simulation models to describe different CIMMYT maize germplasm. In the classification example described for Mexico (see above), a simulation model was linked to the GIS (Chapman 1992). The crop simulation model used monthly temperature data to estimate flowering dates, using the parameters of a tropical intermediate genotype for each of the cells in the GIS. The results were then grouped to produce a map of the extent of adaptation for this genotype. Where flowering dates exceeded 70 days after sowing, the genotype was considered to be not adapted. Repeating the simulation with other genotype parameters or for other planting provides a series of GIS databases that can be combined to refine estimates of the extent of areas suitable for such a genotype.

Modelling of phenology is only a starting point, as other genotype characteristics can be simulated. The use of historical weather data would enable climatic risk to be considered in selecting and defining genotypes (Muchow et al. 1991).

Wheat. The experimental evaluation of GxE interactions in wheat trials is laborious and time consuming because it requires experimentation under a wide range of environmental conditions that are typically impossible to control. Applying explanatory crop-growth simulation models, in which the relations between crop and/or cultivar characteristics and environmental conditions are quantitatively described, could improve the efficiency of research. These models systematically analyze the relative contribution of the various factors to yield level and stability (De Wit and Penning de Vries 1985). Moreover, if combined with GIS-based information on spatial and temporal variability in environmental conditions, they facilitate the extrapolation of results in time and space.

A well-validated crop growth simulation model for the wheat crop is being used to analyze crop-environment interactions in ME 1 (irrigated, low-rainfall temperate areas), where a substantial proportion of wheat in developing countries is produced. The primary purpose of this model is to understand management (especially planting date) by genotype interactions.

NRM research: projects, challenges and opportunities for using systems methods

Projects
The contribution of crop-improvement research to the development of sustainable agriculture is primarily indirect: Averting the degradation of resources by helping alleviate poverty and reducing population growth. Systems methods can help enhance the efficiency of crop improvement and thereby speed up achieving the results desired. CIMMYT also feels, however, that there are opportunities for it to contribute directly through NRM research, defined as research that aims to develop farm- or community-level interventions to slow down or reverse resource-degradation processes, as well as increase system productivity. NRM research at CIMMYT understandably emphasizes farming systems where maize or wheat are important. CIMMYT has been involved for several years in a number of NRM research endeavors, the most prominent of which include:

- Collaborative research on productivity and sustainability issues (soil fertility decline, groundwater depletion, water-induced land degradation, build-up of pests and diseases) associated with the rice-wheat cropping pattern in South Asia. Much of this research is conducted in collaboration with IRRI and national programs from Bangladesh, India, Nepal, and Pakistan. CIMMYT has contributed by participating with national agricultural research systems (NARS) and IRRI colleagues in diagnostic surveys to elicit a users' perspective on rice-wheat issues, and in on-farm experiments.
- Research on conservation tillage and maize-legume combinations to improve maize-based system productivity and reduce land degradation (particularly erosion) in Mexico and Central America. Much of this research is conducted in collaboration with a CIAT-led consortium that includes other international agricultural research centers, national programs, and NGOs. CIMMYT has participated with NGOs in farmer-participatory research and extension. It has also done strategic agronomic research implemented under the supervision of a NARS-led research network.

In addition to these two projects, CIMMYT conducts research on conservation tillage for wheat-soybean systems in the Southern Cone of Latin America and for highland maize-wheat systems in Mexico. In addition, it is developing a proposal for ecoregional research on long-term soil fertility issues in maize-based systems in Eastern and Southern Africa.

Opportunities for collaboration in applying systems methods

In pursuing NRM research, CIMMYT recognizes that systems methods, combined with a GIS, can be extremely valuable to:

■ understand the processes that underlie land and water degradation problems;

■ assess the extent and incidence of these problems to allow better targeting of research, including crop improvement and NRM;

■ explore interactions between possible solutions to these problems and environmental characteristics, including weather variability;

■ identify the specific environments in which certain interventions are likely to be most attractive to farmers; and

■ explore implications for the future of ongoing degradation processes in terms of trends in crop yields and production, system productivity and resource quality, given the different assumptions on farmer adoption of alternative resource-conserving practices.

CIMMYT welcomes collaboration with other institutions to apply systems methods for the kinds of purposes described above in the context of our ongoing and proposed NRM research endeavors.

A word of caution

Systems approaches can help assess alternative technologies that aim to improve farm productivity while contributing to the conservation of agricultural resources. However, recommendations developed with the help of these approaches must take into account the broad range of factors that affect the adoption by farmers. Many of these factors are subtle and frequently ignored (Tripp et al. 1993):

■ Models that compare technical alternatives are based in part on information about costs and returns. Many of the techniques that enhance productivity and conserve resources, however, are new to farmers. In these cases, accurate cost-and-return information simply may not be available, especially since farmers are known to tailor technologies significantly to their own agroclimatic and socioeconomic circumstances.

■ Farming systems vary tremendously—much more so than germplasm adaptation zones—and the ability of farmers to adapt resource-conserving practices to their own circumstances plays a key role in the acceptance of these practices. In addition, technologies developed through NRM research are complex and require a great deal of site-specific adaptation. In some cases, farmers will need to be formally trained in new techniques.

■ The economic assessment of alternative practices for a crop may fail to take into account the broader farming system. System-based analysis often gives a different picture of the profitability of alternative practices than analysis based on one or two commodities (or one cropping sequence) within that system. This is particularly true when crop-livestock interactions are strong.

■ Finally, farmers are often unwilling to invest in a resource-conserving technology with downstream benefits if they are not assured access to the improved land resource. Land tenure, therefore, is another factor that determines whether these technologies will be adopted.

Acronyms

CGIAR	Consultative Group on International Agricultural Research
CIAT	Centro Internacional de Agricultura Tropical
CIMMYT	Centro Internacional de Mejoramiento de Maiz y Trigo
GIS	geographic information system
GxE	genotype by environment
IRRI	International Rice Research Institute
ME	mega-environment
NARS	national agricultural research system
NGO	non-government organization

References

CGIAR (1985) International Agricultural Research Centers A study of achievements and potential Washington D C Consultative Group on International Agricultural Research

Chapman S C (1992) Implementing crop models in spatial analyses Invited review paper, Proceedings of the 22nd Annual Workshop on Crop Simulation, 23-25 March 1992, Corpus Christi, Texas, USA

CIMMYT (1992) CIMMYT in 1992 Poverty, the environment, and population growth The way forward Annual Report International Maize and Wheat Improvement Center Mexico City

Corbett J (1993) Dynamic crop environment classification using interpolated climate surfaces" Proceedings of the Second International Conference/ Workshop on the Integration of GIS and Environmental Modeling, September 1993, Breckinridge, Colorado, USA (to be published)

DeLacy I, Fox P, Corbett J, Crossa J, Rajaram S, Fischer R A, Van Ginkel M (1993) Long-Term association of locations for testing spring bread wheat Euphytica (forthcoming)

FAO (1981) Methodology and results for South and Central America Report on the Agroecological Zones Project, Vol 3 , Food and Agriculture Organization, Rome, Italy

Harrington L (1993) Sustainability in perspective Strengths and limitations of farming systems research in contributing to a sustainable agriculture Journal of Sustainable Agriculture (forthcoming)

Muchow R C, Hammer G L, Carberry P S (1991) Optimising crop and cultivar selection in response to climatic risk Pages 235-262 in Climatic risk in crop production Models and management in the semi-arid tropics and subtropics Proceedings of an International Symposium, 2-6 July 1990, Brisbane, Australia

Tripp R, Buckles D, Van Nieuwkoop M, Harrington L (1993) Land classification, land economics and technical change Awkward issues in farmer adoption of land-conserving technologies Presented at the "Seminario para la Definicion de una Metodologia de Evaluacion de Tierras para un Agricultural Sostenible en Mexico"

De Wit C T, Penning de Vries F W T (1985) Predictive models in agricultural production Phil Trans R Soc London B 310 309-315

CIP's experiences in the use of systems analysis and simulation

J.L. RUEDA, C. LEON-VELARDE, T. WALKER and H. ZANDSTRA
Consorcio para el Desarrollo Sostenible de la Ecorregion Andina (CONDESAN), International Potato Centre (CIP), P O Box 1558, Lima 100, Peru

Key words agroecological zones, natural resource management, participatory research, production system, research planning, simulation models, sustainability indicators

Abstract

A consortium, CONDESAN, was set up recently by NARS and CIP to coordinate research on sustainable management of Andean natural resources Program-planning workshops identified a wide range of specific topics for high-priority research Collaboration with other CG-centers is also sought The broad institutional participation is expected to lead to early impact and to encourage sustainable management of resources Physical, biological, and socioeconomic models are essential tools in these endeavors

Introduction

Two years ago, CIP initiated its activities on Sustainable Management of Andean Natural Resources. CIP's ecoregional activities focus on the cool tropical highlands of the Andes. They respond directly to Agenda 21 and to concerns about deterioration of land and water resources, the weak research infrastructure available to address these problems, and to the extreme poverty that they engender.

CIP hosted an Andean agroecosystem workshop in March 1992, at which time it was proposed that the center help coordinate research on the sustainable management of Andean natural resources. The activities proposed included research on biodiversity, land and water management, agricultural and environmental policy, pastures and livestock, cropping systems, and nutrient cycling.

Following extensive consultation, it was decided to develop a research consortium (CONDESAN) that could integrate the activities of researchers and development workers. CIP was requested to provide support in project planning, monitoring, communications, and information exchange. The center was also asked to take on research tasks for which it had a comparative advantage. To carry out these activities, heritage sites representative of major agroecologies the Andes were selected.

Governance of the consortium is shared, but not representational. Its structure is open and informal, allowing for membership by many stakeholders, including other networks that are willing to share costs. Recently, two such networks—one on Andean pastures and a second on farming systems research methodology—became active members of the CONDESAN.

The consortium's strong participatory orientation provides an effective mechanisms for joint planning, approval, reporting, and monitoring. An advisory council that will include regional research administrators, environmental specialists, donors,

P Goldsworthy and F W T Penning de Vries (eds), Opportunities, use, and transfer of systems research methods in agriculture to developing countries, 289 - 294
© 1994 *Kluwer Academic Publishers*

and representatives from international centers will provide oversight and programme direction. An executive committee, representing all participating groups, is responsible for the day-to-day programming and coordination.

Much of the field work will be conducted at benchmark sites, representative of major ecologies of the Andean ecoregion. Sites have been proposed initially for Colombia, Ecuador, Peru, and Bolivia. At these sites, research will be of a participatory nature, involving local communities and partner institutions. The research will focus on systems modelling to establish priorities for the monitoring and evaluation of impact of existing and alternative land-use systems, maintenance of biodiversity, and in the design and implementation of policies. The work will pursue component technology interventions at the commodity level to maintain productivity gains and achieve sustainable agricultural production without jeopardizing the natural resource base.

Program development in CONDESAN

A program-planning-by-objective (PPO) workshop held in August 1992 provided well-defined outputs for research on eight/nine lesser known Andean root and tuber crops. Work in this area is now well underway. Program activities include in situ and ex situ conservation, germplasm characterization, production of planting materials, and the rescue of traditional knowledge. A total of 52 projects presented by 24 institutions have been initiated in January 1993. A follow-up meeting of the steering committee and donors was held in Quito in July 1993 to discuss implementation and monitoring.

A separate PPO workshop was held in March 1993 to address other issues, among them land and water management, policy and socioeconomics, livestock and pastures, agroforestry, management training, and communications. Forty-three researchers from national, international, and donor agencies participated. The consortium is now soliciting projects that will address research priorities identified at the meeting.

In August 1993, participating non-governmental organizations (NGOs) created a working group on sustainable rural development to contribute to the understanding, development, and implementation of policies that encourage sustainable rural development. The group includes 16 of the most prestigious NGOs from the region. A well-defined project, with activities closely related to the outputs of the March PPO workshop has been prepared. It includes strong farmer and community participation.

INFOANDINA, the communications system for the Andean ecoregion is underway through the active participation of national electronic communication networks.

The participation of other CGIAR centres and international organizations will support work on agroforestry and soil and water management (ICRAF and IFDC), policy (IFPRI), commodity support (CIMMYT), pastures and livestock (ILCA), and mountain agriculture (UNEP). Close collaboration has been established with IPGRI on biodiversity of Andean crops, and with ISNAR in research planning and policy. Recently, CIAT expressed its wish to collaborate in agroecological mapping and the development of research methodologies that encourage farmer participation. Collaboration with CIAT will allow for diagnostic and field research that ranges from the high Andes to mid-altitude hillside areas.

It is expected that broad institutional participation will lead to early impact and will encourage local and regional policies that support sustainable management of Andean natural resources.

Systems modelling in the Andean ecoregion

Modelling the fate of pesticides in the Andes

Assessment of pesticide fate in developing countries is a potentially important but very difficult undertaking. Actual measurement of pesticide residues in soil layers is hindered by difficulties in field sampling and monitoring, storage, and analysis of soil and water samples, and finding technically specialized staff. Therefore, the judicious use of models is needed to arrive at reliable estimates and informative scenarios.

In the Carchi area of Northern Ecuador, models developed by soil scientists on the physical processes of pesticide movement in soils and water are being combined with those of economists featuring information on the timing of pesticide use. The latter feed information into the pesticide movement models. These, in turn, provide outputs which are used in simulations of the Carchi potato/pasture systems. Some of the scenarios that are being examined using these stochastic simulation models, show the probable consequences of the introduction of a uniform tax on pesticides, a toxicity-based tax, an environmental-mobility-based tax, and of the introduction of improved pest management and varietal technologies.

Agroecosystems analysis

CONDESAN activities focus on two main themes: biodiversity and land/water resources. Work on biodiversity provides information on the balance and the diversity of the many Andean crops and animals that are of potential value for food production in environments where risks are high. CIP collaborates with the Graduate School of Production Ecology (WAU) and several Latin American institutions in research on land and water management, to develop methods for identifying the probable constraints inherent in selected land-use options.

At benchmark sites, emphasis is placed on the development and evaluation of tools for the ex ante evaluation of the performance and sustainability of different land-use systems. The work includes the formulation and validation of several types of physical, biological and socioeconomic models. The models include linear programming and principal component analysis, response surfaces, participatory application of decision-making trees, multiple-goal programming, and risk models. The study involves six Latin American institutions and two European participants.

The models are defined using parameters derived from crop and animal models together with data on climate and land-use characteristics. The crop models, particularly the part of them that relates to the responses to different levels of inputs, are validated through field experiments. The animal and farm models are designed to create different scenarios, from which to examine biological and economic relationships at different levels of the systems concerned. The construction of the models takes into account the major current land-use systems, the existing constraints to attaining sustainability, and the possible means of overcoming these constraints. Models will be used to identify where and when land degradation is likely to be most severe in the Andes. For example, we need to be able to target research on those areas where soils are most erodible and rainfall is most erosive. We also need models to assess frost and drought risk. The use of wider interregional models is expected to provide information on spillover effects as the impact of improving technology and income of Andean households on migration and subsequent land degradation in the Amazonian ecoregion.

CONDESAN activities are also directed to identifying measurable and verifiable indicators of the consequences of resource use practices at the farm level. The indicators must reflect whether land-use practices are adequate or inadequate. Limitations of current practices need to be identified and recommendations made on how to correct them. It calls for interdisciplinary team work and appropriate procedures for the transfer of technology for sustainable land management. Sustainability is one of the major concerns that agricultural systems specialists are facing. Because of the complexity of agroecological systems, a combination of several indicators is required in order to measure sustainability. These indicators arise to explain sustainability in a broad and a narrow sense. The broad sense should include a set of indicators from different disciplines considering a quantitative and a non-quantitative approach; the narrow sense represents a quantitative measure on bio-physical or bio-economic response from different actions that have been carried out at producer level within a particular agroecosystem. Table 1 lists some indicators of sustainability to be considered under study.

A CONDESAN team at the benchmark site of Puno, Peru, is using several mathematical models to measure bio-economic sustainability of a specific agricultural production systems. The sustainability is defined in quantitative terms, as the change in the value of bio-economic indicators over time. For positive indicators, a zero or positive slope indicate that the agricultural system concerned is sustainable; the opposite would indicate the system is not sustainable and that some technological intervention is required. The use of mathematical models requires a data base of the values of different indicators over time. These data are sometimes difficult to obtain.

Computer simulation is an alternative approach. An example is a computer simulation model of an alpaca production system which was used to estimate a bio-economic measure (the combined income from fibre and meat production at constant price) of system productivity during 15 years (figure 1). There were six replications of a typical alpaca production system in Puno (80 ha). The alpaca production system had a grazing capacity of 0.9 alpaca ha^{-1} $year^{-1}$. The growth rate of the forage rate was 2.25, and 10.80 kg dry matter ha^{-1} day^{-1} for dry (70 ha) and wet (10 ha) zones of native pastures. The corresponding values for the digestibility of the forage were 57 and 63 percent, respectively.

Table 1. Some indicators of sustainability

Groups of indicators	Indicators of sustainable land-use practices
Rainfed cropping systems, with or without livestock	Agronomic practices Crop yields (trend/variability) Nutrient balance Diversification (type/change) Land-use change (type/rate) Maintenance of soil cover Production per capita Weeds/pests/diseases (kind/intensity)
Ecological indicators	Soil erosion (type/rate of change) Soil organic matter (change) Soil salinity, pH (change) Soil acidity/organic matter (change) Water supply/quality (change) Biodiversity (species/germplasm) Diversity of landscape
Economic indicators	Net farm profitability (level/change) Efficiency of inputs Availability of domestic fuel Policies Off-farm income (level/opportunities) Labor availability (age/distribution)
Social indicators	Quality of life Literacy (level/gender) Human health Farm family health Adoption of conservation practices Market structure/access Land tenure Education (levels/access/gender)

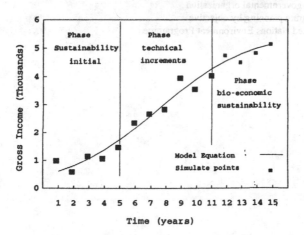

Figure 1. Phases of sustainability estimated by simulation of a alpaca production system.

During the analysis it was feasible to determine three phases:

1. an "initial" phase;
2. a phase of increasing production resulting from technical increases, between four and eight years; and
3. a phase of "stable" production (figure 1).

The changes in the value of the combined income are represented in figure 1 by a logistic function:

$$Y = \frac{B_0}{1 + b_1 e^{-b_2 t}}$$

The parameters that describe the logistic function ($r^2 = 0.91$) were $b_0 = 5553.7$ which represents the level of bio-economic sustainability; $b_1 = 11.34$ represents the technical effort to achieve positive increments; and $b_2 = 0.33$ which indicates the rate of deceleration of the system (diminishing returns). The average of bio-economic response over 15 years of the system was $2,095 \pm 44$ \$ y^{-1}.

Acronyms

CGIAR	Consultative Group on International Agricultural Research
CIAT	Centro Internacional de Agricultura Tropical
CIMMYT	Centro Internacional de Mejoramiento de Maiz y Trigo
CONDESAN	Consorcio para el Desarrollo Sostenible de la Ecorregion Andina
ICRAF	International Centre for Research in Agroforestry
IFDC	International Fertilizer Development Center
IFPRI	International Food Policy Research Institute
ILCA	International Livestock Centre for Africa
IPGRI	International Plant Genetic Resources Institute
ISNAR	International Service for National Agricultural Research
NGO	non-governmental organization
PPO	program planning by objective
UNEP	United Nations Environment Programme

Experience of the use of systems analysis in ICARDA

H.C. HARRIS, T.L. NORDBLOM, A. RODRIGUEZ and P. SMITH
International Center for Agricultural Research in the Dry Areas (ICARDA), P.O. Box 5466, Aleppo, Syria

Key words: linear programming, production function, simulation model, spatial weather generator, systems

Abstract
The International Center for Agricultural Research in the Dry Areas (ICARDA) has an ecoregional mandate and carries out its research within a systems context. One of the requirements of this approach is methods to integrate data to provide information on the functioning of systems within variable environments. The paper briefly describes computer-based tools developed or in use by ICARDA for this purpose. These include: dynamic crop simulation models and production functions linked to a spatial weather generator and a prices generator; models of hydrology and groundwater; and discrete stochastic linear programming models to consider risk. Some applications are mentioned. The paper also considers future needs for models and for training to transfer the tools to new users.

Introduction

The ecoregional nature of its mandate has led ICARDA, from its inception in 1975, to carry out its research in a systems context. The research on systems encompasses a broad range of issues: analysis of environments to define their potential productivity and major constraints; management of soil, natural vegetation, water and nutrients to maintain a strong biophysical resource base for agricultural production; management of crops, pastures and animals and the deployment of human and economic resources in production systems; and the influence of governmental policies on the functioning of systems.

The research methods employed are necessarily diverse. They range from the collection of primary data in on-station classical agronomic trials, when we must have full and long-term control of land and resources, through on-farm trials and demonstrations, surveys of physical, agronomic, social and economic factors, to the use of external primary (weather, imagery, etc.) and secondary data (climate, soils, topography, production statistics, costs, prices, land tenure, water policies, etc.). There is little work directed to the complexity of whole system; rather, individual studies focus on selected components of systems that are perceived as having high priority. An increasing proportion of the research is carried out in, and largely by, national programs with ICARDA staff acting as catalysts, collaborators, and advisors.

The research needs tools to integrate data and provide information on the functioning of systems. We presume that, in the context of this meeting, these tools (models) and their development, application, validity, and utility should be the topic of this paper.

P. Goldsworthy and F.W.T. Penning de Vries (eds.), Opportunities, use, and transfer of systems research methods in agriculture to developing countries, 295 - 302.
© 1994 *Kluwer Academic Publishers.*

There is a wide divergence of thought within ICARDA as to the efficacy of systems analysis. Some regard systems analytical techniques as the only way to rationally approach some of the questions, while others hold that the 'systems approach' has been around a long time, with no obvious effect. In part this divergence is related to how individuals define 'a system'. Some accept the widely quoted definition, 'a limited part of reality with related elements' (e.g,. De Wit and Goudriaan 1978), and agree that, provided the system is judiciously chosen and a clear boundary can be delineated, the behavior of the system can be studied, understood, and eventually modelled by sets of mathematical functions. To others, a 'system' (in the agricultural context) is nothing less than the sum of the physical, biological, economic, social, demographic, cultural, and political factors which impinge on agricultural production. These argue that models are often different in kind from the real world in terms of their dynamics; they are irrelevant in the way they process risk, and totally lacking in any effective mechanism for handling uncertainty.

Perspective of systems analysis in ICARDA

Early in its life, ICARDA espoused systems analysis concepts through the development of a regionally-adapted wheat-growth simulation model, SIMTAG (Stapper 1984), with the objective of using it for agroclimatic analyses of wheat-growing areas. At that time, a proposal to institutionalize the techniques by the appointment of a modeler, whose role would be the integration of research findings, was widely discussed within the Center, but not accepted. The result has been that some work has continued, but it is mainly carried out by scientists who can devote only a small part of their effort to it, and progress has been frustratingly slow. This has been, in part, redressed by collaborative work with advanced institutions in the UK, North America, and Japan.

Systems analysis methods in ICARDA

Some, but not all, of the components of systems analysis in use or being developed at ICARDA are illustrated in figure 1. The figure indicates major components and the linkages among them, but is not a flow chart. Dashed boxes indicate components under development. Most of the components have been discussed elsewhere (Harris et al. 1988), but some have been modified and others added in the interim. We will briefly mention those discussed previously, but will focus most of this section on the modifications and additions.

Research databases
From both on-station and on-farm agronomic trials, ICARDA has accumulated substantial databases on crop yields in relation to the management of cropping systems. Drawn from sites deliberately chosen to span variability, and including eight or more years, these data now demand crop-model development and validation.

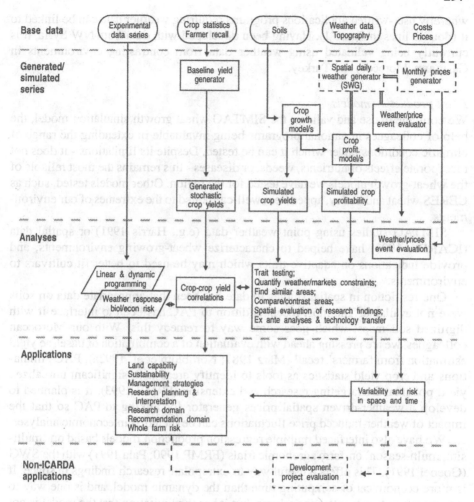

Figure 1. Components of systems analysis methods in use in ICARDA.

Data sets from similar sources relating animal production to quantity and quality of feedstuffs, together with data on feedstuffs on offer through the year, are now also available to begin to formulate integrated models of crop and animal production.

Spatial daily weather generator

The spatial daily weather generator (SWG) serves to both extend weather data series and to allow interpolation between sparse weather-recording sites (Goebel 1990, 1991). Together with crop simulation models and other facilities, it is currently being developed into a Package for Agroecological Classification (PAC). It has an interface whereby user-written applications programs requiring weather data can be linked to it (double lines in figure 1). Having been developed with data from NW Syria, it is currently being validated using weather data from contrasting environments in Germany, Morocco, and Turkey.

Plant production models
We continue to use and validate the SIMTAG wheat growth simulation model, the help of colleagues in national programs being invaluable in extending the range of climatic conditions under which it can be tested. Despite its limitations - it does not incorporate effects of nutrients, weeds, or diseases - this remains the most reliable of the wheat-growth models we have tested for the region. Other models tested, such as CERES wheat and barley, appear less well-calibrated to the extremes of our environments.

SIMTAG studies using point weather data (e.g., Harris 1991) or spatial data (ICARDA 1993) have helped to characterize wheat-growing environments, and provide indications on adaptive traits which may be used to better fit cultivars to environments.

One restriction in spatial studies to date has been that appropriate data on soils were not available. The most recent addition to PAC has been to interface it with digitized soil maps, which goes some way to remedy this. With our Moroccan colleagues, we are pressing ahead with evaluation of a combination of baseline yield estimation from farmers' recall (März 1987; Nordblom et al. 1992a), PAC simulations and crop yield statistics as tools to identify areas with significant unrealized yield potential for targeting research and extension (FRMP 1993). It is planned to develop a weather-driven spatial prices generator for linking to PAC so that the impact of weather-induced price fluctuations can be included in economic analyses.

We have also interfaced multiple regression production models based on 'multi-site, multi-season' on-farm agronomic trials (FRMP 1990; Pala 1991) with the SWG (Goebel 1991). This offers a useful way to extrapolate research findings in space. It is more economical of computing time than the dynamic model, and is one way to incorporate nutrients into our analyses; but it has the limitation that the model is not valid beyond the conditions experienced during the experimental program.

Also of interest in this category are simplified crop growth models for use in optimizing the timing and quantity of supplemental irrigation, and for yield forecasting. Both applications have common requirements for robust soil- and crop-water-use routines, a generalized crop development routine, and facilities for using real-time weather data together with probabilistic expressions of future expectations of weather conditions.

Profitability models
Other developments in hand include risk analysis through crop-profitability models, which will take into account the impact of weather variability not only on crop yields, but also on market prices. Initially, production function models, analogous to the regression models above, have been used (e.g., Saade 1991), but in the longer term

economic models will be linked to PAC. It is anticipated that they will allow of both ex ante and ex post analyses of technological changes, and will provide information of value to planners and policymakers.

Whole farm analysis
Linear programming packages are being applied to model the whole-farm economics of production systems (e.g., Nordblom et al. 1992b). Using as a basis data from trials with six or more years of records and farm and market surveys, ex ante analyses compare the potential profitability of crop-crop and crop-pasture rotations, integrated with livestock, in different environments. Sensitivity analyses illustrate the effect of changes in yield potential to help to identify research goals and potential adoption domains. Recent introduction of considerations of risk into these analyses will increase their utility.

Hydrology models
Studies on water harvesting call for the use of models of the hydrology of small catchments. The catchment may be a field where a proportion of the surface is prepared to increase runoff onto a crop planted in the remaining area, or it may be a natural surface of tens of square kilometers. Currently, work is under way on both scales.

Groundwater model
Concern about the overuse of aquifers in the region has led to the initiation of studies relating to groundwater management. One of us (Smith) expects to begin working with a groundwater model in the immediate future, in a study related to decision making by farmers on the use of ground water.

Geographic information systems
There are currently three PC-based GIS packages in use: ILWIS, IDRISI, and ATLAS-GIS. ICARDA has taken advantage of the offer from Environmental Systems Research Institute (ESRI) to the CG Centers to supply licenses for their ARC/INFO products. We expect to see use of GIS expand, both for visualizing output from spatial studies in PAC, and as we begin to use imagery to evaluate degradation of natural resources (ICARDA 1993; PFLP 1993).

Status of the research

As indicated above, progress is frustratingly slow. ICARDA lacks a critical mass of people with skills in the area of systems analysis, particularly in the physical and biological sciences. We would like to think that, through ICASA, we might be able to find ways to strengthen our capabilities and increase our rate of progress. Other major bottlenecks are:

Data
Access to environmental data to support the software remains a major bottleneck. We look to current initiatives among international institutions to help to redress this situation.

We have begun to compile a database of crop statistics from national records at the smallest available geographical level of aggregation, and to match them with climatic data sets. The next step should be to overlay on these data from surveys which have been, or are being, conducted through the region for many different reasons, in order to make the information more available. Soils data also need to be included, and the digitization of soil maps, currently being undertaken by some national programs through bilateral projects with other international or donor organizations, is a welcome initiative.

Plant models
The cereal models we have been using in agroecological studies need to be upgraded by the inclusion of robust routines for nutrients, pests and diseases, or replaced by existing models that combine an ability to mimic the extremes of our environments with these additional facilities. To date, we have not seen the integration of modelling into plant breeding and physiology programs for hypothesis generation and testing, but that is likely to change in the near future.

We lack models for both crop and forage/pasture legumes, although data sets now exist to support development or calibration. Because of the degraded state of the systems of the region, and the poverty of many farmers, provision of N by legumes is seen as being a key factor in long-term maintenance of production. If we are to address questions of sustainability, we feel that we must incorporate the effect of legumes into our package of analytical tools.

Animal production model
An animal performance model based on the animals' feeding behavior and their nutritional and reproductive physiology is needed to both simulate the performance of the animals under different feeding and management regimes, and integrate the effects of grazing into studies relating to pastures and natural vegetation.

Training

The application of models is fraught with dangers unless users have an adequate knowledge of the principles (statistical, economic, biological, physical) on which they are structured. We are strongly of the opinion that it is not sufficient to train people in the application of complex models, but believe that training needs to begin, not with a model, but with what is modeled. When that is understood, a model becomes a logical statement of what is known and its behavior and projections can be interpreted intelligently. The best attempt to train in this way has unquestionably been the Simulation and Systems Analysis for Rice Production (SARP) program carried out by the Centre for Agrobiological Research and Department of Theoretical Production Ecology, Wageningen, in conjunction with the International Rice Research Institute (Ten Berge 1992). However, we believe the authors of that program would be the first to acknowledge that the effectiveness with which their national program counterparts were able to take up and use the tools of systems analysis varied greatly with educational background. We suggest that training needs to begin earlier in a scientist's life — at the early postgraduate level, when deficiencies in knowledge of basic principles can be identified and most readily remedied.

A need for intensive training means that these methods will not be quickly transferred. It also implies that more resources are needed for training, or that the available resources need to be differently deployed.

Acronyms

ESRI	Environmental Systems Research Institute
GIS	geographic information system
ICARDA	International Center for Agricultural Research in the Dry Areas
PAC	package for agroecological classification
SARP	simulation and systems analysis for rice production
SIMTAG	wheat growth simulation model
SWG	spatial daily weather generator

References

Ten Berge H F M (1992) Building capacity for systems research at national agricultural research centres SARP's experience Pages 515-538 in Penning de Vries F W T, Teng P, Metselaar K (Eds) Systems approaches for agricultural development Kluwer Academic Publishers, Dordrecht, The Netherlands with IRRI, Manilla, Philippines

FRMP (1990) Four year summary of fertilizer research on barley in N Syria (1984-1989) Pages 29-93 in Farm resource management program, Annual Report for 1990 ICARDA, Aleppo, Syria

FRMP (1993) Validation and use of farmer interview methods to define crop yield distributions for yield-gap estimation with crop-growth simulation Pages 26-39 in Farm resource management program, Annual Report for 1992 ICARDA, Aleppo, Syria

Goebel W (1990) Spatial rainfall generation Examples from a case study in the Aleppo area, NW Syria Pages 179-214 Farm resource management program, Annual Report for 1989, ICARDA, Aleppo, Syria

Goebel W (1991) Spatial weather generation More examples from a case study in the Aleppo area, NW Syria Pages 152-183 in Farm resource management program, Annual Report for 1990 ICARDA, Aleppo, Syria

Harris H (1991) The relative impact of water and temperature constraints on wheat productivity on lowland areas of West Asia and North Africa Pages 15-34 in Acevedo E, Conessa A P, Monneveux P, Srivastava J P (Eds) Physiology-breeding of winter cereals for stressed mediterranean environments INRA, Paris, France

Harris H C, Goebel W, Nordblom T L, Jones M J (1988) Defining the impact of variable weather on agricultural production and design of new technology Pages 283-285 in Challenges in dryland agriculture - A global perspective Proceedings of the International Conference on Dryland Farming Amarillo/Bushland, Texas, U S A, 15-18 August 1988

ICARDA (International Center for Agricultural Research in the Dry Areas) 1993 Annual Report for 1992 ICARDA, Aleppo, Syria

Marz U (1987) Methods to simulate distributions of crop yields based on farmer interviews ICARDA, Aleppo, Syria 117 En

Nordblom T, Shomo F, Farihane H, el Mourid M, Boughlala M, Harris H (1992a) Estimating the frequency distribution of farmers' crop yields. Pages 145-158 in Farm resource management program, Annual Report for 1991. ICARDA, Aleppo, Syria.

Nordblom T L, Christiansen S, Nersoyan N, Bahhady F (1992b) A whole-farm model for economic analysis of medic pasture and other dryland crops in two-year rotations with wheat in northwest Syria. ICARDA, Aleppo, Syria. 92 pp.

Pala M (1991) Four year summary of fertilizer research on wheat in NW Syria. Pages 81-105 in Farm resource management program, Annual Report for 1990. ICARDA, Aleppo, Syria.

PFLP (1993) Rangeland resource evaluation for conservation and management. Pages 238-246 in Pasture, forage and livestock program, Annual Report for 1992. ICARDA, Aleppo, Syria.

Saade M (1991) An economic analysis of fertilizer allocation and import policies in Syria. PhD Dissertation, Michigan State University, East Lansing, Michigan. USA.

Stapper M (1984) Simulations assessing the productivity of wheat maturity types in a Mediterranean climate. Unpubl. PhD Thesis, University of New England, NSW, Australia.

De Wit C T, Goudriaan J (1978) Simulation of Ecological Processes. Simulation Monographs. PUDOC, Wageningen, The Netherlands.

Systems analysis for agroforestry research for development: the experience of and future use by ICRAF

D.A. HOEKSTRA
Free-lance consultant, formerly with the International Centre for Research in Agroforestry (ICRAF), P.O. Box 30677, Nairobi, Kenya

Key words: agroforestry, diagnosis, design

Abstract
The experience of the International Centre for Research in Agroforestry (ICRAF) with system analysis is evolving from a qualitative to a more quantitative approach. However, the basic philosophy that knowledge of an existing system (i.e. diagnosis) is essential to plan and evaluate relevant and effective programs in agroforestry research and development (i.e. design), remains the same. It is expected that quantifying the diagnosis of land-use systems and the design of agroforestry technologies will strengthen this basic philosophy and result in a better land use and a more effective use of the scarce research resources. ICRAF's experience in collaborative research with national and international institutions has shown that interinstitutional linkages can help plan and implement agroforestry research and development. Such linkages should be flexible and able to suit a specific need and the prevailing institutional framework. Because of the multi-sectoral nature of agroforestry, these methodological and institutional experiences are useful in developing natural resource management (NRM) research. In addition to interinstitutional linkages, IARCs and NARS will have to deal with their own institutional challenges in designing more efficient research programmes.

Introduction

In the past decade, systems analysis at ICRAF has focused on examining land-use systems to identify or design the potential role of agroforestry in a particular system. This approach has been used extensively in designing research networks in sub-Saharan Africa, which was carried out in collaboration with national and regional research institutes. A major objective of these networks was creating an institutional framework and capability. This paper describes the approach and methods used in system diagnosis and design, as well as the resulting collaborative research programmes. The paper then reviews the approach, methods, institutional arrangements, and some new initiatives, in particular the NRM initiative for the highlands of East and Central Africa, initiated in 1992.

The diagnosis and design methodology

The objective
The Diagnosis and Design (D&D) methodology adopted by ICRAF is based on the philosophy that knowledge of an existing situation (i.e. diagnosis) is essential to plan and to evaluate (i.e. to design) relevant and effective programs of development-oriented agroforestry research. In principle, D&D borrows from established method-

P. Goldsworthy and F.W.T. Penning de Vries (eds.), Opportunities, use, and transfer of systems research methods in agriculture to developing countries, 303 - 311.

ologies such as baseline surveys, sectoral studies, farming-systems surveys, rapid rural appraisals, and feasibility studies. These have been used by many research and development agencies and institutions. D&D, however, is unique in that it has been developed to examine existing land-use systems, identify constraints and their causes, and design appropriate agroforestry technologies to overcome these constraints. The eventual aim of D&D is to improve the performance of land-use systems for the benefit of rural households (Raintree 1987).

The land-use system concept

The unit for D&D can be defined at a macro or a micro level. At the (macro) level of an ecological zone, ICRAF defined land-use systems as a sub-group of a population of land units in which the nature of the structural features, the functional features, and the production constraints of the farming systems within the group, are sufficiently similar for the same agroforestry technology to be appropriate throughout. This is also referred to as an agroforestry recommendation domain area. In addition, if introduced, the technology could be expected to give similar results in all the units of the group (Avila and Minea 1993). At the micro level, a land-use system is defined as a "management unit with a distinctive combination of crops, livestock, trees and/or other production systems on a given unit of land where specific outputs are desired and obtained" (Avila and Minea 1993). This management unit can be a household, a clan, a communal group, cooperatives or a company.

Diagnosis of land-use system constraints and potentials

Every system has to be assessed as to what prevents the household or land user from obtaining optimal outputs from the available resources. Shortcomings can be measured in terms of the extent to which present outputs fall short of potential outputs. Such shortfalls may occur at any level of a land-use system.

The next step is a constraint analysis to identify the underlying causes of the observed constraints. Identifying the main causes is the starting point for designing the agroforestry interventions to remove them.

Design of potential agroforestry technologies

In designing potential agroforestry technologies, researchers may use established technologies or new technologies that are specifically designed to address the issues that cause the system to perform poorly. These new technologies are usually referred to as notional technologies. Established and notional technologies are evaluated in terms of their technical potential, the resources and capability required of the farmer, and the infrastructure and support services required. They are also compared with other, non-agroforestry interventions.

To facilitate the design process, ICRAF developed a computer programme (SCUAF), which models the expected soil changes under agroforestry and the corresponding crop yield responses (Young 1988). ICRAF also commissioned the development of a programme (MULBUD) to model and evaluate multi-period costs and benefits (Etherington 1984).

Review of the D&D methodology

Although D&D has helped focus ICRAF's collaborative research programmes in sub-Saharan Africa, several opportunities for improvement have been identified.

Delineating and quantifying land-use systems

Delineating and quantifying geographic areas as agroforestry recommendation domains is a dynamic exercise. New findings from diagnostic and design research, derived in part from modelling, will lead to the introduction of new technologies and the redefinition of the bio-physical and socioeconomic boundaries for existing technologies. Also, to translate research findings into practical development plans, the area of each technology recommendation domain has to be estimated or measured. A more flexible, quantitative, geo-referenced system to store and retrieve relevant information to delineate and quantify land-use systems will therefore be introduced. ICRAF is now planning to use a geographic information system (GIS) database for its new global work on alternatives to slash-and-burn systems in Latin America, Africa and South East Asia. It will also use GIS for the newly initiated IARC/NARS initiative for NRM in East and Central Africa. Whether this tool will be useful in delineating land-use systems will depend on the availability of relevant secondary data and the interpretation of imageries from remote sensing.

Quantification of problems and constraints

Most of the early diagnostic studies stated problems in terms of "insufficient fuelwood", or "low and declining crop yields", or "insufficient protein in the available animal fodder during the dry season".

Attempts are now being made to quantify some of these statements with the help of follow-up studies. Whole-farm nutrient budgets are used to quantify nutrient gaps and household-level models to determine supply and demand for wood, fuel, and poles. Most of these attempts at modelling use site-specific data. However, the systematic characterization now begun, which goes from the global level to the plot level, will provide the necessary data for extrapolation.

Design of agroforestry technologies

The biggest problem in the D&D approach may be the lack of quantitative data on agroforestry technologies. Technologies were selected or designed because they addressed the problems diagnosed and were perceived to have an impact. Since it was difficult to quantify this impact for many technologies, research programmes were started to measure impact as well as the processes that control it.

ICRAF is now trying to improve the design of agroforestry technologies by developing an expert system and by modelling agroforestry systems. The expert system will be a synthesis of the research and development findings of specific technologies, across different biophysical and socioeconomic environments. It is hoped that such a synthesis of empirical data will help define the critical socioeconomic and biophysical boundaries of different systems. The information can then be used to "match" technologies to land-use systems. Where information on the performance of a given technology is inadequate, ICRAF will adopt a different approach to define the bio-physical limits of the technology. The approach involves standardized field trials at a range of locations that were chosen to represent a wide range of soil and climate conditions. In southern Africa, such an initiative is currently underway for improved fallows, using *Sesbania sesban* and *Tephrosia vogelii*.

The biological modelling of agroforestry will be undertaken by a network of international institutes that is coordinated by the Institute of Terrestrial Ecology (ITE) in Edinburgh, UK. The modelling programme will look at three main priority areas: below-ground processes, modelling tree-crop interactions, and nutrient recycling in agroforestry systems (Agroforestry Modelling Newsletter 1993).

For 1993/94, a number of modelling tasks were planned:

The first task will use a simple tree growth model and a simple crop growth model in succession, to simulate the main pools and fluxes of nitrogen in rotations of tree-fallow and crops in semi-arid regions.

The second task will be to use the MAESTRO (Wang and Jarvis 1990) and HYBRID (under development at ITE) models to define the combined light interception and water use of trees, and to use the output as the environment inputs to drive the PARCH (Bradley and Crout 1993) model for some tropical crops.

The third task will be to use soil models to define the pattern of litter production that is required by tropical soils to meet the demand of crops for nitrogen. For each pattern of input, models will predict the flux of mineralized nitrogen and the nitrogen available to plants. If the pool of nitrogen drops to zero, the nitrogen supply from decomposing litter will be deemed to be insufficient to meet the crop demand. The Hurley soil sub-model will be used, rather than CENTURY (Parton et al. 1992).

The fourth task will be to define the essential elements of a coupled tree-crop model for carbon, nitrogen and water. Although no model currently exists, a simple equation has been developed by ICRAF (1992):

$$I = F - CM + P + L$$

where:

I = the net effect of tree-crop interactions,
F = the benefit of prunings, including the effects of nutrients and mulch (e.g. micro climatic changes at the soil surface such as less evaporation from the soil and modification of soil temperature),
C = the yield reduction caused by inter-specific competion,
M = the consequences of above-ground micro-climatic changes in temperature, light and humidity,
P = the consequence of changes in soil physical properties,
L = the benefit of soil and water conservation.

Each factor can be expressed as a percentage of the sole crop yields in the absence of interaction with trees.

In addition, in 1995, when additional social scientists will be available, ICRAF will initiate integrated economic-biological modelling of land-use systems.

Prioritization of systems and technologies research
In prioritizing research, decisions can be made at various levels, ranging from the global to the farm level. At each level, policymakers and scientists try to influence the decision-making process by highlighting their specific interest or concern. D&D provides an opportunity to identify and systematize these decisions. More guidance can be provided, however, by defining criteria for prioritization between and within ecological zones and between potential technologies in a zone. The decision to select a certain priority zone or land-use system must reflect the potential for agroforestry development. Common criteria used to measure this importance are the size of the area and the size of the population affected by a particular agroforestry technology or set of technologies. But these criteria should be used together with criteria that indicate the impact of the technology in terms of the extent to which it can contribute to solving problems at the farm, national and global levels.

For example, in developing research programmes within agroecological zones in sub-Saharan Africa, higher priority was given to land-use systems and/or technologies that were of greatest common interest. In subsequent research, it was found that some of the soil fertility technologies thus selected had only a limited impact in the zone.

It should be understood, however, that the final word rests with decision makers, who may have different priorities, depending on differences in private and public interests. An economic analysis of the various points of view will highlight these differences and can assist in introducing policy measures that reconcile them.

A summary of the proposed components of the system analysis for the diagnosis and design process for agroforestry are shown in figure 1.

Review of institutional linkages

Because of its complexity, agroforestry requires a multidisciplinary approach. In a NARS that is mainly commodity oriented, agroforestry research may be spread over various institutes, which may result in duplicated efforts and an inefficient use of scarce research resources. Interinstitutional linkages for planning, coordinating and implementing agroforestry research are therefore a key element in developing a proper agroforestry research infrastructure.

Linkages for research planning and coordination
The agroforestry research networks in sub-Saharan Africa have shown that inter-disciplinary and interinstitutional linkages are important. These linkages involve the joint planning, conduct, analysis, and interpretation of research.

Interinstitutional planning activities usually do not require major organizational changes within parent institutions. It is most efficient when the responsibility for

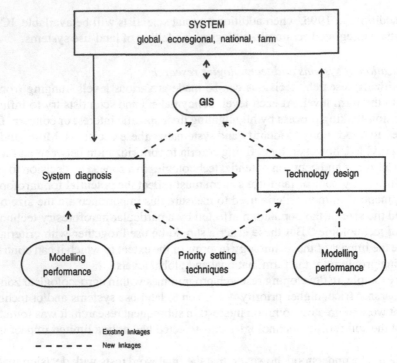

Figure 1. The use of new system analysis tools and techniques in diagnosis and design

national research planning rests with a single institution. A technical committee consisting of the heads of the relevant departments can take charge of the coordination. This is the case, for example, in the recently established National Agricultural Research Organisation (NARO) in Uganda, where an agroforestry committee, chaired by the director of Uganda's Forestry Institute, is responsible. The Institut des Sciences Agronomiques du Burundi (ISABU) and the Institut des Sciences Agronomiques du Rwanda (ISAR) have similar committees. A disadvantage of these "in-house" committees may be that they do not always include potential partners from outside the institutions. When different institutions share the responsibility for research, a secretariat, located in an independent agency, may be responsible for the coordination. In Kenya for example, the National Council for Science and Technology fulfills this coordinating role. Participating institutions may take turns in chairing the secretariat of the agroforestry committee. To institutionalize the coordination function, a committee may be given legal status. This would allow governments and donor agencies to channel funds directly for the implementation of the planning and coordination activities that depend on interinstitutional cooperation.

It is often difficult to institutionalize long-term planning and to coordinate research in regional networks, in particular when there are no established regional bodies or mechanisms to coordinate research activities. Although international agricultural research centers (IARCs) have supported some regional networks, longer-term solutions have to be found to ensure continuity after the IARCs stop their direct involvement in such networks.

Linkages for research implementation
Interinstitutional linkages for research implementation vary between countries. Without an institutional framework for research, the question arises whether it is better to opt for a strategy of joint implementation or for one where institutes carry out coordinated but independent research. To decide on a strategy, decision makers need to consider issues such as the capabilities of the institutions and their policies on collaborating with other institutions.

The research capability of institutes depends on the geographic area they cover, their responsibilities for commodity research, and whether their research is adaptive or process oriented. In developing a national agroforestry research network, the Indian Council for Agriculture Research (ICAR) opted for a strategy of coordinated implementation of research through a network of agricultural institutes and universities with geographically separate mandates. Kenya is an example of a joint implementation strategy, in which two projects were developed within the overall framework of the agroforestry research network for the highlands of East and Central Africa. Both projects are implemented jointly by the Kenya Forestry Research Institute (KEFRI), the Kenya Agriculture Research Institute (KARI), and ICRAF. However, in the first project, KEFRI is the lead institution, while KARI has seconded staff to the team that will implement the research. In the second project, KARI acts as lead institution and KEFRI has seconded scientists. So far, a joint implementation strategy appears to be impractical in countries where the boundaries between institutions, in particular between agriculture and forestry, are much more distinct.

At the national level, separate commodity-oriented research institutes operating in the same geographical area should conduct research on the components of the production system in a way that contributes to the overall productivity of the system. A strategy for the joint implementation of adaptive agroforestry research may be required to ensure this overall productivity goal is achieved. Process-oriented agroforestry research at the national level may not require special arrangements, because such research is more complex and specialized. This means there are fewer research partners, and, hence, fewer opportunities for joint research projects. At an international level, some duplication is likely because of the number of international research institutions involved. It is therefore interesting to see that those working in process-oriented agroforestry research have recognized the need for collaboration, which has already led to the setting up of an international network for agroforestry modelling. Other specialized institutions collaborate on aspects of land and tree tenure, plant protection, germplasm conservation and utilization, biological nitrogen fixation, and documentation (ICRAF 1993).

Integration of agroforestry into NRM

As the NARS and the various CGIAR centers continue to grapple with the problems of improving the productivity of African agriculture, it has become increasingly evident that new agricultural technologies have not made the impact on agricultural production that was expected. One reason may be that the productivity of the resource base has dropped to below the present management systems, and that the resource base is no longer able to support the needs of the rapidly growing population (Loevinsohn and Wangati 1993).

It was against this background and the apparent neglect of research on issues of NRM that the NARS leaders from Burundi, Ethiopia, Kenya, Rwanda, Tanzania, Uganda and Eastern Zaire and the directors of the international centers met at ICRAF in 1992 to explore the possibilities for new regional research initiatives. From this meeting came a proposal for a project titled "Integrated Natural Resource Management Research for the Highlands of East and Central Africa."

System analysis for NRM
This initiative will focus first on compiling an inventory of the natural resource base in the highland region, and on diagnostic studies to identify the potential role of NRM research in three major landforms in the highland areas. Emphasis will be given to the effects of declining soil fertility on the use of the natural resource base and the potentials for reversing this decline by improving systems of land use. Another priority task is an analysis of the effects of evolving farming systems on pests and diseases and the extent to which the problems can be overcome by improved land-use management.

Interinstitutional linkages for NRM research
A committee, consisting of representatives from NARS and IARCs, will govern the programme and liaise with donors. A coordination unit, consisting of country coordinators, will manage the overall research functions of the initiative. Working committees will be responsible for overseeing the research on specific themes. National research institutes will be the home for research and will ensure that the initiative's research priorities remain relevant. National governments will decide on the allocation of responsibilities to national institutions. To begin with, each country will nominate one national center that already has a regional mandate. Such centers must meet the following three criteria: a mandate for research in at least two of the major highland systems, a record of interdisciplinarity in research, and a record of effective collaboration with other institutions and organizations, including universities and NGOs.

IARCs will participate in the initiative as equal partners and in supporting the regional activities initiated by the national centers. This process is part of the evolution of the new role of IARCs in ecoregional NRM research.

Acronyms

CGIAR	Consultative Group for International Agricultural Research
D&D	Diagnosis and Design
GIS	geographic information system
IARC	international agricultural research centre
ICAR	Indian Council for Agriculture Research
ICRAF	International Centre for Research in Agroforestry
ISABU	Institut des Sciences Agronomiques Burundi
ISAR	Institut des Sciences Agronomiques Rwanda
ITE	Institute of Terrestrial Ecology
KARI	Kenya Agriculture Research Institute
KEFRI	Kenya Forestry Research Institute
NARS	national agricultural research system
NGO	non-governmental organization
NRM	natural resource management
SCUAF	soil changes under agroforestry

References

Avila M, Minea S (1993) Personal communication

Agroforestry Modelling Newsletter (1993) Issue 1, Forestry Research Programme, Institute of Terrestrial Ecology

Bradley R C, Crout N M J (1993) The parch model for predicting arable resource capture in hostile environments, User guide University of Nottingham, UK 137 pp

Etherington D M, Matthews P J (1984) MULBUD User's Manual Development Studies Centre, The Australian National University

ICRAF (1992) Annual report 1992, ICRAF, Nairobi, Kenya

ICRAF (1993) The way ahead, Strategic plan ICRAF, Nairobi, Kenya

Loevinsohn M, Wangati F (1993) Integrated Natural Resource Management Research for the Highlands of East and Central Africa Proposal for a collaborative regional initiative

Parton W J, McKeown B M C, Kirchner V, Ojima D (1992) Century user's manual Colorado State University, USA

Raintree J B (1987) compiler and editor, D&D User's manual ICRAF, Nairobi, Kenya

Wang Y P, Jarvis P G (1990) Description and validation of an array model Agricultural and Forest Meteorology, 57 257-280

Young A, Muraya P (1988) Soil changes under agroforestry (SCUAF) a predictive model Pages 655-67 in Rimwanich S (ed) Land conservation for future generations Proceedings of the Fifth International Soil Conservation Conference Department of Land Development, Bangkok, Thailand

The use of systems analysis at international and program levels: IITA's experience

S.S. JAGTAP

International Institute of Tropical Agriculture (IITA), PMB 5320, Ibadan, Nigeria

Key words agroecological zones, decision support systems, GIS, simulation model, systems approach, training

Abstract

A range of methods are available for systematic analysis and evaluation of agricultural production systems The needs of users differ Because of the difficulties of developing adequate models to represent the processes involved, even for a relatively simple, single crop system, it is doubtful whether practical, process-based models for entire cropping systems will be available soon Meantime, the quantitative study of systems must use a combination of process-based modules to represent those parts of the system that are well understood, and empirical models for processes that are less well understood This combination, together with data bases for environmental factors, could be developed in the form of DSS

These DSS must be developed, tested, and applied in close collaboration with NARS and in close collaboration with IARCs and advanced institutions An end user (whether NARS, IARC, or even an advanced institution) would first of all have to develop a conceptual framework that shows how the different components and processes would link up in order to simulate the functioning of a system Such a framework determines the type of data bases and process modules that may be required While NARS have a comparative advantage in adaptive research, IARCs or consortia of NARS can carry out strategic research in regions where national research systems have adequately trained personnel and stable funding Then, the interactive use of DSS and participatory on-farm research methods would provide a powerful tool in improving the farming-systems research by NARS

Research policy environment

The International Institute of Tropical Agriculture (IITA) was founded in 1967 with a mandate for specific food crops, and with ecological and regional responsibilities to develop sustainable production systems in Africa. IITA conducts research, training, and germplasm exchange in partnership with regional and national programs in many parts of sub-Saharan Africa. The research is organized in three substantive divisions: crop improvement, crop protection, and the management of crops and farm resources within a farming-systems framework (IITA 1993). The research program focuses on six major food crops (cassava and yam, plantain, maize, cowpea, and soybean) and smallholder cropping systems (for farms that are generally smaller than three hectares) in the humid savanna (growing period 10-12 months) and moist savanna (growing period 6-9 months) of Africa.

The agricultural potential of African countries varies widely. Central Africa, humid West Africa, and southern Africa have relatively large areas of arable land and relatively low population densities. Most of the Sahel, parts of mountainous East Africa, and dryer parts of the South African nations have large populations and have to import food. Sub-Saharan Africa is experiencing a profound economic and

313

P Goldsworthy and F W T Penning de Vries (eds), Opportunities, use, and transfer of systems research methods in agriculture to developing countries, 313 - 322

environmental crisis. It is the only region in the developing world with a declining per capita food production. Of all the developing regions, it relies most heavily on imported food, it has the highest proportion of land area that suffers from loss of fertility, and the highest percentage of people suffering from malnutrition.

The African farmer works in a uniquely inhospitable environment. About 20 percent of the entire continent is desert, and in another 10 percent the soils are too sandy for agriculture. An area larger than the United States is rendered largely unusable for agriculture by the tsetse fly. Only 19 percent of the soils have no fertility limitations. Most arable soils are coarse and their content of clay is too small to absorb and hold enough moisture for crop plants. A large proportion of the soils are highly susceptible to erosion. African farmers have adapted to their environment by practicing shifting or fallow cultivation. They clear the land, plant crops for a few years until the nutrient level declines and weeds take over, then abandon the land and allow it to return to its natural state. Given time, trees, shrubs, and grasses regrow and slowly restore the fertility of the soils. Shifting cultivation can sustain agriculture indefinitely, even in harsh conditions, when fallow periods are long enough. But when there are too many people and fallow periods are shortened or disappear completely, land productivity declines rapidly.

Added to the pressure on the land caused by agriculture is the need of the people to gather fuel wood and graze their livestock. As fuel-wood supplies become scarcer, farmers have to burn animal dung and crop residues rather than use them to enrich and sustain the soil. As more land looses nutrients, vegetative cover decreases, soil erosion becomes more evident, and the soil's moisture-holding capacity declines. The vulnerability of crops and pasture to uncertain rainfall becomes greater, and in a prolonged dry spell, the entire system may collapse.

IITA has used the major agroecological zones of its mandate region as a starting point for planning and developing its research program.

Systems research methods

Sub-Saharan Africa is very diverse, and IITA faces the problem of a mosaic of agricultural systems, with each part of the mosaic having a different set of constraints and resource endowments. The challenge is to identify within this diversity of agricultural systems the basic principles that are of general importance in determining the success or failure of the systems. Strategic research focusses on these principles. Our conceptual framework of research is under constant review and our methods are adapted and modified to suit particular needs. The systems research framework developed at IITA (figure 1) has several interlinked phases of activity:

- *Resource characterization and diagnosis.* To determine the constraints and problems that smallholder farmers face, the characteristics of the physical, chemical, and biological elements of the natural resources available and the socioeconomic implications of using those resources need to be defined. This characterization is conducted in close collaboration with national agricultural research systems (NARS).

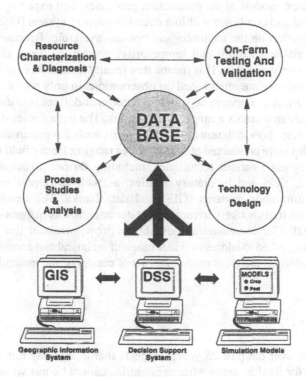

Figure 1. Conceptual research framework integrating research activities related to resource and crop management

- *Process studies and analysis.* This phase includes identification and examination of the factors that determine sustainability and cause degradation of the resource base. It also requires an understanding of the interactions between these factors.
- *Technology design.* In this phase, existing resource management practices are modified or new ones are designed that are capable of sustaining a stable or increasing output while avoiding the degradation of the resource base.
- *On-farm testing, validation and adaptation.* This phase entails the on-farm screening, and evaluation of the technologies generated during experimental station research. It also includes adjusting or adapting existing technology to a particular set of agroecological and socioeconomic environmental conditions.
- *Decision support systems.* Matching appropriate technology to the requirements of farmers across an area of great diversity calls for a comprehensive knowledge

of the range of technologies available and their characteristics in terms of the resources they require and the potential contribution they can make. Environmental databases, models of the production processes, and expert systems need to be developed and combined within a decision support system (DSS) that will provide information on the technological options available. Because rainfall is often very variable spatially and temporarily, production from rainfed crop systems is also very variable. This means that results from field experiments are site specific, and they are often based on observations in only a small number of years. The only way to improve on standard recommendations is to use appropriate models, validated across a range of conditions. The case studies described in this paper illustrate how different systems-analysis tools have been used at IITA. The case studies were conducted at spatial scales ranging from a field to an entire continent. They used various techniques, including farmer participation, rapid rural appraisal, field and laboratory studies, as well as remote sensing and geographic information systems (GIS) and data banks. The development of systems analysis tools is likely to receive greater attention as budgets continue to tighten and calls for increased accountability grow, while at the same time research is being asked to address a wider range of technical and economic issues. This paper highlights progress made in terms of concepts and methods.

GIS

Rationale
IITA's experience over the past 25 years suggests that in the long run, the rapidly growing demand for food in many African countries cannot be met without considering the capacity of the natural environment. It has become clear that promoting sustainable agriculture involves the coordination of information from various sectors, so that the impact of a chosen strategy on a region may be better understood. Agricultural production and the protection of natural resources both require better use of information about existing resources. GIS are used extensively at IITA to capture, store, and analyze quantitative and descriptive data for this purpose in collaboration with its partners in the national agricultural research systems. IITA uses GIS to describe the environment where research is conducted and in which the results are to be used. It is used to define agroecological boundaries, which helps research programs focus on the most important issues for research. GIS play a major role in defining key research sites within an agroecological area where representative research for the entire area can be done (Jagtap 1993a).

Applications
The Agroecological Studies Unit of IITA has developed a raster-based Resource Information System (RIS) and over 150 digitized geo-referenced databases covering elevation, rainfall, temperature, soils, vegetation, farming systems, population density, and communication networks, on a grid of 10 minutes of arc for Africa. Staff

from 14 African national agricultural research programs, including Guinea, Sierra Leone, Côte d'Ivoire, Mali, Burkina Fasso, Ghana, Togo, the Republic of Benin, Nigeria, Cameroon, the Central African Republic, Zaire, Congo, and Gabon have been trained in the use of RIS. These countries, in partnership with IITA, will develop biophysical and socioeconomic data bases to categorize their agricultural systems. Several research projects in IITA have used RIS to identify potential areas for research or data collection. Examples include the identification of sites ecologically and economically suitable for developing technologies for inland valleys (Izac et al. 1994) and sites for sampling banana pests throughout Uganda (Jagtap 1993b).

The geographically referenced data bases have been linked with a water-balance model and with agroclimatic crop suitability models of FAO (1978) in the form of an agroclimatic atlas for Africa. From data on a 10-minute by 10-minute grid, the computerized atlas provides information on monthly rainfall, estimated potential evapotranspiration, maximum and minimum temperatures, radiation, day length, length of growing period (based on water balance), soil type, soil constraints, natural vegetation, estimated population density, and crop suitability. The RIS is used by breeders to select crop varieties for dispatch to different locations, to identify similarities and differences between research sites, and as an environmental informa-tion storage and retrieval system for Africa. The information retrieved is presented both in tables and graphically as illustrated in figure 2.

In another application of systems methods, the potential of environments in northern Cameroon was assessed for sorghum production. To do this, a combination of data bases, a sorghum crop model, and RIS were used. This region of Cameroon has an arid and a semi-arid climate with a single rainfall season. The area contains 38 unique soil types with considerable variations in the depth of the root zone, water-holding capacity, and drainage characteristics. Sorghum, millet, cowpea, maize, cotton, and groundnut are grown, bullock-drawn ploughs are used to prepare the land. Crops are sown at the beginning of the rain and may have to be re-seeded or transplanted if there is too little rainfall immediately after sowing for the crop to establish. Since there is abundant solar radiation, the region has a high production potential. There is very little development of irrigation and only subsistence agricul-ture is practiced, with little or no external inputs. Using daily weather data (precipi-tation, radiation, and temperatures) from between 1960 and 1991, simulations were run for an early and a late maturing of sorghums cultivar with 0 and 30 kg N ha^{-1}. The maps in figure 3 are an example of an output which shows how risks of production losses can change when planting is delayed by 0, 15, and 30 days after the onset of rains. The results showed that risks of production losses are lower when farmers wait for the rains to become established and if they apply fertilizer (Jagtap and Carsky 1993).

Constraints
A shortage of suitable computer software is currently a constraint. The software available is often expensive, not suitable for agricultural applications, and it requires extensive specialized training. One of the major efforts and costs involved in a GIS

318

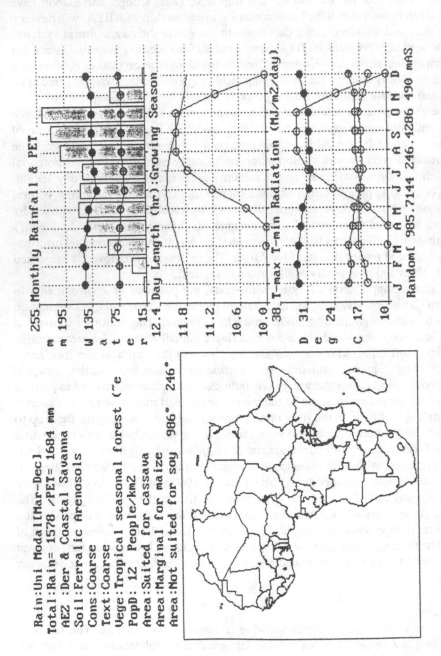

Figure 2. Agroclimatic information retrieval system, displaying climatic, soils, demographic, and crop-suitability data throughout Africa, using a 10-minute by 10-minute georeferenced data base

319

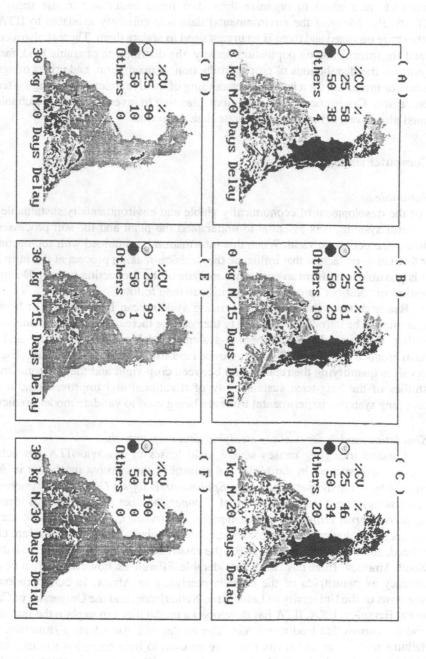

Figure 3 Maps showing risks measured as coefficient of variation (CV, percent) associated with year-to-year variations in the yield of rainfed sorghum production in northern Cameroon under different fertilizer application rates and dates of sowing, using weather data from 1960-1991

project is the collecting of data. Much of the data that is collected is not fully used. The capabilities of GIS techniques to analyze and display results will encourage scientists and others to organize their data bases better and to use them more effectively. Many of the environmental data sets currently available to IITA and others are outdated and there is an urgent need to update them. There is also a critical need for information on population density, the distribution of arable land, farming systems, the distributions of crops, production, consumption, and prices of agricultural commodities, for a better understanding of the dynamics of farming systems in the region. Governments and institutions planning to invest in systems technologies must also invest in timely and accurate data collection.

Computer simulation models

Rationale
For the development of economically viable and environmentally sustainable agricultural systems, it is essential to understand the plant and the soil processes that determine economic yield. When this information is combined with socioeconomic information on factors that influence the decision-making process at the farm level, it is possible to perform an economic analysis of the production system. Design and testing of "best-bet" production systems can then follow.

Research must show how the existing systems can be improved or how new systems can be introduced that satisfy the need for increased production and sustainability. IITA intends to use cropping-systems models to study the short- and long-term productivity of current and improved land-use and cropping systems. The work involves quantifying the relationship between crop yield and the environment, and studies of the long-term sustainability of traditional and improved land-use and cropping systems. Experimental trials are being used to validate model predictions.

Simulation models for cassava mealybug and crop growth.
Cassava mealybug causes severe yield losses in cassava. IITA has achieved considerable success in the biological control of the cassava mealybug in Africa, mainly by using the parasitic wasp, *Epidinocarsis lopezi De Santis (Hymenoptera, Encyrtidae)*. Since the introduction of *E. lopezi*, another parasitoide was identified, *E. diversicornis (Howard) (Hymenoptera, Encyrtidae)*, and found to be useful in Africa for biological control of the cassava mealybug. The two species are closely related, and both are known to attack the cassava mealybug in their native habitat in South America. There have been considerable differences, however, in their observed efficacy as parasitoids of the cassava mealybug in Africa. In collaboration with scientists of the University of Leiden, the Netherlands, and the University of California at Berkeley, USA, IITA has developed a model that can explain the interactions and the causes that lead to success. The model will also identify situations where failures may occur and it can therefore be used to help scientists to make rational choices for biological control efforts in the future (Gutierrez et al. 1988a and 1988b).

It will be used to develop strategies for further research on the control of pests in Africa. The models have been modified to study cowpea and maize pests. The approach that has been used is an important contribution to the science of biological control, and it is likely to be a useful tool in other situations for the control of crop pests.

Cropping system models

The CERES Maize model developed by International Benchmark Sites Network for Agrotechnology Transfer (IBSNAT) was tested during the 1990 and 1991 growing season at Ibadan in southwestern Nigeria (Jagtap et al. 1993). The measurements of the crop and estimates derived from the simulation model were compared and found to agree within 10 percent. The calibrated model was then used to simulate on-farm trials (G. Weber, unpublished). The grain yields predicted by the model were closer to the observed values for fields where productivity was high than where productivity was poor. Additional on-station experiments with different fertilizer inputs are currently underway to test the model. Experiments are also being conducted to test soybean simulation models as well as maize-soybean intercropping models.

Challenges

Many crop models, representing varying degrees of physiological detail, have been developed during the past 20 years. As more physiological details are added, the models become less transparent to the user. The risk is that they will become "black boxes" for users who are unfamiliar with the underlying principles in the models. This would reduce the usefulness of such models as tools in applied research. IITA is committed to improve the performance of cropping systems that have received little research attention from the research institutions in developed countries, where the productivity of agriculture is already high. Little has been done to develop methodologies for research on complex, multiple crop systems in sub-Saharan Africa. Little is known about how the growth of component crops determines their performance in associations. Developing process-based models for mixed populations of a mixture of species is likely to be more difficult than for crops of a single species. The immediate need is to analyze the utilization of resources in existing cropping systems, to identify the inefficiencies. Factors such as the duration of the cropping season, the amount of solar radiation, and the supply of water and nutrient from the soil affect the performance of crops in the mixture. The models must therefore be capable of simulating competition among the component crops.

The role of training

IITA's training programs ensure that technologies developed through research reach those who can use them. The potential beneficiaries include national research programs and, ultimately, the consumers of agricultural produce. IITA conducts training programs at its headquarters, but there is now an increased emphasis on

322

in-country training programs. It expects that national programs and research networks will eventually be able to conduct their own training. IITA has strengthened the capability of national research systems by training more than 7000 African scientists and technicians. All training courses at IITA have a farming-systems orientation. In 1994, a course on GIS methods was conducted for the first time to train representatives from countries in West and Central Africa in the use of GIS methods in farming-systems research and agricultural-research planning.

Acronyms

DSS	decision support system
FAO	Food and Agriculture Organization
IARC	international agricultural research center
IBSNAT	International Benchmark Sites Network for Agrotechnology Transfer
IITA	International Institute of Tropical Agriculture
NARS	national agricultural research system
RIS	resource information system

References

IITA (1993) Unlocking Africa's Potential, IITA Medium-term Plan, 1994-98. International Institute of Tropical Agriculture, Ibadan, Nigeria.

FAO (1978) Agroecological Zones Project. Vol. I Methodology and Results for Africa. FAO, Via delle Terme di Caracalla, 00100 Rome, Italy.

Gutierrez A P, Wermelinger B, Schulthess F, Baumgartner J U, Herren H R, Ellis C K, Yaninek J S (1988a) Analysis of biological control of cassava pests in Africa. I. Simulation of carbon, nitrogen and water dynamics in cassava. Journal of Applied Ecology, 25:901-920.

Gutierrez A P, Neuenschwander P, Schulthess F, Herren H R, Baumgartner J U, Wermelinger B (1988b) Analysis of biological control of cassava pests in Africa. II. Cassava Mealybug *Phenacoccus Manihoti*, Journal of Applied Ecology, 25:921-940.

Weber G. (personal communication). On-farm performance of maize data.

Izac A M N, Jagtap S S, Mokadem A, Thenkabail P S (1993) Agroecosystems Characterization in International Agricultural Research: A case study of Inland Valley Systems in West and Central Africa. For Agriculture, Ecosystems and Environment.

Jagtap S S, Mornu M, Kang B T (1993) Simulation of growth, development and yield of maize in the transition zone of Nigeria. Agricultural Systems 41,215-229.

Jagtap S S (1993a) Resource Information System (RIS) User's Guide, Agroecological Studies Unit, International Institute of Tropical Agriculture, Ibadan, Nigeria.

Jagtap S S (1993) Diagnostic survey site selection using GIS for effective biological and integrated control of highland banana pests. Proceedings of "Biological and Integrated Control of Highland Banana and Plantain Pests and Diseases". International Institute of Tropical Agriculture, Ibadan, Nigeria.

Jagtap S S, Carsky R J (1993) Quantitative assessment of sorghum production environment in Northern Cameroon: Integrating data bases, crop models and GIS, ASAE paper #933562, ASAE, St. Joseph Michigan 49085-9659, USA.

Capacity building and human resource development for applying systems analysis in rice research

M.J. KROPFF[1], F.W.T PENNING DE VRIES[2] and P.S. TENG[1]

[1] International Rice Research Institute (IRRI), P.O. Box 933, Manila 1099, Philippines.
[2] DLO Research Institute for Agrobiology and Soil Fertility, P.O. Box 14, 6700 AA Wageningen, The Netherlands.

Key words: reorganization, research priorities, simulation, systems approach, team building, training

Abstract

Systems approaches have been used for a long time in IRRI's research. In the beginning, such activities were based on the interests of individual scientists. More recently, when IRRI reorganized its research, it developed ecosystem-based programs. The new structure facilitates interdisciplinary research on the complex issues that we face today, such as the need to increase productivity and resource-use efficiency in both favorable and unfavorable rice ecosystems. Systems approaches are essential to tackle such complex problems. In ongoing projects, systems approaches are used successfully as research tools at different levels of integration and have been identified as important tools in IRRI's mid-term plan for 1994-98. Over time, IRRI's relationship with the national agricultural research systems (NARS) has evolved into new modes of partnership that are based on collaborative research. Systems research is being integrated into these collaborative projects. Much attention has been given and will continue to be given to capacity building in using systems approaches at IRRI as well as at collaborating NARS.

Introduction

The International Rice Research Institute (IRRI) was founded in the early 1960s. Since then, rice-growing environments have greatly changed. More than 70 percent of the rice worldwide is currently being produced in high-yielding irrigated rice ecosystems, while only 17 percent comes from rainfed lowland systems. The onset of the Green Revolution, in the early 1960s, was marked by changes in the characteristics of the rice plant for irrigated rice ecosystems, which led to an unprecedented leap in rice production. This breakthrough, however, had unforeseen social and environmental costs (IRRI 1993). The small genetic basis of modern high-yielding varieties and the misuse of pesticides became important issues and changed countries' research agendas. Interdisciplinary work started to improve the use and evaluation of rice germplasm. New varieties with pest resistance and shorter duration followed IR8, the first semidwarf variety that was launched in the 1960s.

The research approach followed in the early days was to develop genotype-based technologies, which allowed new genotypes to be cultivated successfully across a broad range of environments. Thus, improved genotypes and improved crop management practices were developed simultaneously so that genotype-by-environment interactions were minimized.

During the 1970s, IRRI's work was extended to include research for the unfavorable rice ecosystems—the rainfed lowlands, the uplands, and the deepwater and tidal

323

P. Goldsworthy and F.W.T. Penning de Vries (eds.), Opportunities, use, and transfer of systems research methods in agriculture to developing countries, 323 - 339.
© 1994 Kluwer Academic Publishers.

324

wetlands. In the 1980s and early 1990s, the research program was further expanded to deal with more complex systems problems. IRRI and the national agricultural research centers (NARCs) and systems (NARS) developed new forms of collaboration such as research consortia. In the irrigated rice systems, knowledge-based technologies are being developed to optimize resource use and efficiency. In the unfavorable and highly variable rice ecosystems, genotype-by-environment interactions are crucial. The changes in the research agenda has transformed the research approach for the 1990s; it is knowledge based, system focused, and based on the "think globally, act locally" principle (K.S. Fischer, pers. commun.). As a result, simulation and systems analysis (SSA) has become an important tool at IRRI.

In 1989, IRRI published its long-term strategy (IRRI 1989a) and its work plan for 1990-94 (IRRI 1989b). In the light of the rapidly expanding world population and a greater emphasis on social equity, enhanced sustainability, and environmental protection, these plans aim to meet the challenges in increasing the agricultural yields to meet the demand for food. Systems approaches will be very important in IRRI's future research and have been identified as such in IRRI's midterm plan for 1994-98 (IRRI 1993).

The systems approach can be described as the systematic and quantitative analysis of agricultural systems and the synthesis of comprehensive, functional concepts of these systems. Agricultural systems are defined as well-delineated parts of the real world, consisting of many interacting elements, while their environments have only a one-way effect on them. For instance, in some studies, a field crop may be considered a system in which carbon and nitrogen metabolism interacts with phenology; it responds to climate but does not affect it. The systems approach uses many specific techniques, such as simulation modelling, expert systems with data bases, linear programming, and geographic information systems (GIS). Systems research employs systems approaches and some of the techniques appropriate for its purposes.

This paper describes and discusses the role that systems approaches have played and will play in IRRI's research, with special emphasis on institutional capacity building.

Systems research at IRRI

A historical perspective
IRRI scientists started using SSA soon after it was developed. Models that were developed and used at IRRI ranged from simple descriptive to complex mechanistic, and generally focused on specific issues. They were used mainly by the scientists who developed them. SSA has been used in a wide range of disciplines, from plant physiology to economics (Penning de Vries et al. 1991b). This section gives a brief summary of past systems research at IRRI; Herrera-Reyes (1991) provided a more detailed historical sketch.

One of the first studies in which SSA was used was by Wickham (1973), who predicted the performance of lowland rice through a water-balance model. Later,

Angus and Zandstra (1979) developed IRRIMOD, a multiple-field, multiple-crop, toposequence model for rainfed rice. IRRIMOD was used to predict yields of drought-prone second rainfed wetland rice in Iloilo, Philippines, and to define rice areas and landscape positions suitable for double-cropping of rice. Zandstra et al. (1982) used the same model to determine the agronomic feasibility of major rice-based cropping patterns. An evapotranspiration model that derives stress-day indices to project the productivity of cropping patterns and to account for effects of weather, management, and soil factors was developed (Morris 1985). Garrity and Flinn (1987) used this model to analyze the long-term biological and economic productivity and stability of *Sesbania rostrata*. In the 1980s, RICEMOD, a simulation model for the growth and yield of irrigated rice, was developed (McMennamy and O'Toole 1983). Whisler (1983) validated and compared IRRIMOD and RICEMOD in sensitivity analyses.

Penning de Vries et al. (1989) introduced a modular approach to ecophysiological crop modelling through a series of modules called MACROS. The approach taken was largely based on the "Wageningen models". Aspects of this model have been validated and applied to different varieties, crops and situations (Herrera-Reyes and Penning de Vries 1990). Woodhead et al. (1991) used the water balance developed for MACROS to study upland rice hydrology. The CERES rice model has been used to simulate soil nitrogen transformations and crop response to nitrogen (Buresh et al. 1991). Kirk and Rachpal-Singh (1991) conducted detailed modelling studies on processes related to soil N dynamics. These models improved the understanding of NH_3 losses from paddy rice systems.

In crop protection, models have been used extensively to study the population dynamics of pests as well as the interactions of the pests with the crop (Calvero and Teng 1988; Heong and Fabellar 1989; Rubia and Penning de Vries 1990). In the Agricultural Engineering and in the Social Sciences divisions, modelling activities have been used at different levels of integration (Huisman 1983; Rosegrant 1977, 1992). Models and optimization techniques were developed and used to optimize economic returns from rainfed ricelands with limited water from farm reservoirs (Galang and Bhuiyan 1991). Penning de Vries et al. (1991b) gives a summary of these activities. To present and review systems research in a wider context, IRRI organized a symposium on Systems Approaches for Agricultural Development in 1991 (Penning de Vries et al. 1993; Teng and Penning de Vries 1992).

IRRI's strategic plan emphasizes approaches to balance IRRI's concern for adequate food production with strong concerns for the sustainable use of resources, environmental health, and increased employment. The plan explicitly states that systems approaches will have to play an important role in IRRI's future research. It shifts research "upstream" toward more strategic research, because the national systems had developed a relatively strong capacity to conduct adaptive and applied research (IRRI 1989a). It also refocuses the research agenda by developing interdisciplinary, ecosystem-based research programs to balance commodity research on rice with research on sustainability and environmental security. Five programs were established for the different rice ecosystems: irrigated, rainfed lowland, upland, and tidal wetlands and deepwater. A cross-ecosystems program was established to generate knowledge needed for technology development in more than one ecosystem. The 1990-94 workplan defines about 50 projects in the five programs. About 14 of these projects specified models as tools to meet their objectives (Penning de Vries and Kropff 1991). Model development is concentrated in the cross-ecosystems program, whereas applications of the tools are generally found in the ecosystem-based programs.

IRRI's cross-ecosystems work bridges the basic research done at other advanced institutions, the institute's ecosystem-specific programs, and national rice research systems. One objective is to use and/or develop modern scientific tools, methods, and knowledge to address current and anticipated rice-production problems common to several ecosystems at different levels of integration. Another objective is to develop promising technologies that may be derived from more basic research available to national programs. The research aims to significantly improve rice science and technology by linking research activities at different levels of integration, from gene to plant to ecosystem and, ultimately, to the global community. This goal requires systems approaches. The scope and focus of the program can be demonstrated best in a graphic manner, with maps at different levels of integration (figure 1). The research being conducted can be characterized by several levels of integration: the genome level, the crop level, the farm level, and the community level. At all these levels, characterization is an important activity. The genome maps characterize varietal traits that have implications at the crop level. Special attention is paid to the quantitative links between genome characteristics, physiological traits, and crop production possibilities in a given environment. Again, SSA will be used as a tool. Quantitative knowledge on the performance of rice crops in the ecosystems is essential to identify constraints, and thus to set research priorities and develop and evaluate the impact of new technology at the community level. At the highest level of integration, the long-term socioeconomic and environmental impacts of new rice technologies are analyzed. The focus is on ex ante impact assessment for effective research planning.

Current programs

IRRI's midterm plan for 1994-98 (IRRI 1993) is designed to make progress toward resolving researchable questions in several areas: IRRI's responsibilities regarding people (providing sufficient and affordable rice, improving the welfare of rice communities), permanency (preserving and using genetic resources, sustaining the resource base of intensive rice systems, sustaining the resources of the rainfed lowlands, stabilizing the uplands), productivity (increasing resource use efficiency), and protection (reducing the use of pesticides, global climate change and rice) (IRRI 1993). New projects were defined, and the focus of the ecosystem-based programs was sharpened. In most objectives listed in the plan, a systems approach will have to resolve the problems. These objectives are the following:

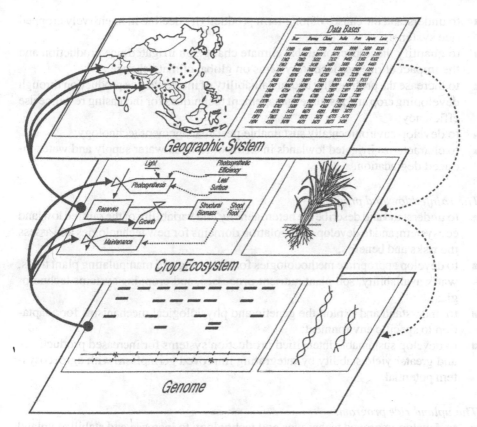

Figure 1. Interrelationships among major levels of research complexity. Molecular biology tools and systems science and simulation enable a better understanding and broad application of new knowledge in agroecosystems (IRRI 1993).

The cross-ecosystems program:

- to improve the understanding of pest diversity, the pest-plant coevolution process, and the biology and ecology of pests, and to apply management strategies for different ecosystems;
- to develop systems approaches and models that integrate knowledge on favorable and unfavorable environments and predict ecosystem behavior for different scenarios;
- to generate knowledge on the interactive effects of agroecology and socioeconomic environments on people and the natural resource base.

The irrigated-rice program:

- to increase the yield potential in tropical environments by 50 percent;

- to understand the causes of long-term productivity decline in intensively cropped rice systems;
- to quantify the impact of global climate change on irrigated rice production and the impact of rice production systems on global climate change;
- to increase the profitability and sustainability of intensive rice production through developing crop and resource management techniques for increasing resource use efficiency;
- to develop environmentally sustainable pest-management technology;
- to characterize irrigated lowlands in terms of reliable water supply and water-induced degradation.

The rainfed lowland program:
- to understand and describe the heterogeneity and variability of the rainfed lowland ecosystem, and to develop extrapolation domains for new technologies and assess the risks and benefits;
- to develop appropriate methodologies for analyzing and manipulating plant traits, water availability, soil plant nutrient processes, and cropping systems technologies;
- to understand and exploit the genetic and physiological mechanisms for adaptation to specific environment;
- to develop sustainable intensified production systems for increased productivity and greater yield stability by integrating improved germplasm with agroecosystem potential.

The upland rice program:
- to develop improved techniques and technology to increase and stabilize upland rice yields;
- to develop a range of upland rice production practices that will help rehabilitate degraded uplands and transform them into sustainable agroecosystems.

The floodprone rice ecosystem program:
- to characterize the changing production environments for possible shifts in research on land use and environmental protection;
- to develop improved crop and resource management practices.

Current projects
The following examples illustrate what systems research at IRRI means.

Declining yields. A decline in rice productivity has been observed in long-term experiments on intensive rice cropping systems (Flinn et al. 1982). Whereas in the late sixties, the first semidwarf variety (IR8) yielded 10 ton ha^{-1} on IRRI's experimental farm (IRRI 1968), recent dry-season yields generally did not exceed 6-7 ton ha^{-1} at the same fertilizer rates. Possible causes include reduced genetic yield potential, buildup of soilborne diseases, accumulated boron, altered soil fertility, and climate change. We focused on the hypothesis that the current low yields at IRRI's farm were related partly to a change in the N supply environment, causing low N concentrations in leaf tissue, especially during the grain-filling period, resulting in early senescence of leaves and low rates of photosynthesis (Kropff et al. 1994; Cassman et al. 1994a).

In the 1991 wet-season (WS) experiments, which were conducted with adapted N management, yields of 6 ton ha^{-1} were achieved (similar to WS yields in the 1960s). The ecophysiological rice model ORYZA1 (based on SUCROS and MACROS and evaluated with a large number of data sets [Kropff et al. 1993b]) was used to predict a dry-season (DS) yield in the beginning of the 1992 DS, using varietal parameters derived from the WS experiments. It predicted yields of 8 ton ha^{-1} in typical DS weather, and a further increase to 9.3 ton ha^{-1} when leaf N was increased by 20 percent. In the subsequent DS experiment, yields of 9.5 ton ha^{-1} were indeed achieved with increased N input (Kropff et al. 1993a,c) (figure 2). This analysis showed that greater fertilizer N inputs than presently recommended were required to achieve yields of 9-10 ton ha^{-1}. These yield levels were achieved on the IRRI farm in the late 1960s and early 1970s with 50 percent less fertilizer N inputs (IRRI 1968, 1973). These findings support the hypothesis that the effective N supply from the soil may decrease in the long term when soil is cropped under submerged conditions in continuous double- or triple-cropped rice systems, despite the conservation of soil organic carbon and total N (Cassman et al. 1994b).

Future research, with experimental and modelling components, will be directed toward the quantitative understanding of the soil N supply capacity (Cassman et al. 1994b). This knowledge, integrated in a model, will allow us to derive knowledge-based N management decision support systems. This can help optimize the crop N supply and fertilizer-use efficiency, and minimize N losses and pollution of the environment. The model will also be used to predict the suitability of varieties for

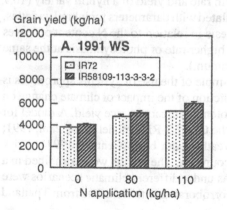
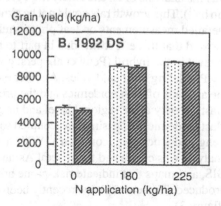

Figure 2. Grain yield (14 percent moisture) of IR72 and IR58109-113-3-3-2 at different rates of N application in the 1991 wet season (A) and the 1992 dry season (B)

specific environments. This will help make germplasm testing programs more efficient.

Increasing the yield potential. With the use of models in several studies, the physiological characteristics needed to develop new rice varieties with an increased yield potential have been determined (Penning de Vries 1991; Dingkuhn et al. 1991; Kropff et al. 1994). Since we do not foresee any large changes in photosynthetic efficiency and respiration costs, increased yield potential must come from an increased allocation of stem reserves, from a prolonged grain-filling period, from an increased growth rate during grain filling, or from a combination of these sources. Penning de Vries (1991), Dingkuhn et al. (1991), and Kropff et al. (1994) emphasized the lengthening of grain filling duration as the main option to increase the yield plateau, combined with an increased sink size. To achieve 15 ton ha^{-1}, 38 days of effective grain filling would be needed.

For the same reason, the yield potential of a rice variety at higher latitudes is greater than in the tropics; the grain-filling period is longer because of a lower mean temperature. The ORYZA1 model simulates an increase in rice yield potential for Los Baños, Philippines, of 2 ton ha^{-1} when the mean temperature is 3^{0}C lower. The model was run for weather situations in Yanco, Australia, and it correctly pointed out that yields of more than 14 ton ha^{-1} could be achieved. The research group in Yanco has achieved such yields. IRRI's breeding programs already uses indications of specific plant characteristics that are important to raise the yield ceiling.

It has become almost a tradition for ecophysiological studies to use models. INTERCOM, a model for interplant competition (Kropff and Van Laar 1993), demonstrated that crop yield can be increased when stems are short and panicles do not shade leaves. Physiological research verified this in experiments before it was used as a characteristic in breeding programs (T.L. Setter, pers. commun.).

Another example of a feedback between modelling and experimental research was the analysis of an extremely high growth rate and yield of a hybrid variety (10.7 ton ha^{-1}). This growth rate could not be simulated with parameters for inbred varieties. Detailed measurements on leaf photosynthesis in relation to the N content of leaves showed that this effect was due in part to a higher rate of photosynthesis at the same N level in the hybrid (Peng et al., pers. commun.).

Predicting blast epidemics. Another example of the use of systems approaches is the analysis of blast epidemics and the prediction of the impact of climate change on blast severity at the regional level and its potential impact on rice yield. A model for blast epidemics (Blastsim) was coupled to the CERES-Rice model (Teng et al. 1991; Teng and Luo, pers. commun.). With a weather data base (Centeno 1991), a risk analysis was conducted for several Asian countries. The results were combined in a GIS, and maps that indicate risk-prone areas under different climatic scenarios were produced. These areas have recently been corroborated with field data from Thailand (figure 3).

Climate-change effects on rice production. The increased concentration of CO_2 and other greenhouse gases is expected to change the earth's climate over the next century.

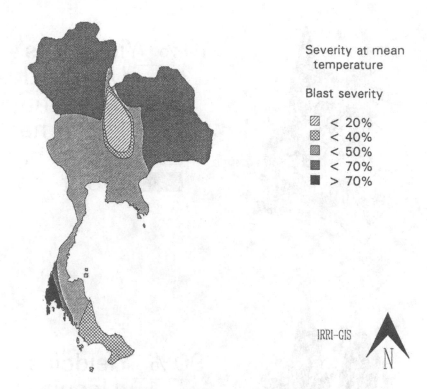

Severity at mean
temperature

Blast severity

☒ < 20%
▨ < 40%
▨ < 50%
■ < 70%
■ > 70%

IRRI–GIS

N

Figure 3. Simulated area under the disease progress curve for blast in Thailand (Teng et al. unpubl.).

Because of the complexity of the effects of temperature and CO_2 on physiological and morphological processes and their interactions, ecophysiological models are needed to predict the impact of expected climate-change scenarios. Preliminary studies using several models have been carried out (Penning de Vries 1993; Kropff et al. 1993a; Singh and Padilla 1993). In cooperation with scientists from five NARS in Asia, several rice crop models are being used to predict the effect of different climate-change scenarios on rice production in Asia (Matthews et al., pers. commun.).

Predicting risk of rice production in rainfed lowland systems. Rainfed lowland systems are highly variable, both spatially (soils) and temporally (precipitation only, no irrigation). This implies that results from experiments are site-specific and based on only a small number of years. The only way to improve standard recommendations is to use appropriate models, validated across a reasonable range of conditions. IRRI is now using such a model and GIS to analyze the effect of soil type and weather on rice yield loss at the regional level. A case study, conducted in the province of Tarlac in the Philippines, demonstrated the wide range in yield loss due to expected drought in a wet season (figure 4; Wopereis et al. 1993; Wopereis 1993). This model and the

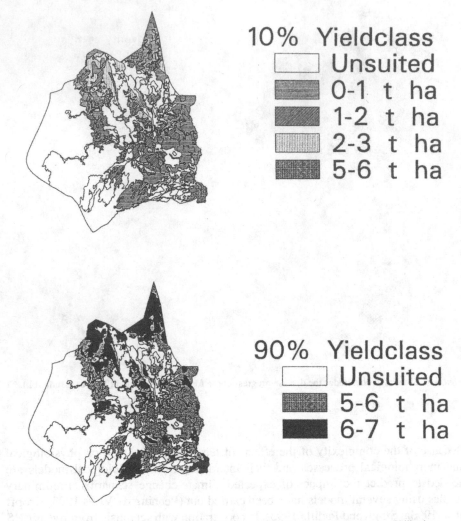

Figure 4 Maps of simulated rainfed yields at different probabilities in Tarlac province, Philippines

CERES-Rice model are being used to extrapolate research findings of new technologies for water and nutrient management in the rainfed lowland consortia.

Qualitative systems modelling in diagnostic surveys. Teams of IRRI and NARS rice researchers have conducted diagnostic surveys in different rice ecosystems in many rice-growing developing countries. Surveys are conducted to better understand farm- and community-level constraints and opportunities associated with overall rice systems or to analyze particular farmer systems (e.g., direct, dry-seeded, rainfed, lowland rice in India or Indonesia). The outputs of diagnostic surveys were a better understanding of farmers' systems and the identification and prioritization of problems and of problem-solving research.

Qualitative systems modelling has been a key tool in this process. Problems are identified (e.g., long-term declining soil fertility in the rice-wheat systems of South Asia), interacting causes are identified and diagrammed (figure 5), and research to address the causes is identified and prioritized. Problems are met in order of priority, as they occur over time and space and in terms of yield loss (or cost increases). Research prioritization is based on the priority of problems and on the probability of success, given the research resources (including human resources). Examples of such systems modelling and diagramming include work on irrigated rice farmers' N management (Fujisaka 1993) and on the rice-wheat system in South Asia (Fujisaka et al. 1994).

Spatial analyses using GIS and models. As complementary tools for extrapolation, GIS and simulation models have become especially important now that more of our work focuses on extrapolation. International collaborative projects such as the germplasm evaluation project, the research consortia, and the resource management network includes a few key locations where more intensive experiments and observations are conducted. At IRRI, a task force developed a plan to integrate existing

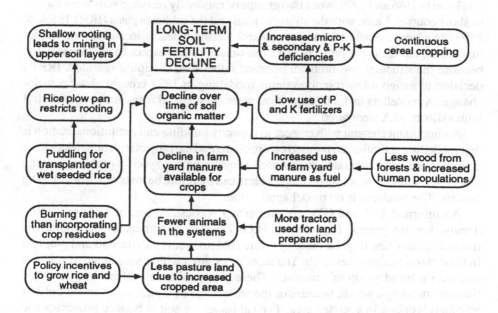

Figure 5. Conceptual model for the causes of the long-term problem of declining soil fertility in the rice-wheat system in Southeast Asia. (Fujisaka et al. 1994).

data bases into a relational data base, also for use with GIS. This data base will be accessible through the local area computer network at IRRI and the wide area computer network with NARS, IRRI liaison offices, and other institutions. The germplasm data base and the rice statistics data base have been included in the system. A weather data base, containing long-term daily weather data, has been developed by IRRI's Climate Unit using the data collected by Oldeman et al. (1986) with NARS and data collected by IRRI and several NARCs in a project with the US Environmental Protection Agency.

Individual skills and institutional capacity

IRRI

Before 1989, when IRRI's research was structured by the various disciplines and by department, systems research was conducted by individual scientists. Models were developed by visiting scientists, but there was no concerted effort. The models were not strongly interdisciplinary either (Herrera-Reyes 1991). In 1986, an interdisciplinary team of IRRI's nationally recruited staff was trained in a Systems Analysis and Simulation in Rice Production (SARP) project (for details, see p. 335). After the course, however, the researchers continued to operate individually because they were working in different projects. The team approach could neither be implemented at this level in 1988 and 1990, when two groups of nationally recruited staff were trained in short courses. Later, with the strategic plan and the midterm plans (IRRI 1989a,b, 1993) in place, the role of systems research became clearly identified and implemented in the different projects. The multidisciplinary aspect came in naturally, because the structure was problem oriented rather than discipline oriented. IRRI's decision to assign a key role to systems modelling in 1989 brought about a major change. A modellers and a GIS position were opened and many international staff with skills in SSA were hired.

An important element with respect to capacity building and institutionalization is that modelling activities are focused on problems and the project structure. Modelling activities are fully integrated in research projects with a specific problem-solving objective. Scientists working on this problem can therefore be involved in developing models. The emphasis is on model application.

An informal SSA and GIS working group was established in 1990 by IRRI's deputy director general for research H.G. Zandstra to stimulate and coordinate modelling activities. Regular meetings were held to discuss the concepts and progress in the different areas of research. The meetings started with a wide range of subjects and with a broad group of scientists. These meetings have evolved into so-called thematic meetings, which, because of the wide range of topics, are attended only by scientists working in a related area. Typical issues are soil N balance processes and modelling, designing new plant types, modelling the performance of rainfed lowland rice systems, combining modelling and GIS, and developing GIS applications.

When SSA was identified as an important tool in IRRI's program-related research, a two-day training course was developed for interested internationally recruited staff (most of whom are project leaders). The course addressed the general concepts used in SSA, simulation modelling techniques, hands-on modelling with a simple simulation programming language using simple crop models, and hands-on work with the newly developed rice models. The objective of the course was to familiarize participants with the tools.

This course will be followed by a five-week modelling course for nationally recruited staff. This course will make intensive use of the courseware program developed by the Wageningen Agricultural University (WAU) and the Courseware Midden Nederland company, both of The Netherlands. The course will consist of quantitative crop ecology (two weeks), SSA concepts and crop simulation (one week), and extensive examples of uses of crop models (two weeks). The first participants will be trained as trainers for extra courses within IRRI and for new teams in the NARCs. An important aspect of this approach is that the models are not presented as final or unalterable products. Rather, the idea is to understand the principles of modelling and to allow the user to adapt the models to his specific needs. These courses will be developed in such a way that they can be used also by NARCs to train colleagues.

NARS

With partnership as one of IRRI's major themes (IRRI 1993) and with the enhancement of national rice research capacity as one of IRRI's missions, IRRI considers NARS as its most important partners. Although several national programs continue to need support in developing their research programs, others have gained considerable strength in the past decades and are now capable of conducting adaptive and applied research and some strategic research (IRRI 1993). These systems need access to tools and methodologies. In strong NARS, such as those collaborating in research consortia, systems tools have become necessary for the strategic research components. In 1985, through SARP, IRRI and the Wageningen modelling groups at the Centre for Agrobiological Research (now DLO-Research Institute for Agrobiology and Soil Fertility) and the Department of Theoretical Production Ecology of WAU started a major training program on implementing and using SSA at the NARCs involved in rice research (Penning de Vries et al. 1988; Ten Berge 1993; Kropff et al. 1992). This project was funded partly by the Dutch Ministry for Development Cooperation and partly by the participating organizations. Sixteen interdisciplinary teams in nine countries were trained in a program that consisted of an eight-week training course, a case study of about one year at the home institute, and a workshop to present and evaluate the results.

These training activities were well appreciated, and the teams continued to work on the project after the training phase. Gradually, the emphasis moved from technology transfer to collaborative research. The research activities of the teams are fully integrated with the other research activities in the institutes. In 1992, SARP was extended with renewed funding in the form of a collaborative research network. The

336

results of the various case studies were documented by Penning de Vries et al. (1991a). Research is conducted in four areas: agroecosystems, potential production, crop and soil management, and crop protection. The approach has been successful; most scientists are still active in the field, and the Indian and Chinese teams have organized training courses for secondary teams in their countries (Sinha et al. 1993). These courses were funded by the NARS. The research emphasis of the teams is changing as well—from parameterizing models for their own environment to developing and applying models for the specific needs in their research.

NARCs continue to need support in building institutional capacity because they are generally structured along disciplinary lines. In the new interdisciplinary approach, teams work outside the common research and project structures. Already, SARP is planning to continue collaborating with its teams at the NARCs and to design an effective organizational structure. Figure 6 shows how this might evolve.

We view SSA capacity building as taking place in two manners: incorporating SSA into the research planning process, and using SSA as a tool in research to solve problems. In incorporating SSA into the research planning process, a systems approach allows the targeting of issues with a potential impact. Using SSA as a tool in research to solve problems results in developing and using models for specific purposes, such as designing new plant types and quantifying pest effects on rice yield.

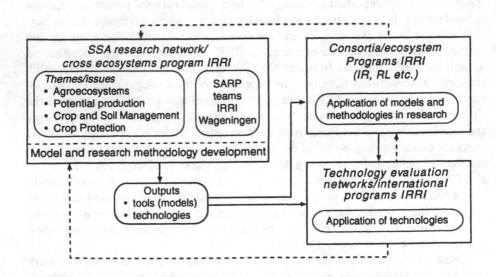

Figure 6. An international research network on the use and application of simulation and systems approaches for rice-based cropping systems. Broken lines indicate research needs.

Conclusions

The complex issues in rice research call for a stronger need for systems approaches in research. The introduction of ecosystem-based and multidisciplinary problem-oriented programs stimulated systems research at IRRI substantially. Further developments will be needed to cover the whole area from problem definition in farming systems via strategic research to practical application of the results. Building individual capacity can be done in various ways, for example through training programs and exchanges with visiting scientists. However, to build the institutional capacity, the research organization has to be structured along problem-oriented programs, and the organization must stimulate interdisciplinary systems research.

Acronyms

DS	dry season
GIS	geographic information system
IRRI	international rice research institute
NARC	national agricultural research center
NARS	national agricultural research system
SARP	systems analysis and simulation in rice production
SSA	simulation and systems analysis
WS	wet season

References

Angus J F, Zandstra H G (1979) Climatic factors and the modeling of rice growth and yield. Pages 189-199 in Agroclimatology of the rice crop. IRRI, Manila, Philippines.

Ten Berge H F M (1993) Building capacity for systems research at national agricultural centers: SARP's experience. Pages 515-538 in Penning de Vries F W T, Teng P S, Metselaar K (Eds.) Systems approaches for agricultural development. Kluwer Academic Publishers and IRRI. Kluwer Academic Publishers, Dordrecht, The Netherlands.

Buresh R J, Singh U, Godwin D C, Ritchie J T, De Datta S K (1991) Simulating soil nitrogen transformations and crop response to nitrogen using the CERES-RICE model. Pages 43-46 in Penning de Vries F W T, Kropff M J, Teng P S, Kirk G J D (Eds.) Systems simulation at IRRI. IRRI Res. Pap. Ser. 151.

Cassman K G, Kropff M J, Zhende Yan (1994a) A conceptual framework for nitrogen management of irrigated rice in high yield environments. In Proceedings of the 1992 International Rice Research Conference. IRRI, Manila, Philippines. (in press).

Cassman K G, De Datta S K, Olk D C, Alcantara J M, Samson M I, Descalsota J P, Dizon M A (1994b) Yield decline and the nitrogen economy of long-term experiments on continuous, irrigated rice systems in the tropics. Adv. Soil Sci. (in press).

Calvero S B, Teng P S (1988) Interfacing diseases severity to the IBSNAT CERES rice model: a computer simulation on rice yield loss. Poster paper presented at the 19th Annual Convention of the Pest Control Council of the Philippines, 3-7 May 1988, Cebu City.

Centeno G (1991) Glicom: A comprehensive climate data base system. Pages 65-67 in Penning de Vries F W T, Kropff M J, Teng P S, Kirk G J D (Eds.) Systems simulation at IRRI. IRRI Res. Pap. Ser. 151.

Dingkuhn M, Penning de Vries F W T, De Datta S K, Van Laar H H (1991) Concept for a new plant type for direct seeded flooded tropical rice. Pages 17-38 in Direct seeded flooded rice in the tropics. Selected papers from the International Rice Research Conference, 27-31 August 1990, Seoul, Korea. IRRI, Manila, Philippines.

Flinn J C, De Datta S K, Labadan E (1982) An analysis of long-term rice yields in a wetland soil. Field Crops Res. 5:201-216.

Fujisaka S (1993) Were farmers wrong in rejecting a recommendation? The case of nitrogen at transplanting for irrigated rice. Agric. Syst. 43:271-286.

Fujisaka S, Harrington L, Hobbs P (1994) Rice-wheat in South Asia: Systems and long-term priorities established through diagnostic research. Agric. Syst. (in press).

Galang A L, Bhuiyan S I (1990) Simulation model use for analyzing decision variables in multipump rice irrigation system. Agric. Water Manage. 17(4):339-350.

Galang A L, Bhuiyan S I (1991) Optimizing economic returns from rainfed ricelands with limited water in farm reservoirs. Pages 48-51 in Penning de Vries F W T, Kropff M J, Teng P S, Kirk G J D (Eds.) Systems simulation at IRRI. IRRI Res. Pap. Ser. 151.

Garrity D P, Flinn J C (1987) Farm-level management systems for green manure crops in different rice environments. Paper presented at the Symposium on Sustainable Agriculture — The role of green manure crops in rice farming systems, 25-29 May 1987. IRRI, Manila, Philippines.

Heong K L, Fabellar L T (1989) Feeding model of the rice leaffolder. Paper presented at the Workshop on Simulation of Insect Damage on Rice, 1-5 May 1989, Khon Kaen, Thailand.

Herrera-Reyes C G (1991) History of modeling at IRRI. Pages 5-9 in Penning de Vries F W T, Kropff M J, Teng P S, Kirk G J D (Eds.) Systems simulation at IRRI. IRRI Res. Pap. Ser. 151.

Herrera-Reyes C G, Penning de Vries F W T (1990) Evaluation of a model for simulating the potential production of rice. Philipp. J. Crop Sci. 14:1-32.

Huisman A (1983) Estimating risk of fertilizer use in rainfed rice production. IRRI Res. Pap. Ser. 93.

IRRI (1968) Annual Report. Manila 1099, Philippines.

IRRI (1973) Annual Report. Manila, Philippines.

IRRI (1989a) IRRI towards 2000 and beyond. Manila, Philippines. 66 p.

IRRI (1989b) Implementing the strategy. Workplan for 1990-1994. Manila, Philippines. 93 p.

IRRI (1993) Rice research in a time of change. IRRI's medium term plan for 1994-1998. Manila, Philippines.

Kirk G J D, Rachpal-Singh (1991) Modeling soil processes with N dynamics as an example. Pages 40-42 in Penning de Vries F W T, Kropff M J, Teng P S, Kirk G J D (Eds.) Systems simulation at IRRI. IRRI Res. Pap. Ser. 151.

Kropff M J, Van Laar H H (Eds.) (1993) Modeling crop-weed interactions. CAB International, Wallingford, UK and IRRI, Manila, Philippines. 274 p.

Kropff M J, Ten Berge H F M, Wopereis M C S (1992) SARP: Research plans for 1991-1996. IRRI, Manila, Philippines.

Kropff M J, Cassman K G, Penning de Vries F W T, Van Laar H H (1993a) Increasing the yield plateau in rice and the role of global climate change. J. Agric. Meteorol. 48:795-798.

Kropff M J, Van Laar H H, Ten Berge H F M (1993b) ORYZA1, a basic model for irrigated rice production. IRRI, Manila, Philippines. 89 p.

Kropff M J, Cassman K G, Van Laar H H, Peng S (1993c) Nitrogen and yield potential of irrigated rice. Plant and Soil. (in press). Also pages 533-536 in Barrow N J (Ed.) Plant nutrition-from genetic engineering to field practice. Kluwer Academic Publishers, Dordrecht, The Netherlands.

Kropff M J, Cassman K G, Van Laar H H (1994) Quantitative understanding of the irrigated rice ecosystem for increased yield potential. In Proceedings of the 1992 International Rice Research Conference. IRRI, Manila, Philippines (in press).

McMennamy J A, O'Toole J C (1983) RICEMOD: a physiologically based rice growth and yield model. IRRI Res. Pap. Ser. 87.

Morris R A (1985) Simulation models in crop research. Paper presented at the Department of Statistics training. Philippines. IRRI, Los Baños, Philippines.

Oldeman L R, Seshu D V, Cady F B (1986) Response of rice to weather variables. IRRI, Manila, Philippines.

Penning de Vries F W T (1991) Improving yields: Designing and testing VHYVs. Pages 13-19 in Penning de Vries F W T, Kropff M J, Teng P S, Kirk G J D (Eds.) Systems simulation at IRRI. IRRI Res. Pap. Ser. 151.

Penning de Vries F W T (1993) Rice production and climate change Pages 175-89 in Penning de Vries F W T, Teng P S, Metselaar K (Eds) Systems approaches for agricultural development Kluwer Academic Publishers and IRRI Kluwer Academic Publishers, Dordrecht, The Netherlands

Penning de Vries F W T, Kropff M J (1991) Modeling in IRRIs work plan for 1990-1994 Page 11 in Penning de Vries F W T, Kropff M J, Teng P S, Kirk G J D (Eds) Systems simulation at IRRI IRRI Res Pap Ser 151

Penning de Vries F W T, Jansen D M, Ten Berge H F M, Bakema A (1989) Simulation of ecophysiological processes of growth of several annual crops Simulation Monograph 29 PUDOC, Wageningen and IRRI, Philippines 271 p

Penning de Vries F W T, Van Laar H H, Kropff M J (Eds) (1991a) Simulation and systems analysis for rice production (SARP) Selected papers presented at workshops PUDOC, Wageningen, The Netherlands, 369 p

Penning de Vries F W T, Kropff M J, Teng P S, Kirk G J D (Eds) (1991b) Systems simulation at IRRI IRRI Res Pap Ser 151 67 p

Penning de Vries F W T, Rabbinge R, Jansen D M, Bakema A (1988) Transfer of systems analysis and simulation in agriculture to developing countries Agric Admin Ext 29 85-96

Penning de Vries F W T, Teng P S, Metselaar K (Eds) (1993) Systems approaches for agricultural development Kluwer Academic Publishers and IRRI Kluwer Academic Publishers, Dordrecht, The Netherlands 542 pp

Rosegrant M W (1977) Risk and farmer decisionmaking A model for policy analysis Paper presented at the IRRI Saturday Seminar 23 April 1977 IRRI, Los Baños, Laguna

Rosegrant M W (1992) The impact of irrigation on production and income variability Simulation of diversion irrigation in the Philippines Agric Syst 40 283-303

Rubia E G, Penning de Vries F W T (1990) Simulation of yield reduction caused by stemborers in rice J Plant Prot Trop 7 (2) 87-102

Sinha S K, Aggarwal P K, Kalra N, Singh A K (Eds) (1993) Systems analysis and crop simulation in agriculture Case study reports Proceedings of the Workshop on Systems analysis and crop simulation in agriculture 1-4 March 1993, IARI, New Delhi, India

Singh U, Padilla J L (1993) Simulating rice response to climatic change Submitted to Amer Soc Agron Spec Publn on Climatic Change and International Impacts

Teng P S, Calvero S, Pinnschmidt H (1991) Simulation of rice pathosystems and disease losses Pages 27-33 in Penning de Vries F W T, Kropff M J, Teng P S, Kirk G J D (Eds) Systems simulation at IRRI IRRI Res Pap Ser 151

Teng P S, Penning de Vries F W T (Eds) (1992) Systems approaches for agricultural development Agric Syst 40 1-3 309 p

Whisler F D (1983) Comparison of rice crop simulation models RICEMOD and IRRIMOD Paper presented at the IRRI Saturday Seminar 30 April 1983 IRRI, Los Baños, Laguna

Wopereis M C S (1993) Quantifying the impact of soil and climate variability on rainfed rice production PhD thesis, Agricultural University, Wageningen, The Netherlands 188 p

Wopereis M C S, Kropff M J, Hunt E D, Sanidad W, Bouma J (1993) Case study on regional application of crop growth simulation models to predict rainfed rice yields Tarlac province Pages 27-46 in Bouman B A M, Van Laar H H, Wang Zhaoqian (Eds) Agro-ecology of rice-based systems SARP Research proceedings, October 1993 CABO-DLO, TPE-WAU and IRRI CABO, Wageningen, The Netherlands

Woodhead T, Ten Berge H F M, de San Agustin E M (1991) Modeling upland rice hydrology Pages 53-60 in Penning de Vries F W T, Kropff M J, Teng P S, Kirk G J D (Eds) Systems simulation at IRRI IRRI Res Pap Ser 15

Wickham T (1973) Predicting yield benefits in lowland rice through a water balance model In Water management in Philippine irrigation systems Research and operations IRRI, Manila, Philippines

Zandstra H G, Samarita D E, Pontipedra A N (1982) Growing season analyses for rainfed wetland fields IRRI Res Pap Ser 73

Systems research at WARDA

M. DINGKUHN

West Africa Rice Development Association (WARDA), 01 B P 2551, Bouake 01, Côte d'Ivoire

Keywords agroecological zoning, crop simulation models, research consortia, rice, Sahel, sustainability, systems research, WARDA, West Africa

Abstract

WARDA was restructured and became a CGIAR center in 1986 Research is organized by ecosystems, including the upland-inland valley swamp continuum, the Sahel irrigated rice ecosystem, and the mangrove rice ecosystem programs This paper discusses the institutional and collaborative requirements for systems research in West Africa It also proposes a symmetry in complexity and integration between researched system and research system An open-center concept was developed at WARDA as a means to achieve a critical mass of scientists and to encourage a more interdisciplinary and interinstitutional integration for systems research Examples of the concept are the task-force system that integrates the effects of NARS and WARDA, the consortium program that focuses on inland valleys in the humid zone, and the Sahel diagnostic network that emphasizes crop-climate interactions in the arid zone A need to intensify training programs to achieve full NARS-IARC partnership in systems research has been identified WARDA's systems research on Sahel irrigated rice is presented in more detail, with an emphasis on cropping calendar x genotype x climate interactions The regional variations in crop calendars was analyzed and explained using simulation methods Improvements are now sought through breeding and organizational studies at the level of the village or irrigation scheme

Introduction

Unlike the more usual reductionist research approach (Dillon 1976), systems research is research on the elements of a system, which also acknowledges the relationships among the elements. An initial conceptual effort defines the borders of the system concerned and the links to the environment outside. It also chooses a level of detail that is adequate for the research objectives (Anderson and Hardaker 1992).

In the past two decades, a remarkable shift in international agricultural research has taken place from discipline-centered towards systems research. This shift is probably driven by an increased awareness of ecological concerns and by the frustrating experience with past attempts to transfer technologies from one environment to another, without sufficient attention to important differences between them. Another element that undoubtedly contributed to the change in approach has been the continued reduction in research budgets, including the budgets of the centers of the Consultative Group on International Agricultural Research (CGIAR). Peer reviews and strategic planning exercises have more than ever emphasized the need for efficiency and measurable impact (Ozgediz 1990). Since research organizations cannot afford to generate technologies that are inappropriate, more use is being made of systems methods to ensure that research is relevant.

The impact that IRRI's initial emphasis on high-yielding rice varieties (HYV) has had on tropical rice production has for some time masked the importance of systems

P Goldsworthy and F W T Penning de Vries (eds), Opportunities, use, and transfer of systems research methods in agriculture to developing countries, 341 - 350

research in rice science. The more favorable tropical Asian rice environments were uniquely prepared for the adoption of HYV in the 1960s and '70s, because intensive and highly sophisticated cultural systems were already the established tradition. No comparable success was achieved in West Africa, partly because existing rice production systems were extensive and partly because technologies that were successful in Asia were initially transferred to African environments, with little regard for the nature of existing systems. It is now evident in both Africa and Asia that research for less-favorable ecosystems, or those where intensification of production is alien to local culture, requires new concepts based on systems analysis.

Since joining the CGIAR system, the West Africa Rice Development Association (WARDA) has made a fresh effort to understand and improve rice-based production systems in West Africa. The present analysis provides an update on this effort and seeks to determine the structural and institutional requirements for systems research.

A brief portrait of WARDA

WARDA is a both a member of the CGIAR system and an association of West-African rice-producing states. Its role as a pluri-national research organization with a regional commodity mandate for rice dates back to 1970. On joining the CGIAR in 1986, WARDA redefined its strategy (WARDA 1988a), its organizational structure, and its interaction with national agricultural research centers (NARS) (WARDA 1988b). Committed to strengthening rice research and development, particularly in its member states, WARDA has a privileged and—by CGIAR standards—close relationship with NARS in the region.

WARDA's research is structured to cater to the needs of different rice-production ecosystems. Individual research programs address rice-based cropping systems in the upland/inland valley continuum, the Sahel irrigated rice ecosystem, and the mangrove rice ecosystem. Each program (except the mangrove rice program, which currently operates as a NARS-based network for varietal evaluation) is made up of objective-oriented, interdisciplinary projects (WARDA 1992). Project designs reflect the interactive nature of factors influencing rice production. The ecosystem-based research programs each include an agroecological characterization project serving as a mechanism to join discipline-oriented projects.

Integrated research systems for meaningful systems research

Although a small institute, WARDA is responsible for addressing rice production in very diverse physical and socioeconomic environments in West Africa. The resource base of most rice production systems is ecologically and economically fragile. It requires a holistic approach to understand component interactions within these systems. Most regional research in the past was component oriented (germplasm, crop response to fertilizer, etc.). However, the results of some past research, which classifies ecosystems by hydrology, topography, and soil taxonomy for parts of West

Africa, forms the basis for WARDA's present, ecosystem-oriented research structure (Andriesse 1985; Andriesse and Fresco 1991; Hekstra et al. 1983; Oosterbaan 1986; Windmeijer and Andriesse 1993).

Research on rice-based production systems as diverse as those in West Africa requires both in-depth studies for prototype situations and more generic studies, the results of which can be extrapolated to different situations. Disciplinary, component studies are essential but must not result in disconnected subrealities, based on perceptions from individual physical or social sciences. The integration of approaches requires critical mass both in terms of scientists (for disciplinary coverage) and hard data (for the necessary levels of confidence needed for integrated models and weighting factors). Systems research, therefore, requires research structures that reflect the complexity of the systems being studied.

Like other CGIAR centers with an ecoregional mandate (ODI 1992), WARDA's research emphasizes technology development and resource management, based on a vision of how current production systems might (or should) evolve. Creating this vision and translating it into concrete development scenarios is probably the most difficult task and can only be accomplished through close interaction with others in the region. Given its small size, WARDA cannot achieve its regional mandate without institutional collaboration with other research organizations working in or concerned with the region. To achieve critical mass and full regional coverage for its projects, WARDA has embarked on an innovative approach of interorganizational and disciplinary integration, called the "open-center" approach.

The open-center approach is basically a guiding philosophy, but it has found (or is currently finding) its expression in several organizational frameworks. In the future they will represent the main avenues for collaboration in systems research. Among the structures reflecting the open-center approach are the task-force system, the consortium program for inland valley swamps, and the Sahel diagnostic network.

The task-force system (WARDA 1988b) is a mechanism designed to strengthen regional, task-oriented NARS research while providing NARS with effective logistical and scientific backup. WARDA, in turn, obtains improved access to its mandate region. All task-force members benefit from the mobilization of resources (expertise and facilities) in the region to solve common problems. Since 1991, a number of task forces have been created to focus either on a region or an ecosystem (e.g., irrigated rice in the Sahel) or a group of closely linked disciplines (e.g., integrated pest management [IPM]). Breeding-oriented task forces work in close coordination with the International Network for Germplasm Evaluation for Rice (INGER). Task forces consist of NARS and WARDA scientists, and through a steering committee, they manage "catalytic" budgets to promote key research activities of NARS that are of regional relevance. Task forces also represent a forum for NARS and WARDA scientists to meet, discuss, and develop collaborative research.

The consortium program (WARDA 1993a) is a concerted program of action by research and development institutions to develop options for the productive and sustainable use of inland valleys in sub-Saharan Africa. The consortium program is still establishing itself, but it will eventually come under WARDA's technical leadership with several international agricultural research centers (IARCs), NARS, and research centers in Europe. In consultation with policymakers, the program will help prepare and implement a master plan for the development of inland valleys, based on their agroecological and socioeconomic characteristics.

The Sahel diagnostic network is a component of WARDA's agroecological characterization project for Sahel irrigated rice. A network approach was chosen because of the vastness of the Sahel area and the scattered nature of the irrigation systems within it. Climatic stresses are important determinants of production. Crop simulation in combination with continuous, multilocation experiments, and automatic weather stations are used to aid agroecological characterization and varietal research, which are currently the two main projects of the Sahel irrigated rice program. The network, comprising NARS and WARDA teams operating at key sites, is coordinated by WARDA and closely associated with the Sahel task force, though independently funded. The main objectives are to generate comprehensive data sets to describe key environments, to develop and test systems analysis tools, and to facilitate the multilocation testing of new emerging technologies.

The organizational arrangements (task forces, the consortium program, and the Sahel diagnostic network) represent structures for collaboration born from the need to collect and integrate information from large, diverse regions. Systems analysis methods like crop, watershed, and economic modelling fit in naturally and will be used systematically. The following sections discuss our experience with systems analysis tools in a specific ecosystem, that of irrigated rice in the Sahel.

Systems research for irrigated rice in the Sahel

A portrait of irrigated rice in the Sahel

The Sahel is the transition zone from the Sahara desert to the Guinea savanna. Erratic rainfall in past decades has rendered irrigated rice the only staple food production system having stable yields, which average about 5 ton ha^{-1} in the major irrigation schemes of Senegal, Burkina Faso, and Niger, and about 2.5 ton ha^{-1} in less intensive systems. Irrigation is based on major permanent rivers.

For rice, the Sahel is also the transition zone from the subtropics, where a pronounced cool season permits only a single summer crop, and the humid tropics, where two rice crops can be grown in a year. The production of two crops of rice in a year is one of the main objectives of NARS in the Sahel.

Irrigated rice culture in the Sahel is of recent origin, most farmers being first- or second-generation rice growers. No traditional varietal or cultural background for irrigated rice exists in the region. Ever since the introduction of irrigation, the gap between actual rice yields (today about 3.5 ton ha^{-1} paddy on 130,000 ha) and potential yields (on more than 3 million ha of irrigable land) has led to large investments in government-managed schemes (Dingkuhn and Miezan 1992). However, production has fallen short of projections and the sustainability of existing systems has become controversial. Meanwhile, the growth of urban population and rising per capita consumption force the mainly agriculture-based states of the Sahel to import 70 percent of their rice.

Since the late 1980s, government intervention in Sahelian rice production has decreased and the private sector has become more active. The changing production systems now require more research to meet the ecological and economic challenges in the Sahel. Lack of research is a major factor contributing to low rice production. Technologies were transferred directly from Asia to the Sahel without evaluating their compatibility with local socioeconomic and cultural realities. Because the cost of production is high, yield potential was the main criterion of the local parastatal for selecting new varieties. The result was a narrow, ecologically vulnerable germplasm base.

WARDA is currently placing its limited resources into two regional projects: agroecological characterization, and rice germplasm improvement and diversification.

Methods for characterization of the ecosystem

In 1991, a multidisciplinary characterization project was initiated. Its objectives were the qualitative and quantitative analysis of constraints to rice production in the Sahel, their principal interactions, and their incidence in the region. The project began with a regional inventory of production and yield data, the area cropped, cultural practices and crop calendars, the rice varieties used, and the biotic and abiotic stresses encountered. A regional study is being conducted on the cost of rice production and how this is related to the type of production system and the policy environment, using the policy analysis matrix (PAM) approach.

The main elements in the experimental studies on crop-environment interactions are the continuous "rice-garden" growth and yield trials with key varieties planted at monthly intervals at six sites in four countries (Alocilja et al. 1981; Dingkuhn 1992). Each rice-garden trial is associated with micrometeorological stations and yield trials that contain local check varieties and advanced selections from the germplasm improvement project. The main biotic and abiotic stresses are studied in associated experiments depending on the resources available at individual sites.

Crop duration and yield were extremely variable but followed a distinct, annual pattern. Simulation models were developed to explain variations based on photoperiod, weather, water temperature, and crop canopy properties. The largest yield variations were due to chilling during the reproductive stage, a stress that was also simulated. Comprehensive models of potential yield, which are sensitive to the major climatic stresses in the Sahel, are being developed on the basis of rice models L3QT (Dingkuhn et al. 1991; Dingkuhn and Penning de Vries 1993) and ORYZA (Kropff, Van Laar and Ten Berge 1993).

Determinations of the gap between simulated and observed yields are being conducted for key environments in the Sahel. Two types of yield-gap studies are planned: analyses of the contributions of different factors to the yield gaps and an evaluation of varietal performance across sites and seasons using potential yield as a reference level.

Targeting key constraints using simulation
Climatic and organizational factors were identified as the main constraints to double cropping with rice. These factors, in turn, gave rise to other constraints. For example, the long duration of the dry-season rice crop, due to low temperatures, frequently delays the planting of the wet-season crop. This in turn causes cold-induced spikelet sterility, resulting in low yields or even total crop failure (WARDA 1993a, b). Organizational constraints are the indirect result of climatic constraints: tight cropping calendars lead to seasonal peaks in farmers' activity, which leads to bottlenecks in the availability of shared equipment (Le Gal 1992a, b, 1993; Dingkuhn, Le Gal and Poussin 1993). The inflexibility in the crop calendar makes irrigation systems more expensive because it requires a large pump and canal capacity that is underused between periods of peak demand (Raes and Sy 1993).

Solutions to the problems associated with the crop calendar can be sought at various levels. Different varieties, modified crop calendars, organizational improvements, or different successions of crops can be combined as the basis of more sustainable practices for individual situations. To develop and test such combinations, component models are needed that can be used alone or in combination to simulate successions of field operations, crop ontogeny, the effects of major stresses, and the demand and supply of human and material resources.

A key component for such a comprehensive, cropping-system-based analysis was developed and validated in 1992-93 in the Senegal delta and valley: RIDEV, a rice ontogeny model which predicts the time taken to maturity and the probability of temperature-induced spikelet sterility (Dingkuhn, Le Gal and Poussin 1993; Dingkuhn et al. 1993). The model can be used for transplanted or direct-seeded rice. A meteorological database of 30 years of daily records from 38 sites in the Sahel was established for regional simulation studies. The model was calibrated at Ndiaye in Senegal for 49 diverse rice genotypes potentially suitable for the Sahel and was validated for the Niono site in Mali (Dingkuhn and Miezan 1993). Derived versions of the model to simulate field-level water use (in collaboration with the Catholic University Leuven, Belgium) and potential yield (with the SARP project at IRRI) are in preparation.

WARDA is collaborating with CIRAD, ORSTOM, and ISRA to simulate scenarios of cropping systems at the level of a village or irrigation scheme. The guiding principle in this collaborative work is that models should accommodate quantitative, physical, or economic parameters determined by the researcher and the criteria that govern decisions in farming communities. Examples of parameters set by the researcher are crop parameters (e.g., weather- and genotype-dependent crop duration and the effects of major stresses), quantifiable organizational constraints (e.g., the supply of manpower or machinery), soil trafficability, and the water requirements in planted plots. Parameters based on farmers' decision-making criteria might include diverse issues such as religious holidays, competing activities, or varietal and cultural preferences.

The modelling software currently being adapted for cropping system simulation is OTELO (Attonaty et al. 1990; Dingkuhn, Le Gal and Poussin 1993), a model originally developed to guide agricultural production systems in France. Agronomic and socioeconomic case studies for individual village communities in the Senegal delta were conducted by CIRAD/ISRA (Le Gal 1992a, b, 1993). These case studies serve as a basis for model adaptation and testing.

Clients and collaborators in systems research for the Sahel
Like other CGIAR institutes, WARDA's immediate clients are NARS, though farmers are the ultimate beneficiaries. This hierarchy is not ideal for WARDA's role in West Africa, and the Sahel in particular, for two reasons.

First, WARDA needs the collaboration of NARS to achieve a critical mass to achieve any impact in rice-cropping systems research, in which case the client becomes a partner in research. At WARDA, we believe that this is also the situation in other ecoregional IARCs. NARS have information on natural resource use and the socioeconomics of local rural production systems. Without the participation of NARS as partners to the IARCs, it is unlikely that systems research at the regional level could succeed.

Second, irrigated rice production in the Sahel was managed in the past by parastatal organizations equipped with their own research units. Some important parastatals in the Sahel have a high level of autonomy and very limited interaction with NARS. Although the role of parastatals is currently changing, a partnership between parastatal organizations and WARDA is essential to facilitate technology transfer and, thus, impact. It is easier to apply simulation-aided systems research to rice production in the Sahel than to the traditional, rainfed rice-production systems in the humid zone of West Africa. Production systems in the Sahel commonly have full control of water, they are severely limited by a comparatively small number of physical and socioeconomic constraints, their potential for production is very large, and the characteristics of the systems are comparatively well known.

Regional and interinstitutional efforts in systems analysis are generally promising in the Sahel, although limited by restricted resources available to both NARS and WARDA. WARDA's effort comes at a crucial time: government-managed rice-production systems are rapidly giving way to a largely inexperienced private sector (Le Gal 1992a), resulting in both opportunity and environmental risk.

The new level of partnership in systems research between NARS and IARCs will require continuing adjustment in research and training programs. The SARP project in Asia has demonstrated how training can stimulate research and how systems research can benefit from training, particularly on a regional and an interinstitutional level. The concept of clients becoming full partners in research depends on increased investments in training.

Prospects for systems research for rice in West Africa

Will systems research on rice be more successful than the earlier, reductionist focus on system components? The two approaches are complementary. Disciplinary research with a focus on specific problems will continue to be essential. Systems research will have its greatest impact as an aid to the identification of constraints and the determination of their relative importance, and as a means to establish a rational consensus on priorities. Systems research can generate information that cannot be obtained readily by other approaches (except sometimes by intuition). It therefore improves the efficiency of the disciplinary research that it guides, always provided that communication and coordination within the research system are also effective.

WARDA's current approach of combining systems research, agroecological characterization, and in-depth studies is likely to have a measurable impact in the irrigated environments of the Sahel in the medium term. The upland-inland valley continuum in the humid zone, however, is much more diverse and complex than the Sahelian irrigated rice environments. Its complexity makes the use of systems-research methods a necessary part of any study of production systems in the zone. Because of this heterogeneity, it is necessary to identify subsystems within the zone that are sufficiently homogenous for systems research. The greatest impact of new technology can be expected to occur in production systems that are amenable to both research and change. For this reason, WARDA has initiated the consortium program for inland valley bottoms and associated watersheds, an environment with great potential, particularly in the largely unused wetlands.

Conclusion

An open-center, multilateral research system may at present be the most promising institutional approach to rice system research in West Africa. It is, however, difficult to focus and manage. Systems research tools shared by partners across a region or ecosystem might provide additional incentives to establish common databases, nomenclatures, and even approaches to research issues. Umbrella projects such as the consortium program for sub-Saharan Africa or SARP in Asia, in conjunction with training programs, provide some of the necessary, interorganizational coordination and communication.

Interinstitutional collaboration that involves the NARS and the IARCs as partners will be a necessary feature of research at the level of a region or ecosystem. This development requires investments in training, particularly in the use of systems research tools and in the generation and management of databases.

Acronyms

CGIAR	Consultative Group on International Agricultural Research
CIRAD	Centre de Coopération Internationale en Recherche Agronomique pour le Développement

HYV   high-yielding variety
IARC   international agricultural research center
IRRI   International Rice Research Institute
ISRA   Institut Sénégalais de Recherches Agricoles
NARS   national agricultural research system
SARP   simulation and systems analysis for rice production
WARDA   West Africa Rice Development Association

References

Alocilja R B, Cervantes E B, Haws L D (1981) A continuous rice production system The rice garden IRRI, Los Baños, Philippines

Anderson J R, Hardaker J B (1992) Efficacy and efficiency in agricultural research A systems view Pages 105-124 in Teng P S, Penning de Vries F W T (Eds) Systems approaches for agricultural development Elsevier Science Publishers, London - New York

Andriesse W (1985) Area and distribution Pages 15-30 in The conference of wetland utilization for rice production in Subsaharan Africa IITA, Ibadan

Andriesse W, Fresco L O (1991) A characterization of rice-growing environments in West Africa Agriculture, Ecosystems and Environment 33 377-395

Attonaty J M, Chatelin M H, Poussin P C, Soler L G (1990) Un simulateur à base de connaissance pour raisonner equipement et organisation du travail en agriculture Pages 291-297 in Bougine P, Walliser B (Eds) Economics and artificial intelligence Paris, France

Dillon J L (1976) The economics of systems research Agricultural Systems 1(1) 5-22

Dingkuhn M (1992) Physiological and ecological basis of varietal rice crop duration in the Sahel Pages 12-22 in WARDA Annual Report for 1991 WARDA, Bouake, Côte d'Ivoire

Dingkuhn M, Miezan K M (1992) Temperature-related problems in Sahel irrigated rice WARDA State of The Arts Paper WARDA, Bouake, Côte d'Ivoire

Dingkuhn M, Miezan K M (1993) Climatic determinants of irrigated rice performance in the Sahel II Model validation and simulation-based characterization of the varietal spectrum present in the Sahel Agricultural Systems (submitted)

Dingkuhn M, Penning de Vries F W T (1993) Improvement of rice plant type concepts Systems research enables interaction of physiology and breeding Pages 19-35 in Penning de Vries F W T, Teng P, Metselaar K (Eds) Systems approaches for agricultural development Kluwer Academic Publishers, Dordrecht, The Netherlands

Dingkuhn M, Le Gal P Y, Poussin J C (1993) RIDEV Un modèle de développement du riz pour le choix des variétés et calendriers culturaux Atelier ORSTOM/CIRAD "Nianga, Laboratoire de la Culture Irriguée," St Louis 19-21 October 1993 (proceedings in press)

Dingkuhn M, Penning de Vries F W T, De Datta S K, van Laar H H (1991) Concepts for a new plant type for direct seeded flooded tropical rice Pages 17-38 in Direct seeded flooded rice in the tropics International Rice Research Institute, Los Baños

Dingkuhn M, Sow A, Samb A, Asch F (1993) Climatic determinants of irrigated rice performance in the Sahel I Photothermal response of flowering and interactions with microclimate Agricultural Systems (submitted)

Hekstra P, Andriesse W, Bus G, De Vries C A (1983) Wetland utilization research project, West Africa Phase I, the Inventory (4 vols) ILRI, Wageningen, The Netherlands

Kropff M J, Van Laar H H, Ten Berge H F M (1993) Oryza1, a basic model for irrigated lowland rice production IRRI, Los Baños, Philippines

Le Gal P Y (1992a) Le delta du fleuve Sénégal Une région en pleine mutation Département des Systèmes Agroalimentaires et Ruraux CIRAD-SAR CIRAD/SAR Doc No 70/92, BP 5035 Montpellier, France

Le Gal P Y (1992b) Informal irrigation A solution for Sahelian countries? Some remarks from case studies in the Senegal river delta Pages 779-788 in Feyen J, Mwendera E, Badji M (Eds) Advances in planning, design and management of irrigation systems as related to sustainable land use Vol 2 Center for Irrigation Engineering, K U Leuven, Leuven, Belgium

Le Gal P Y (1993) Rapport annuel d'activités 1992. I. Organisation dur travail et double riziculture. II. Organisations paysannes et irrigation privée. CIRAD/SAR and ISRA-DRCSI, BP 240 St. Louis, Senegal.

ODI (1992) The CGIAR in transition. Implications for the poor, sustainability and the national research systems. Network Paper 31, June 1992. Centre for Development Research, Copenhagen, Denmark.

Oosterbaan R J (1986) Wetland utilization research project: Suggestions for hydrologic and water management research. ILRI, Wageningen, The Netherlands.

Ozgediz S (1990) Strategic planning concepts and issues. Pages 267-282 in Echeverria R G (Ed.) Methods for diagnosing research system constraints and assessing the impact of agricultural research. Vol. I, Diagnosing agricultural research system constraints. ISNAR, The Hague.

Raes D, Sy B (1993) Bilan d'eau et coût d'énergie de périmètres rizicoles. Delta et Vallée du Fleuve Senegal: Campagnes de 1991 et 1992. Bulletin Technique No. 6. SAED - Centre de Ndiaye, BP 74, Saint Louis, Senegal.

WARDA (1988a). WARDA's strategic plan: 1990-2000. WARDA, Bouake, Côte d'Ivoire.

WARDA (1988b) Un programme de partenariat. Nouvelle vision et approche de la collaboration entre l'ADRAO et les services nationaux de recherche agricole. WARDA, Bouake, Côte d'Ivoire.

WARDA (1992) Research project, sub-project and activity summaries 1992-1994. WARDA, Bouake, Côte d'Ivoire.

WARDA (1993a) Sustainable use of inland valley agro-ecosystems in sub-Saharan Africa. A Proposal for a consortium program. WARDA, Bouake, Côte d'Ivoire.

WARDA (1993b) Annual report for 1992. WARDA, Bouake, Côte d'Ivoire.

Windmeijer P N, Andriesse W (Eds.) (1993) Inland valleys in West Africa. An agroecological characterization of rice-growing environments in West Africa. ILRI Publication 52. ILRI, Wageningen, The Netherlands.

Discussion on Section F: The use of systems methods in research at an international level

Evolution of systems approaches in CGIAR centers

Because of the number of IARCs and the constraints on time at this workshop, there was no opportunity to discuss the accounts by the centers of their use of systems methods in agricultural research. The organizers sincerely regret this, not only because their papers are an important contribution to the workshop, but also because it is some time since there was an occasion for the centers to review collectively their progress and their needs in this field of activity. For the NARS, the centers continue to be a window on agricultural research, and their experiences are therefore likely to serve as a useful guide to NARS in their efforts to develop their own capacity to apply systems methods in agricultural research.

Systems applications

Over the past 20 years, the original focus of the CGIAR centers on increasing food production has moved progressively to one that aims for a balance between production-oriented objectives and research to improve the efficiency and ensure the sustainability of resource use in agricultural production. Some of the IARCs (CIAT, ICARDA, ICRISAT, and IITA) had a regional mandate when they were established, but in all the centers the agroecological aspects of their work have grown in importance. All are now concerned in different ways with the systematic, quantitative analysis of agricultural and land-use systems. The centers are employing a range of systems approaches and methods for various purposes, which include ex ante evaluation of the productivity and sustainability of land-use systems, and analysis for research planning. Information is the key for this knowledge-based agroecological research.

The IARCs, like NARS, are the users rather than the developers of the various methods and models. CIMMYT, IRRI, and ICARDA, among others, use them as tools to make predictions and support decisions in their own research. Their experience and that of other users has helped focus further research on critical areas, and to refine the use of systems methods for a variety of applications.

The papers in this section illustrate some of the applications for which these methods are being used. They also demonstrate a very substantial growth and development in the use of systems methods since the last time the centers met formally to discuss their application in international agricultural research in Rome, in 1986 (Bunting 1987).

Compared with the more traditional commodity- and production-oriented research, it is the greater complexity of the research on resource management in which IARCs have become increasingly engaged that has led them to use systems approaches more widely. These approaches are now used to characterize existing agroecological systems, analyze land-use patterns and options, understand soil, water

351

and plant-nutrition relations, and study the distribution of crops and cropping systems. CIAT is planning to seek the collaboration and expertise of those from outside the CGIAR system who have developed simulation models to deal with the complexities of pasture- and livestock-management systems, to adapt some of their models for use in the environments where CIAT works (Torres). ICARDA also requires simulation models for pasture and livestock research. They require models to represent the role that self-regenerating annual legume pastures could play in the sustainability of crop livestock systems in mediterranean environments.

Between them, the centers cover all of the world's main agricultural environments. CIP is concerned with the cool Andean highlands and conducts research at sites that represent the diversity in the region (Zandstra). WARDA uses systems methods to explore the feasibility of improving the adaptation of West-African rice production systems, which are of comparatively recent origin and located in a transition zone for rice production, between the lowland tropics and subtropical climates to the north (Dingkuhn). Although IRRI has a global responsibility for rice research, including research on lowland swamp, upland, and deep-water flood rice, most of its work concerns the irrigated rice production systems of Asia (Kropff et al).

CIAT used a natural resource database coupled with GIS, initially to select and describe research sites for its crop improvement research on cassava, beans, and rice, but more recently to define and locate research sites that are representative of a selection of the most important and vulnerable tropical environments in Latin America, where it intends to concentrate its research into the problems of natural resource management (Torres). At ICRAF, systems methods have been employed to characterize African land-use patterns in which agroforestry is or could become integrated into the resource management system. The East African Highlands are of special interest because of their vulnerability to soil erosion (Hoekstra).

Not all of the work is devoted to whole systems. CIMMYT employs systems methods to increase the efficiency of crop-improvement research, thereby contributing to sustainability objectives. With its responsibility for maize and wheat improvement, CIMMYT has been seen as one of the principal commodity research centers, but it too now allocates some of its research resources to system-level research. In collaboration with IRRI, CIMMYT scientists are investigating the probable causes of declining yields in wheat-rice based systems in South Asia. It is also studying conservation tillage methods and mixed maize-legume crop systems on hillsides in Central America.

System levels
Agroecological systems consist of a hierarchy of different systems levels. An ascending agricultural hierarchy might be: gene, plant, plant community (crop), cropping system, farming system, and agroecological system. A corresponding spatial hierarchy might be represented by: a soil unit, a field, a farm unit, farming landscape, a catchment area, and an agroecological region. There is no agreed taxonomy for the purposes of system analysis, and from the schematic representation

that some of the centers have included in their paper (e.g., see Harris, Kropff, Torres), it appears that their perceptions of the hierarchy differ, depending on their purpose.

The centers use systems methods for various purposes that correspond with different levels within the hierarchy. Several centers use systems methods at the level of a sub-system for soil management studies to study, for example, the dynamics of soil nitrogen (CIAT and IRRI), the nutrient levels in soils (ICARDA), and at CIAT as part of the research to alleviate problems on acid soils. At another level, CIAT uses a whole-farm analysis to compare land-use technologies and the probable economic consequences of applying them.

All the centers are now using systems methods of one kind or another to characterize environments of special interest to them. ICARDA for example uses historical weather data and simulation methods to estimate the probability of any given outcome from the use of a technology or practice, in an environment where the spatial and temporal variation in rainfall is such that field trials alone cannot provide satisfactory answers (Harris et al).

CIMMYT, IRRI, and WARDA use simulation models and GIS to characterize the environments of major wheat-, maize-, and rice-producing areas, with the aim of improving the interpretation of their international trials and the efficiency of their crop-improvement research.

Simulation methods are used to investigate the dynamics of pest populations and, at IRRI, of the frequency and intensity of disease pressures as part of the crop-protection research and, at IITA, in the program of biological control of the cassava mealy bug. CIP is conducting studies in the Andean region on the movement and persistence of pesticide residues and on the probable effects of different tax options to discourage the indiscriminate use of pesticides. This is one of the few examples of the application of systems methods to compare policy options.

Because the tools are not yet available, there is little work at present that attempts to integrate the complexity of whole system (Harris), though the aim is to do so eventually (CIAT, ICARDA, IITA, and ICRAF). CIAT plans, in time, to develop a framework that would enable studies conducted at different levels of the land-management systems in tropical America to be linked. Studies on the long-term socio-economic effects of rice technologies represent the highest level of system aggregation that IRRI anticipates it will be concerned with.

Methods
Most of the centers are now using GIS and it is clear they have found it to be a powerful tool for dealing with the spatial dimension of complex agroecological problems. It is often being used to complement simulation methods for geographic extrapolation of research findings (e.g., CIAT, CIMMYT, ICARDA, IRRI, and WARDA). This combination of methods offers completely new opportunities for the centers, but also for the more advanced and better endowed NARS, in targeting their research. In the past, to improve varieties that would perform well over wide areas, the international crop-improvement programs had to minimize the effects of genotype-by-environment (GxE) interactions, and sacrifice the advantages of specific

adaptation that is a feature of many varieties of crops that have evolved locally. Research on resource management faces a similar problem because of the local-specific nature of many resource management technologies. The specific adaptation of the rice cropping systems that WARDA is helping to develop for different areas in West Africa illustrates this point (Dingkuhn). The combination of GIS and simulation methods, with the facility this offers to bring more information to bear on a problem, makes it possible to target research, whether it is for varietal improvement or resource management, in a way that was impossible until very recently.

This is a particularly clear example of how one method often complements another. Another example is the way in which CIMMYT uses varietal performance data with multiple regression and cluster analysis to further refine the results from its GIS procedures to characterize its main environments and target varietal development to meet their specific needs (Harrington et al).

Other systems methods being used by the centers include:

- expert decision-support systems used by CIAT for dealing with problems on acid soils and for management of soil nitrogen. Also used by ICRAF for examining options in the design of agroforestry systems;
- a modified linear programming procedure that WARDA uses to determine rice cropping systems options in West Africa (Dingkuhn);
- stochastic rainfall models used by CIAT;
- a spatial weather generator used by ICARDA to supplement the information it obtains from incomplete sets of daily weather records.

More standardization, first among IARCs and then between them and NARS, on systems methods and their application would be helpful. Some methods may be easier to standardize than others. For example, the only way to deal with some of the limitations in hardware-software compatibility in GIS may be by agreement among users to stick to certain options.

IRRI and ICARDA have adopted a modular approach to the development of systems analysis methods that others, in time, may follow.

Availability of data

IARCs use data for different purposes. It seems that obtaining the data they require is still a problem for at least some of the international centers, just as, according to the NARS, it is for the national systems (see Section C). ICARDA has valuable data from crop, pasture, and livestock experiments and from surveys, but has experienced difficulty in obtaining the environmental data it needs for the geographic region it serves. IITA apparently also has difficulty in obtaining up-to-date environmental data.

Some other centers seem to have more success in obtaining the data they need. IRRI, for example, has a weather database with long-term daily data and arrangements for updating this information. WARDA has been able to benefit from the historical records kept in francophone West Africa, and they make use of 30 years of daily weather data from 38 sites.

It seems that to some extent the centers still collect data on environmental elements independently. In the light of the discussion in this section and in Section C, it would be interesting to review the extent to which it has been possible to implement the recommendations of the inter-center workshop on agricultural environments in Rome in 1986, for improving the collection and accessibility of environmental data, for both the centers and the NARS. The recommendations included the following (Bunting 1987):

- Since many environmental data elements are common to all centers, data collection should be divided amongst centers, so that national services are not asked to provide the same data for each of several centers.
- Collectively, harmonize the activities that centers carry out in one nation with what they do in others, so as to operate effectively as a system.
- Assist nations in collecting fuller and more accurate environmental data, to assemble them from their different sources in the nation, and to store and manage them in internationally compatible ways; where necessary support national services for collection of weather data, and help arrest the decline in quality of services for lack of funds.
- Assist national programs in establishing their own geographical databases and information systems for environmental information, and link them to the worldwide environmental data-management network that is envisaged.

Institutional arrangements
In their reorganization to achieve a balance between production-oriented objectives and sustainability objectives, IARCs are clearly taking account of the need to harmonize their operations with those of the NARS. Many of the centers (e.g., CIAT, CIP, IRRI, and ICRAF) have extensive collaborative arrangements with other IARCs, NARS, institutes, and NGOs, in the form of both consortia and informal networks. CIP includes extension services in some of its institutional arrangements. All of them place great emphasis on maintaining close relations with NARS partners, and WARDA in particular underlines the dependence of the centers on the local knowledge that NARS bring to the partnership (Dingkuhn).

The ecoregional work that the centers are planning includes broad geographic areas of considerable diversity—too broad for many individual NARS (e.g., East African Highlands Initiative; Slash and Burn Initiative [ICRAF]; Desertification Initiative [ICRISAT], Agricultural Sustainability on Cento-American Hillside [CIAT, CIMMYT, CATIE, IICA]). Even within this diversity, lower levels of ecological units, such as major watersheds and river basins or specific ecosystems, may extend across the national boundaries of several countries. Concern was expressed by NARS representatives that on the one hand there was the danger that they may be called on to devote a significant part of their capacity to transboundary issues, and on the other that their direct contact with international centers on matters of national concern will suffer. Clearly, a balanced approach to national and international dimensions is required. The recommendation made in the discussion on Section D—to consider establishing specific benchmark sites representative of the agroecological regions of importance to a national program as a means to demonstrate to policy makers the value of systems methods—could serve also as a way to overcome this difference in spatial scale. It would enable individual NARS to contribute to a larger ecoregional objective while focussing on issues of direct relevance to their national interests.

It will clearly be necessary to establish effective mechanisms for providing and allocating resources for joint programs. (IBSRAM, for example, has proposed the formation of a research consortia or some formalized linkage to manage a new agenda of activities on soil-, water- and nutrient-management research. NARS would form linkages to the consortia rather than to IARCs and other NARS.) It was suggested that for ecoregional studies, cooperation between IARCs and NARS could be improved by introducing joint steering committees with clear administrative responsibilities (see also Dingkuhn; and Rueda et al.). At the same time, it was recognized that the results of ecoregional research may have little influence on policy decisions affecting the use of natural resources in individual countries, unless policymakers as well as scientists are involved in the planning and implementation of ecoregional activities. It will be important to ensure that they are involved. WARDA is already accustomed to doing this, for unlike most of the other centers it has a mandate that specifically includes advising its member governments.

Interdisciplinary research
The IARCs have experienced some of the institutional changes that the NARS also face. The shift from disciplinary to more problem-oriented, systems-based research, and the need to deal with complex issues, has required a much more interdisciplinary approach to plan, conduct, and interprete research (Dingkuhn, Harris et al., Kropff et al.).

At IRRI, the use of simulation methods was initially not strongly interdisciplinary. Later, an interdisciplinary organization developed naturally because of the problem-oriented character of the research (Kropff et al.).

In some of the IARCs, systems methods are now used widely (Kropff et al.). In others, it is still only a very small group of people who are using and promoting systems methods in research. There are differences of view, and not all scientists are convinced of the value of the methods (Harris et al.). The institutional constraints to the introduction of systems methods are likely to be more marked in NARS, most of which are organized along disciplinary or departmental lines (see discussions on Sections C and D). The different experiences of the IARCs may therefore serve as useful indicators to help NARS identify the factors that influence institutional integration of systems methods, and so enable them to avoid some of the problems.

Staffing levels
There was no opportunity during the workshop to survey the total staff commitment to the development and assimilation of systems methods for the applications that are being used, the system must have spent considerably more than the estimated seven man-years that were committed in 1986 (Bunting 1987). However, the resources

committed to this kind of work vary markedly between centers. At some, there are still only a few scientists who devote part of their time to this work, and progress has therefore been correspondingly slow (Harris et al.). All the IARCs have experienced severe budget constraints during the past few years. At the same time, they, like NARS, have to undertake significantly more complex tasks without additional resources, and sometimes with fewer resources than they had previously. The impression gained was that the total number of scientists involved is still small compared with the magnitude of the task, and it is to be hoped that past levels of support wil at least be maintained or even improved.

Summary of recommendations

Section B

Integration of economic and ecological approaches
Economic and ecological approaches are complementary in systems research. Both are needed, and a much stronger link between the two approaches needs to be sought so that the strengths of one overcome the limitations of the other. The workshop recommends more vigorous efforts on the part of the national and international organizations to promote an interdisciplinary dialogue on ways to integrate biophysical and economic modelling with particular attention to the needs of NARS for research and development planning.

System level
By specifying their needs, model users in national and international organizations should encourage modellers to devote special attention to the hierarchical linkages between systems levels that would permit easier movement from one level of analysis to another. There is a particular need for better hierarchical integration to link the different decision-making levels for purposes of agricultural research planning (e.g., at strategic and tactical levels).

Improved methods for determining priorities
When production-oriented objectives and research to improve the efficiency and to ensure the sustainability of resource use in agricultural production are combined, the issues to be addressed become more complex. The workshop recognized that because of this additional complexity, IARCs and the NARS urgently need improved procedures for determining priorities and making decisions about the allocation of research resources. Procedures are needed that include appropriate measures of the costs and benefits of the environmental and natural resource consequences of different resource-use options. The workshop recommends that the development of such procedures should be one of the main objectives of the integration of economic and ecological approaches recommended above.

Section C

Supporting interinstitutional and interdisciplinary research
Institutional reorganization can be disruptive and often encounters resistance, which takes time to overcome. The workshop therefore recommends that while such change is taking place, NARS should be encouraged and assisted by IARCs and donors to introduce funding mechanisms that would encourage more institutional collaboration and interdisciplinary research. By allocating funds to programs and projects, rather

359

than to organizations or departments, organizations can be persuaded to focus more on joint efforts to solve complex resource management problems.

Strengthening linkages between research and policy
The workshop recommends that more attention be given by NARS and IARCs to making policymakers aware of the opportunities for improving the overall productivity of land, water, and biological resources that can result from the application of systems methods and NRM approaches. NARS and IARCs have an obligation to participate more directly in the policy dialogue. High priority should be given by NARS to strengthening linkages between natural resource users, scientists, planners, and policymakers.

Supporting long-term goals
The workshop recognizes that the vulnerability of NARS to externally-funded priorities makes it imperative that they develop a capability to determine their own priorities to negotiate the allocation of investments with donors.

Natural resources and socioeconomic databases
The workshop endorses the continuing efforts by the CGIAR and UNEP to harmonize the environmental-information-management activities throughout the CGIAR, and to integrate natural resources and socioeconomic information required for sustainable agricultural development at the international, regional, and national level. As noted above, little progress was made on the last item at two previous workshops where it was discussed. But the issue arises with increasing frequency, as it did at this workshop, as attention to NRM and work at a systems level expands and the technology for information management improves. For this reason, it should be high on the list of current priorities in the UNEP/GRID and CGIAR project. The workshop urges that in the course of these efforts, early attention is given to the role and the needs of the national services, which are the traditional sources of much of this information.

Section D

Institutional change
The workshop recommended that research managers in IARCs and NARS exercise caution to ensure that the work of groups of systems scientists is integrated into the programs of the remainder of the institute in which they are based. The need to guard against any tendency to become isolated is particularly important in institutional environments without an established culture of systems research. The workshop recognizes that apart from the institutional changes required, the increased complexity of NRM research calls for more investment in diagnosis and research planning.

Strengthen research-policy links

The workshop recommends that IARCs and NARS consider a proposal to establish well-defined benchmark sites and data sets representing key agroecological zones as a way to show practical applications of systems research, and so gain greater policy support. The sites would not be "dry labs", but key locations within a broader agroecological context, which would allow individual NARS to contribute to a wider regional objective while focussing on agroecological systems and issues of direct relevance to national interests. Further reference to this idea is made in Section F, in relation to the ecoregional work of IARCs.

Adapting models to NARS needs

The workshop wishes to draw attention to the plea from NARS that those who develop models should validate them under a wider range of conditions. NARS have also requested better documentation of the models.

Section E

The workshop recognized that a shortage of scientists trained in systems approaches is currently one of the most acute constraints to the introduction and adoption of systems methods in developing countries. It recommends that high priority be given to training, and recommends to the CGIAR, NARS, and donor agencies the following:

- Introduce an understanding of the principles involved in systems approaches into the educational background, starting at the universities and at post-graduate level. The CGIAR, NARS, and donors should seek to create more opportunities for graduate and post-graduate training in systems methods and in subjects related to agroecology in the developed nations and in developing-country universities.
- Reinforce efforts to build the capacity of the national programs by providing additional short-term training courses in systems approaches, similar to those that have been provided for rice scientists by the SARP project, and by IBSNAT, but with particular emphasis on the needs of rainfed agricultural systems. Where appropriate, train groups of scientists who on return will continue to operate as a team. Include short courses that are designed for managers and project leaders.
- CGIAR centers should continue to promote competence in this field by providing, in cooperation with the universities, opportunities for graduate research and other forms of training for their national cooperators. These training sessions should be conducted in an interdisciplinary environment at the centers.

Acronyms

AEZ	agroecological zone
AFE	agricultural and food engineering
AIT	Asian Institute of Technology
APSRU	agricultural production systems research unit
BARC	Bangladesh Agricultural Research Council
BLB	bacterial leaf blight
CABO-DLO	Research Institute for Agrobiology and Soil Fertility
CGIAR	Consultative Group on International Agricultural Research
CIAT	Centro Internacional de Agricultura Tropical
CIMMYT	Centro Internacional de Mejoramiento de Maiz y Trigo
CIP	Centro Internacional de la Papa
CIRAD	Centre de Coopération Internationale en Recherche Agronomique pour le Développement
CONDESAN	Consorcio para el Desarrollo Sostenible de la Ecorregion Andina
D&D	Diagnosis and Design
DAT	days after transplanting
DRSAEA	Department for Research on Agrarian and Agricultural Economics
DS	dry season
DSS	decision support system
DWR	deep water rice
EC	European Community
EMBRAPA	Empresa Brasilera de Pesquisa Agropecuaria
ESRI	Environmental Systems Research Institute
FAO	Food and Agriculture Organization
FSR	farming-systems research
FSR&D	farming-systems research and development
GDP	gross domestic product
GIS	geographic information systems
GOAL	general optimal allocation of land use
GRID	Global Resource Information Database
GxE	genotype by environment
HUMUS	Hydrologic Unit Model for the US
HYV	high-yielding variety
IARC	international agricultural research center
IBSNAT	International Benchmark Sites Network for Agrotechnology Transfer
ICAR	Indian Council of Agricultural Research
ICARDA	International Center for Agricultural Research in the Dry Areas
ICASA	International Consortium for Application of System Approaches to Agriculture
ICRAF	International Centre for Research in Agroforestry
IFDC	International Fertilizer Development Center
IFPRI	International Food Policy Research Institute
IITA	International Institute of Tropical Agriculture
ILCA	International Livestock Centre for Africa
IMGP	interactive multiple-goal programming
IPGRI	International Plant Genetic Resources Institute
IPM	integrated pest management
IRM	integrated resource management
IRR	internal rate of return
IRRI	International Rice Research Institute
ISABU	Institut des Sciences Agronomiques Burundi
ISAR	Institut des Sciences Agronomiques Rwanda

ISNAR	International Service for National Agricultural Research
ISRA	Institut Sénégalais de Recherches Agricoles
ISRIC	International Soil Reference and Information Centre
ITE	Institute of Terrestrial Ecology
IWM	integrated weed management
KARI	Kenya Agriculture Research Institute
KEFRI	Kenya Forestry Research Institute
KIT	Koninklijk Instituut voor de Tropen
LEISA	low external-input sustainable agricultural
LP	linear programming
ME	mega-environment
MIS	management information system
MOA	Ministry of Agriculture
NARC	national agricultural research center
NARS	national agricultural research system
NCA	net cropped area
NGO	non-governmental organization
NPK	nitrogen-phosphorus-potassium
NRC	natural resource conservation
NRI	Natural Resources Institute
NRM	natural resource management
OR	operational research
ORSTOM	Office de la Recherche Scientifique et Technique Outre Mer
PAC	package for agroecological classification
PCARRD	Philippine Council for Agriculture and Resources Research and Development
PPO	program planning by objective
R&D	research and development
RCC	Regional Computer Center
RDBMS	relational database management system
RIS	resource information system
RNR	renewable natural resources
SANREM CRSP	Sustainable Agriculture and Natural Resource Management Collaborative Research Support Program
SARP	simulation and systems analysis for rice production
SCUAF	soil changes under agroforestry
SD	systems diagnosis
SIMTAG	wheat growth simulation model
SOTER	soils and terrain digital database
SSA	simulation and systems analysis
STW	shallow tubewells
SWG	spatial daily weather generator
UNCED	United Nations Conference on the Environment and Development
UNEP	United Nations Environment Programme
USAID	United States Agency for International Development
WARDA	West Africa Rice Development Association
WAU	Wageningen Agricultural University
WS	wet season

Index

Systems Approaches for Sustainable Agricultural Development

1. Th. Alberda, H. van Keulen, N.G. Seligman and C.T. de Wit (eds.): *Food from Dry Lands*. An Integrated Approach to Planning of Agricultural Development. 1992   ISBN 0-7923-1877-3

2. F.W.T. Penning de Vries, P.S. Teng and K. Metselaar (eds.): *Systems Approaches for Agricultural Development*. Proceedings of the International Symposium (Bangkok, Thailand, December 1991). 1993
                              ISBN 0-7923-1880-3; Pb 0-7923-1881-1

3. P. Goldsworthy and F.W.T. Penning de Vries (eds.): *Opportunities, Use, and Transfer Of Systems Research Methods in Agriculture to Developing Countries*. 1994   ISBN 0-7923-3205-9; Pb: ISBN 0-7923-7923-3206-7

KLUWER ACADEMIC PUBLISHERS – DORDRECHT / BOSTON / LONDON